AF616356

WITHDRAWN FROM
STOCK

Risk and Failure Analysis for Improved Performance and Reliability

SAGAMORE ARMY MATERIALS RESEARCH CONFERENCE PROCEEDINGS

Available from Plenum Press

9th: Fundamentals of Deformation Processing
Edited by Walter A. Backofen, John J. Burke, Louis F. Coffin, Jr., Norman L. Reed, and Volker Weiss

10th: Fatigue: An Interdisciplinary Approach
Edited by John J. Burke, Norman L. Reed, and Volker Weiss

12th: Strengthening Mechanisms: Metals and Ceramics
Edited by John J. Burke, Norman L. Reed, and Volker Weiss

13th: Surfaces and Interfaces I: Chemical and Physical Characteristics
Edited by John J. Burke, Norman L. Reed, and Volker Weiss

14th: Surfaces and Interfaces II: Physical and Mechanical Properties
Edited by John J. Burke, Norman L. Reed, and Volker Weiss

15th: Ultrafine-Grain Ceramics
Edited by John J. Burke, Norman L. Reed, and Volker Weiss

16th: Ultrafine-Grain Metals
Edited by John J. Burke and Volker Weiss

17th: Shock Waves
Edited by John J. Burke and Volker Weiss

18th: Powder Metallurgy for High-Performance Applications
Edited by John J. Burke and Volker Weiss

19th: Block and Graft Copolymers
Edited by John J. Burke and Volker Weiss

20th: Characterization of Materials in Research: Ceramics and Polymers
Edited by John J. Burke and Volker Weiss

21st: Advances in Deformation Processing
Edited by John J. Burke and Volker Weiss

22nd: Application of Fracture Mechanics to Design
Edited by John J. Burke and Volker Weiss

23rd: Nondestructive Evaluation of Materials
Edited by John J. Burke and Volker Weiss

24th: Risk and Failure Analysis for Improved Performance and Reliability
Edited by John J. Burke and Volker Weiss

Risk and Failure Analysis for Improved Performance and Reliability

Edited by

John J. Burke

Army Materials and Mechanics Research Center
Watertown, Massachusetts

and

Volker Weiss

Syracuse University
Syracuse, New York

PLENUM PRESS • NEW YORK AND LONDON

Library of Congress Cataloging in Publication Data

Sagamore Army Materials Research Conference, 24th, Raquette Lake, N.Y., 1977.
Risk and failure analysis for improved performance and reliability.

Includes index.
1. Materials–Testing–Congresses. 2. Reliability (Engineering)–Congresses. 3. Munitions–Materials–Testing–Congresses. I. Burke John J. II. Weiss, Volker, 1930-
III. Title.
TA410.S17 1977 620.1'1 80-12346
ISBN 0-306-40446-X

Proceedings of the Twenty-fourth Sagamore Army Materials Research Conference on Risk and Failure Analysis for Improved Performance and Reliability, held at Bolton Landing, Lake George, New York, August 21–26, 1977.

A Division of Plenum Publishing Corporation
227 West 17th Street, New York, N.Y. 10011

Printed in the United States of America

SAGAMORE CONFERENCE COMMITTEE

Chairman
JOHN J. BURKE
Army Materials and Mechanics Research Center

Program Director
VOLKER WEISS
Syracuse University

Secretary
ARAM TARPINIAN
Army Materials and Mechanics Research Center

Conference Coordinator
JOSEPH A. BERNIER
Army Materials and Mechanics Research Center

PROGRAM COMMITTEE

JOHN J. BURKE
Army Materials and Mechanics Research Center

M. J. BUCKLEY
Air Force Materials Laboratory

G. DARCY
Army Materials and Mechanics Research Center

H. HERGLOTZ
E. I. du Pont de Nemours & Company

GEORGE MAYER
Army Research Office

R. B. THOMPSON
Rockwell International Science Center

VOLKER WEISS
Syracuse University

S. WEISSMANN
Rutgers University

Arrangements at Sagamore Conference Center
Helen B. DeMascio
Syracuse University

Preface

The Army Materials and Mechanics Research Center of Watertown, Massachusetts in cooperation with the Materials Science Group of the Department of Chemical Engineering and Materials Science of Syracuse University has conducted the Sagamore Army Materials Research Conference since 1954. The main purpose of these conferences has been to gather together over 150 scientists and engineers from academic institutions, industry and government who are uniquely qualified to explore in depth a subject of importance to the Department of Defense, the Army and the scientific community.

This volume, RISK AND FAILURE ANALYSIS FOR IMPROVED PERFORMANCE AND RELIABILITY, addresses the areas of Techniques of Failure Analysis, Risk and Failure Analysis for Design Against Fracture, Risk and Failure Analysis for Design Against Fatigue, Elevated Temperature Effects, Environmental Effects, Systems Approach to Production Reliability Integration and Outlook - Emerging Needs and Techniques.

We wish to acknowledge the dedicated assistance of Joseph M. Bernier of the Army Materials and Mechanics Research Center and Helen Brown DeMascio of Syracuse University throughout the stages of the conference planning and finally the publication of this book is deeply appreciated.

Syracuse University
Syracuse, New York

The Editors

Contents

SESSION IV

RISK AND FAILURE ANALYSIS FOR DESIGN AGAINST FATIGUE
T. E. Davidson, Moderator

SESSION V

ELEVATED TEMPERATURE EFFECTS
P. Fopiano, Moderator

SESSION VI

ENVIRONMENTAL EFFECTS
A. McEvily, Moderator

SESSION VII

SYSTEMS APPROACH TO PRODUCTION RELIABILITY INTEGRATION
D. Morlock, Moderator

SESSION VIII

OUTLOOK - EMERGING NEEDS AND TECHNIQUES
F. W. Schmiedeshoff, Moderator

CHAPTER 1

RISK AND FAILURE ANALYSIS FOR IMPROVED PERFORMANCE AND RELIABILITY

Thomas J. Dolan

Professor Emeritus, University of Illinois

Urbana, Illinois

ABSTRACT

The care and philosophy employed in material selection, design analyses, fabrication, and maintenance must be sufficient to limit the risk of failure. Failure analysis requires careful sorting of a variety of information to determine how and why a metal part failed, and to prevent a recurrence. To improve safety and reliability, a philosophy of design and prototype evaluation based on the risk of failure is more sound than the stereotyped application of empiricisms, codes, specifications and factors of safety commonly used. Designers must document all conceivable failures in a system, determine by analyses the effect on system operation, and rank the risk of each potential failure according to its combined influence of severity and probability of occurrence. Design codes based upon handbook values for properties of materials are often misleading. A probability of failure exists due to the many uncertainties or variability of the basic structural reactions of a metal; significant changes in mechanical behavior occur due to processing operations, field repairs, adverse or unforeseen loadings and environment, or deterioration with time, temperature, or operating conditions. Consideration must be given to man-machine interactions to prevent accidents in complex systems. Considerable latitude in use and misuse of equipment must be foreseen in order to predict and evaluate the resistance to each possible mode of failure. Careful consideration of the complete life cycle is necessary for selecting optimum materials that will withstand the modifications due to processing and service history, yet provide minimum risk of failure with improved safety and reliability.

INTRODUCTION

The title for this 24th Sagamore Conference covers a broad and comprehensive field. We cannot cover all these factors in complete detail, but only scan the surface as to where we now are and where we hope to go. Risk is always present in the development of any product; failure analysis implies that failures do occur in products in service. We must improve the performance for better reliability and better product function. Another interpretation of "performance" implies safety, which is becoming vitally important in the operation of every piece of mechanical equipment. "Reliability" involves many interpretations, including long term satisfactory service, minimum of maintenance, low operating costs, and availability to perform all of the functions for which the operator feels a need in the area of its warranty and its possible modes of operation. Thus, "risk and reliability" are somewhat opposed. Risk implies a probability of a failure that is at variance with normal coes of design which usually infer that if one applies specifications and a "safety factor" conscientiously, no failures will occur.

In accordance with "Murphy's" law, if there is a remote possibility of anything going wrong at any time, it will go wrong. This every aspect of the design, development, fabrication, service history, operator errors and service induced defects must be studied in order to minimize risk of failure to the lowest possible level consistent with the functions to be performed and the economics of the market [1].

The next chapter of this book concerns techniques of failure anaysis; the experiences of the past must be accumulated and documented to determine what possible modes of failure have occurred and what remedies could be used to restrict undesirable events to a minimum [1-24]. Chapters on risk and failure analysis for design against fracture (either of a ductile or brittle nature), and design to prevent fatigue from cyclic loadings will be considered. Subsequent chapters will deal with the important aspects of elevated temperature in specialized types of equipment, and the severe environmental effects that one may encounter in various products over long periods of service.

Most of the techniques of analysis are concerned with individual components. It is also important to use a systems approach to study the effects on reliability due to the interactions that occur in a product comprised of many elements. The book will conclude with a disucssion of outlooks for the future. What are the emerging needs and techniques that can be developed for improvement of performance, reliability, safety and minimum risk of failures? By intermingling of experts from many fields of endeavor, we hope that the integrated effect will be stimulating, and that you will find a new approach to the avoidance of (or minimum risk of) failures for the future.

The young engineering gradute is confronted with many handbooks of data from standard specimen tests with which to initiate his analysis for a particular design. It should be realized and emphasized that any standard data of this type doe not yeild numerical values indicative of the probably service performance. They merely serve as an index to determine whether or not the givenmaterial has mechanical properties approximately the same as (or widely different from) those materials that have been used more or less satisfactorily in the past. There is no assurance that a new material having the same (or better) properties than that which served satisfactorily in the past will of itself guarantee satisfactory service performance in a specific application.

A striking example of this was evident in the drastic failure of the heat treated wire used for construction of the Mount Hope and Ambassador suspension bridges a half century ago [11,22]. By all standard laboratory rests the new heat treated wire showed mechanical strength qualities somewhat better than that of the cold drawn wire that had given satisfactory performance in many suspension bridges. However, before erection of these tow bridges was completed, it was found that many strands had broken before service loads were applied, making it necessary to dismantle both bridges and replace the suspension cables with the old type cold drawn wire. Apparently some galvanizing and stress corrosion problems developed which had not been anticipated in the higher strength heat treated wire.

While the desirability of developing a new engineering project is often obvious, there may also be intermediate considerations which must be weighed. Occasionally a proposed development is demonstrated impossible on the basis of physical principles, but a more usual situation is that the technical objective is achievable in principle, but not feasible because of the lack of complete technology available. For example, Leonardo da Vinci understood the principles of flight and might well have mastered the problem of stability, but he was not supported by a technology providing high strength-to-weight ratio structures or adequate propulsion. There are many instances of inventions feasible in principle which could not be realized because the technology of the period did not provide the materials, the precision of manufacture or other essential support. In some instances, the monetary criterion may control, because a development may measure up to all criterial and meet an obvious need, but be unsound for reason that the volume of the demand may be insufficient. Judging novel engineering projects for their technical and economic feasibility is a high level of engineering activity, and success requires effective collaboration with other groups such as businessmen, bankers, public officials, if the objectives agreed upon are to represent an economic or social gain sufficient to justify the effort. Skill in the choice of technical objectives is often the highest level of the art of engineering. Any industrial or engineering organization can generate a tremendous volume of work by simply analyzing in detail all the proposals made for new plants or products. To reduce the work to reasonable proportions someone with good judgment and the ability to analyze the risk involved must sift them out.

When it comes to developing or improving a product in the research laboratory there is always the tendency to reduce a complex problem to a series of related analytical and experimental studies. These studies apply inductive reasoning to eliminate unimportant variables and evaluate quantitatively the relationships between shapes, materials, conditions and behavior. But this means that the progress is slow and the possibility of an eventual solution may be in the dim and distant future. In general, industry cannot, will not, and should not, wait that long. Without awaiting the ultimate answers to its problems industry will forge ahead by trial and error, if no other way is available. It will test and use many variables en masse and try to determine their integrated rather than their individual effects and will also utilize basic discoveries as they are made known. The fact is that the highest expression of engineering is in the use of incomplete information in designing and making useful, economical, safe, and reliable structures. It is in these developments that principles of risk and failure analyses become vitally important.

SOURCES OF FAILURE

In studying a failure we must understand "what are the causes leading to failure" and "what are the modes of failure that might be anticipated" in the proposed service environment. The reasons for failure may be classified into three categories:

I. Faulty Processing: Inclusions, voids, laps, delaminations, burns, gouges, fins, porosity, lack of penetration, undercuts, cracks, excess plastic deformation, decarburization, residual stress, dissolved hydrogen, mis-match in fitting parts.

II. Design Considerations: Improper assumptions of loading types, underestimating magnitudes, cyclic forces inadequately known or underestimated, lack of proper design details to resist localized stresses, neglect of thermal, corrosive, or fretting conditions, inadequacy or impossibility of rational stress analysis, error in selecting and evaluating significant material property that measures resistance in each possible failure mode.

Figure 1a. Seventeen foot long longitudinal crack in cylindrical shell of a Yankee dryer. (a) Crack originated in flange at bolted-on head.

III. Deterioration and Unexpected Service Conditions: Unforeseen loadings, wear, corrosion, improper maintenance, disintegration, radiation damage, abnormal or accidental operating conditions, vibrations, impacts, ablation, thermal shock, improper repairs, chemical attack.

Some years ago a survey of 470 failures in industrial plants indicated that about one-third of the failures were due to service conditions, one-third to design consideration, and one-third to shop and metallurgical practices involved in the production. Thus the inferences from failure analyses require that a complete risk analysis be conducted involving all possible modes of failure, and all probabilities of a defective condition that may exist. Methods of fault tree analysis [12,13] provide a quantitative means of making such an assessment and can be valuable for situations of complex systems.

At this point several fractures will be discussed to illustrate some of the complexities that must be considered in sources of failure. The photo in Figure 1(a) shows the large cast iron container called a Yankee dryer used in a paper mill to dry tissue paper. It is heated with steam internally and ruptured longitudinally with a 17 foot crack on startup. Removing a segment at the end showed the origin to be a severe gouge in the flange [Figure 1(b)] caused by erosion from steam leakage through a bolt hole. Failure

Figure 1(b). Origin of crack in flange; steam leakage eroded gouge at the bolt hole.

developed from a combination of thermal and pressure stresses, and local bolt loads.

Figure 2 illustrates the fatigue failure of a large aluminum forging in an engine part. The surface was damaged by pitting in a poorly lubricated interface under moderate cyclic stresses with the presence of high residual tensile stresses due to quenching.

A complex sudden rupture is shown in the views of a track pin from a crawler type tractor in Figure 3. Analysis indicated the pin was quenched to high hardness, and then surface induction hardened, leaving high residual tensile stresses in a three-dimensional pattern inside the pin. Small flaws developed in one end from harsh abrasion. Because of low toughness a small flaw

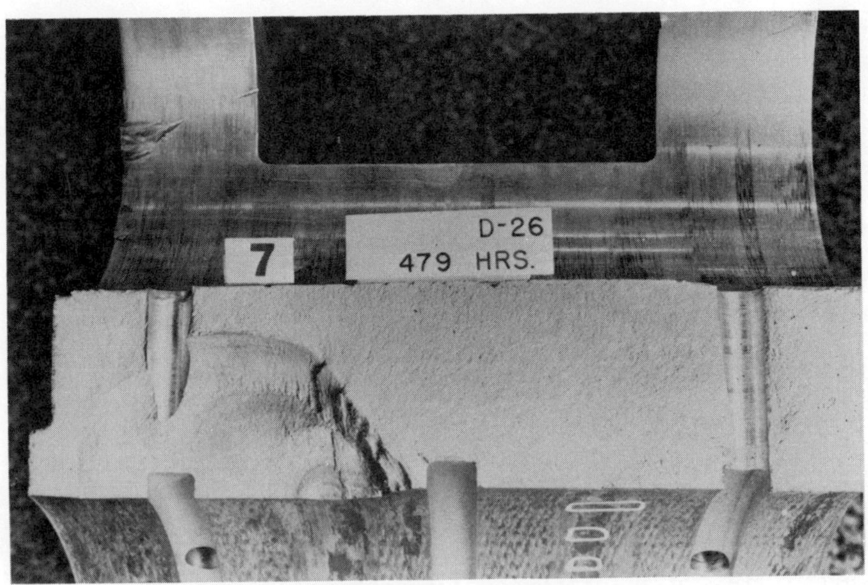

Figure 2. Failure of a 463 pound aluminum forging in an engine part. Fatigue initiated from surface damage by pitting.

was critial and developed a catastrophic fracture from the high residual stresses. When the crack reached the far end, it was inhibited by a radial compression from the high press fit into the track links. This caused the crack to turn 90° and travel in a circumferential direction where axial tensions of high magnitude still existed.

Accidents to equipment in service may lead to controversy as to the cause of a fracture. The smooth textures of the fracture in Figure 4(a) confused investigators who were not knowledgeable; they suggested this was a fatigue fracture. The smearing in the circumferential direction indicates an impact torque that sheared off the shaft when the front wheel of the truck hit a culvert. Further evidence of the severe overload was evident in the twisting of the splines on the shaft [Figure 4(b)], and by impact indentations in the worm of the steering box.

ANALYSIS OF FAILURES

It is important that the investigator of a failure not approach the study by preconceiving an answer before making the detailed investigation [15]. In studying the resonance of sound

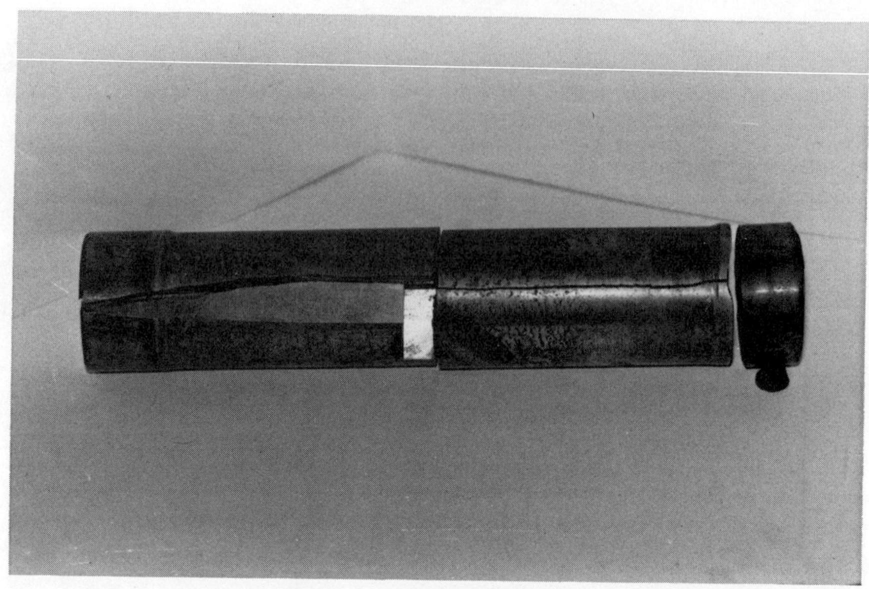

Figure 3(a). Unique failure of track pin on crawler type tractor. Fractures in 2-1/2 inch diameter pin.

Figure 3(b). Origin from grinding abrasion against gravel.

Figure 3(c). Longitudinal radial fracture from origin "O".

Figure 3(d). Radial crack turns 90° at far end separating end portion.

Figure 4(a). Torsional overload fracture of Pitman arm shaft on a truck. Fracture face.

in pipes and cavities, Lord Rayleigh in his book "Theory of Sound" remarks, "When the theoretical result is known, it is almost impossible to arrive at an independent opinion by experiment." Do not approach the analysis of a failure by starting with a hypothesis before making the detailed investigation. Some failures such as those encountered years ago in the British Comet airplanes [8] illustrate the occasional complexity developed by interactions between the materials selection, the design details, new types of service loadings, and a final failure consisting of a rapidly running crack initiated from a very small fatigue crack which had developed to critical size. In other instances, subtle localized zones of chemical attack or stress corrosion cracking may initiate failures that are difficult to categorize as sudden brittle fractures or fatigue problems; they might be initiated by the microscopic actions occurring prior to the development of visual cracking [16]. Embrittlement, diffusion, and localized corrosion often require

Figure 4(b). Bent splines indicate severe overload deformation of shaft.

careful documentation of the service history (time, temperature, loading, and environment) supplemented by chemical analysis or even electron micrographs.

The location and condition of all adjacent parts after the incident should be studied to determine where and how the forces arose that were transmitted to the part in question to result in the failure observed. These are sometimes necessary to confirm the analysis beyond a reasonable doubt, and the complex interaction of several modes developing to final fracture may require additional laboratory testing.

As an illustration of problems in failure analysis, the following illustrations show some of the misinterpretations arrived at by self-appointed "experts". In Figure 5 the arrow points to the zone claimed to be the initiation of a corrosion fatigue failure in the water pump shaft. The fracture is through the root of a ground groove that serves as the inner race of a ball bearing, and the shaft is induction hardened to Rockwell C-60. This opinion was developed from a preconceived hypothesis in analyzing the fracture. Another metallurgist claimed the hardened surface contained too much retained austenite (about 15%), that transformed to brittle martensite under the rolling balls and caused the shaft to disintegrate! The shaft actually was broken by an impact during a collision which hit the fan blade on the front of the shaft. Telltale brinell marks from the balls in the shaft grooves gave evidence of the severity of the blow which ruptured the shaft.

There has been extensive litigation resulting from fractures of automobile axles. Figure 6 shows two views of an axle ruptured at the zone where it emerges from the differential housing. Plaintiffs claimed the shaft ruptured in torsion because it was too weak; the manufacturer proved the shaft fractured in bending. Both agreed the origina was at the location of the toothpick in Figure 6(b). A head-on collision shoved the engine and drive shaft rearward about one foot, snapping the piece of axle off at the exit from the differential. The slight torque from the wheel rolling forward at this instant caused the crack to develop in a diagonal manner as it progressed away from the origin.

Figure 5. Fracture face of broken automotive water pump shaft. (Surface hardened to R_c 60.)

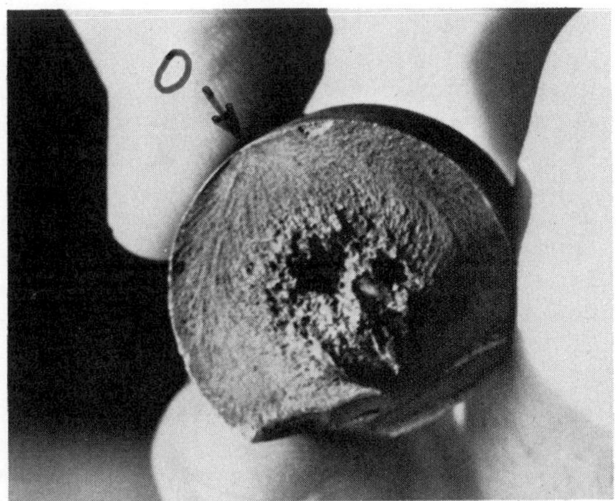

Figure 6(a). Bending impact fracture of automotive drive axle. Face showing origin of fracture.

Figure 6(b). Side view of diagonal development from origin at "O".

In Figure 7 the herringbone markings in the induction hardened case point to the origin and are indicative of a catastrophic impact fracture. The coarse core condition is not metallurgically defective but represents the sudden termination of the rapidly developed separation.

In analyzing impact failures it is important to study adjacent parts to track down sources of the severe loading which resulted in the fracture. Figure 8 shows evidence of the deformation and crack in the central housing indicating the axle was slammed downward during an impact, breaking it off at the bearing support. Deformation of the brake drum backing plate shows the effect of a lateral downward force delivered from the wheel.

In making an analysis of a failed part, care must be used in detailed visual, optical, metallurgical examination, chemical and hardness tests, etc., without careless handling that may destroy important evidence.

Deterioration during service in a specific environment needs special consideration. Many types of surface disintegration, chemical activity, or metal transfer affect stability of the component; these are influenced by the time, temperature, and dosage of the critical factors in the service environment. Not only must the failed part itself be examined in great detail, but background

Figure 7. Impact fracture of induction hardened automotive axle shaft.

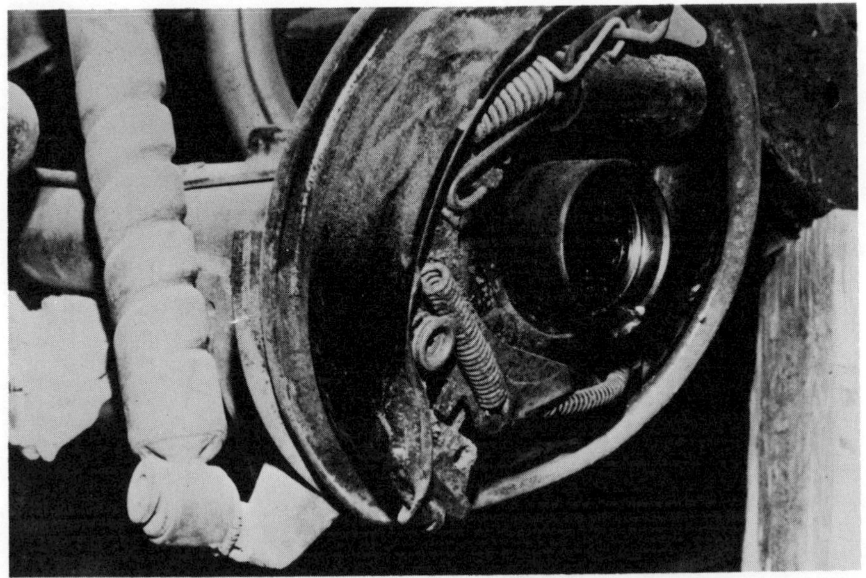

Figure 8. Right rear portion of auto after fracture of axle. (Deformations indicate severe impact.)

information on its chemistry, processing and fabrication, service history, and environment, etc., need to be correlated. A rational and complete analysis must be based on positive supporting evidence (rather than the absence of contrary evidence).

MATERIALS SELECTION

For equipment to operate under unique and severe environmental conditions, a designer is confronted with many variables in the selection and evaluation of the optimum material from the wide variety available. Components such as gas turbines, nuclear reactors, space missiles, submarines and cryogenic equipment are subjected to combinations of severe environments which may involve extremely high or low temperature, corrosive liquids, high vacuum, progressive deterioration due to radiation, surface wear, etc. In spite of the many "standard" mechanical tests available today and the large number of simulated-service experiments being conducted, there is confusion in the interpretation of existing data as far as its application to a particular design is concerned. The selection must often be confined to a small group of candidate materials due to the necessity to design for outstanding resistance in one characteristic (for example, high-temperature resistance in the case of a gas-turbine blade, or inertness to the chemical environment in some forms of chemical-processing equipment). There

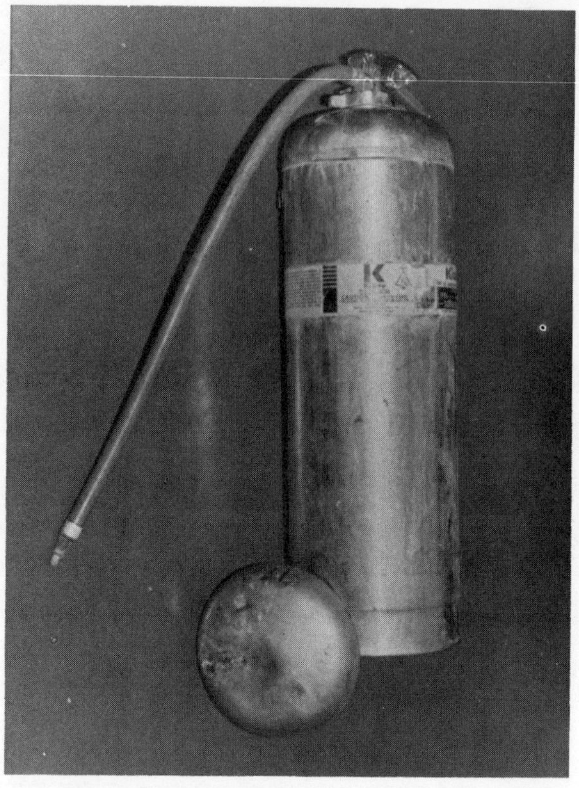

Figure 9(a). Stainless steel fire extinguisher in which lower head blew out during refilling with air pressure. Pieces after rupture.

are many other factors which must be considered such as resistance to brittle fracture, fabricability, wear resistance, ductility, etc., before optimum selection can be made of the most suitable material. Confusion often results because of the lack of significance attributable to many "standard" mechanical properties obtained in laboratory tests or listed in handbooks.

Consideration of the possible modes of failure sets up the criteria under which the part must operate and determines the specific types of mechanical resistance required in the material. For materials evaluation, we must then select those methods of mechanical testing that most nearly evaluate the specific resistance required in this particular service condition. There are, of course, some gaps in which new methods need to be evolved for better determination of quantitative values that can be used for final design purposes. Table 1 presents in preliminary form an organized

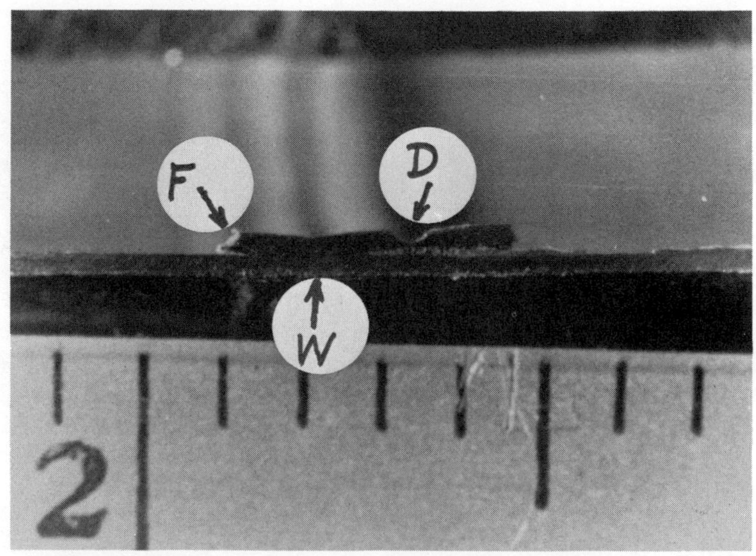

Figure 9(b) Section of shell at zone of fracture F with piece of head attached.

approach for selection of material and design stresses. More detail, of course, needs to be added in the last column to arrive at quantitative values.

An example of improper materials selection is shown in Figure 9 where the lower head of a stainless steel fire extinguisher blew out during repressurizing with air. A small zone about 3/16" on either side of the weld was subjected to "weld decay". That is, the chromium combined with carbon and was precipitated out in grain boundaries at the intermediate temperature of this zone during welding of the head. Corrosion in local ditch-like zones decreased the thickness from 0.030 inches to 0.006 inches when fracture took place. A stabilized grade of stainless would have prevented this depletion of chromium in zones of the head.

Complex interactions of metal are shown by the fracture of the stainless steel pressure vessel in Figure 10. After 400 pressurizations a small corrosion fatigue crack originated at "A" at the threaded end next to the bronze plug. Failure was initiated by too high a hardness, embrittlement from hydrogen and local corrosion with dissimilar metal in a moist atmosphere and propagated catastrophically because of low toughness at the high strength.

For failures due to faulty processing or fabrication, there are few standard tests that can be used for evaluation to cover all

Table 1. Relation of failure to operating conditions and mechanical properties of the material.

Mode of Failure	Loading Mode			Stress Type			Operating Temperature			Material Type		Significant Mechanical Resistance of the Material Measured by:
	Static	Repeated	Impact	Tension	Compression	Shear	Low	Room Temp.	High	Brittle	Ductile	
Brittle fracture	X	X	X	X			X	X		X	X	Charpy V-notch transition temperature. Notch toughness. K_{IC} toughness measurements.
Ductile fracture	X			X		X		X	X		X	Tensile strength. Shearing yield strength.
Fatigue (millions of cycles)		X		X		X	X	X	X	X	X	Fatigue strength for expected life, with typical stress raisers present.
Low-cycle fatigue		X		X		X	X	X	X	X	X	Static ductility available and the peak cyclic plastic strain expected at stress raisers during prescribed life.
Corrosion fatigue		X		X		X		X	X	X	X	Corrosion fatigue strength for the metal and contaminant and for similar time.*
Buckling	X		X		X		X	X	X	X	X	Modulus of elasticity and compressive yield strength.
Gross yielding	X			X	X	X	X	X	X		X	Yield strength.
Creep	X			X	X	X			X	X	X	Creep rate or sustained stress-rupture strength for the temperature and expected life.*
Caustic or hydrogen embrittlement	X			X				X	X	X	X	Stability under simultaneous stress and H_2 or other chemical environment.*
Stress-corrosion cracking	X			X		X		X	X	X	X	Residual or imposed stress and corrosion resistance to the environment. K_{Iscc} measurements.*

* Items strongly dependent upon elapsed time.

Table 2. Failure due to processing methods or to deterioration.

	Category	Method	Effect
I.	Processing and Fabrication		
1.	Mechanical	Cold forming, stretching, bending machining, polishing, grinding, etc.	Each of the processing operations will alter gross or local mechanical properties and may result in micro or macro cracks, or depletion of ductility in localized zones. Surface effects and metallurgical changes from processing may have significant influence on fatigue strength, brittle fracture resistance, and corrosion resistance. Anisotropic properties, zones of dissimilar material, and orientation of principal stresses with respect to unfavorable structural characteristics should be given detailed study in evaluating the resistance to failure in the final product. This will require detailed research to appraise the changes in resistance caused by each specific processing or fabrication operation.
2.	Thermal	Heat treating	
		Welding, brazing, etc.	
3.	Chemical	Processing base material	
		Cleaning	
		Plating	
		Chemical coatings	
III.	Deterioration		Each specific environment or operation needs unique analysis of the significant structural action that limits the usefulness in the service intended.
1.	Mechanical		Specialized abrasion, galling, cavitation, wear, cyclic or slow flaw growth, etc.
2.	Chemical		Stability and activity dependent upon temperature and severity of environment. Oxidation, intergranular attack, diffusion and alloying from foreign elements uniquely determined by the chemical agents, time, and temperatures of operation.
3.	Thermal		Metallurgical changes, grain growth, ablation, melting, etc. dependent upon melting point and stability in the time and temperature for prescribed service.
4.	Corrosion		Time, temperature, simultaneous stressing, frequency of wetting and composition of the corrosive agent as well as the chemical composition and processing of the structural member and its mating parts.
5.	Radiation damage		Influenced by time, temperature, and intensity of the dosage.

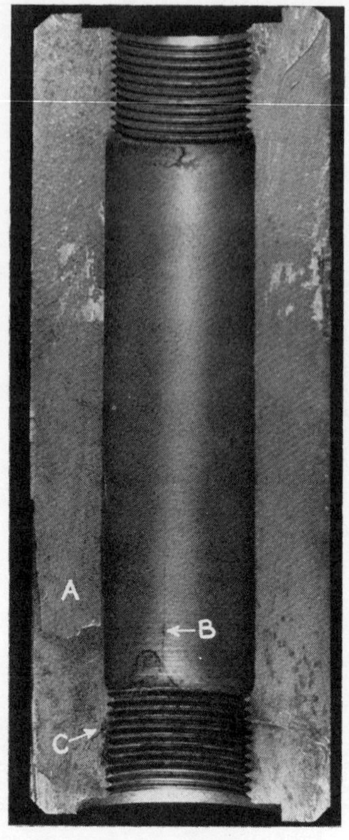

Figure 10. Fracture of martensitic stainless steel pressure vessel after 400 pressurizations [1].
A Brittle fracture face
B Secondary crack
C Zone of contact with bronze plug where corrosion fatigue crack initiated

of the possible inherent flaws that may be induced by such operations as casting, forging, welding, machining, grinding, heat treating, plating, chemical diffusion, or careless assembly operations. As outlined in Table 2, each processing operation may induce residual stresses and modify the mechanical properties by severe cold work in local zones, under-bead cracking, local heating, porosity, hydrogen embrittlement, nonmetallic inclusions, and a multitude of other localized effects which might be categorized as "defects or flaws". In some applications, it is the progression of small flaws which drastically affect the resistance of the member and determine the nearness to failure. The material evaluation should include samples that have been processed by the method intended in the final structural component so as to include normally expected processing "flaws" in the determination of the mechanical resistance.

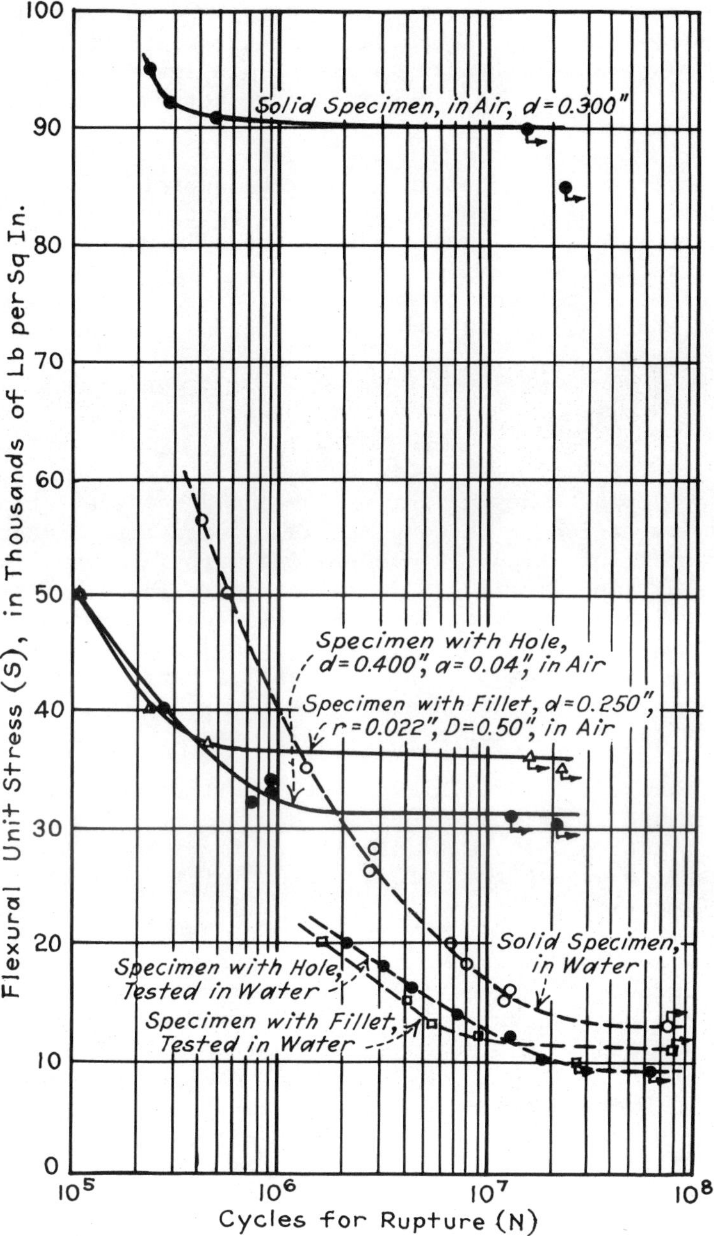

Figure 11. Bending fatigue test data for 3140 steel. (Cycled 1750 cy/sec; simultaneous corrosion with tap water) [9]

The seriousness of an environment that combines the simultaneous effects of corrosion and cyclic stressing is emphasized by the laboratory test data of Figure 11. These results show the severe reduction of fatigue strength in neutral tap water at 1750 cycles/minute [9] in addition to the strength reduction caused by stress raisers.

The environmental conditions must be appraised as major factors that may alter the mechanical properties, or develop defects that lead to failure. Strong emphasis must be placed on critically examining every operation done on the material as comprising the vital parameters in the environment and service conditions that must be included in selecting the material for a given component.

ZERO DEFECTS

In developing a design philosophy and procedure for nondestructive inspection or testing, it is important to recognize that in theory or in practice all engineering materials contain faults, blemishes, or imperfections. In recent years there has been a great emphasis on the production of mechanical components with "zero defects". Philosophically, one should recognize that all materials contain flaws or imperfections of various types; depending on their size, orientation and distribution they might be classified as "defects" when located in critically stressed zones.

In high strength steels even very small flaws of the type shown in Figure 12 in weld zones may be of critical size to result in sudden fracture. In this instance radiographic inspection was not adequate to detect the defects which initiated rupture during proof testing [18].

Many flaws may be too small to be detected by currently available methods of nondestructive examination. Techniques such as X-ray, dye penetrants, Magnaflux, eddy currents, ultrasonics, etc., are limited in the sizes and shapes of defects that can be detected. Usually only cracks of macro or visible size can be discovered. What is essential is that the potentially dangerous triggering defects by located and appraised quantitatively (size, shape, location, etc.) in view of the type of service intended.

For given stresses in certain types of service a critical flaw size exists for the material; larger flaws will stimulate sudden catastrophic fracture. In adverse environments flaws may continue to grow if stresses are maintained above a threshold value. Some micro size flaws may grow by a progressive fracturing mechanism under repeated stressing and develop "fatigue" failure.

Figure 12. Small weld defects caused catastrophic rupture of spherical vessel. (Inconel 718 at R_c 40 to 44) [1]

In view of the statistical nature of the sizes, dispersion, and locations of defects, the probability of finding a potentially dangerous triggering defect is difficult to estimate on a quantitative basis. Many parts may contain flaws which are located in zones of low service stress and hence do not cause problems. On the other hand, the detection of flaws that lead to a probably life that is satisfactory gives no guarantee against premature or catastrophic failure of some small percentage of the components produced. Though every possible precaution is taken in quality control, only gross defects are detected; because inhomogeneities always exist no part is fabricated with "zero defects". One should regard this philosophically as the inevitability of the improbable. In other words, chance effects may cause a very small proportion of the parts produced to fail because of the impossibility of adequately locating and measuring all microflaws present in material for every component produced.

DESIGN CONSIDERATIONS

Many design codes for equipment such as aircraft have, in the past, been committed to design on the basis of a series of static loadings involving: (a) "design load" as the nominal load for which the component is designed to operate, (b) "limit load factor" applied to the design load to compensate for uncertainty regarding exact

TABLE 3. GENERAL APPROACH FOR MECHANICAL DESIGN

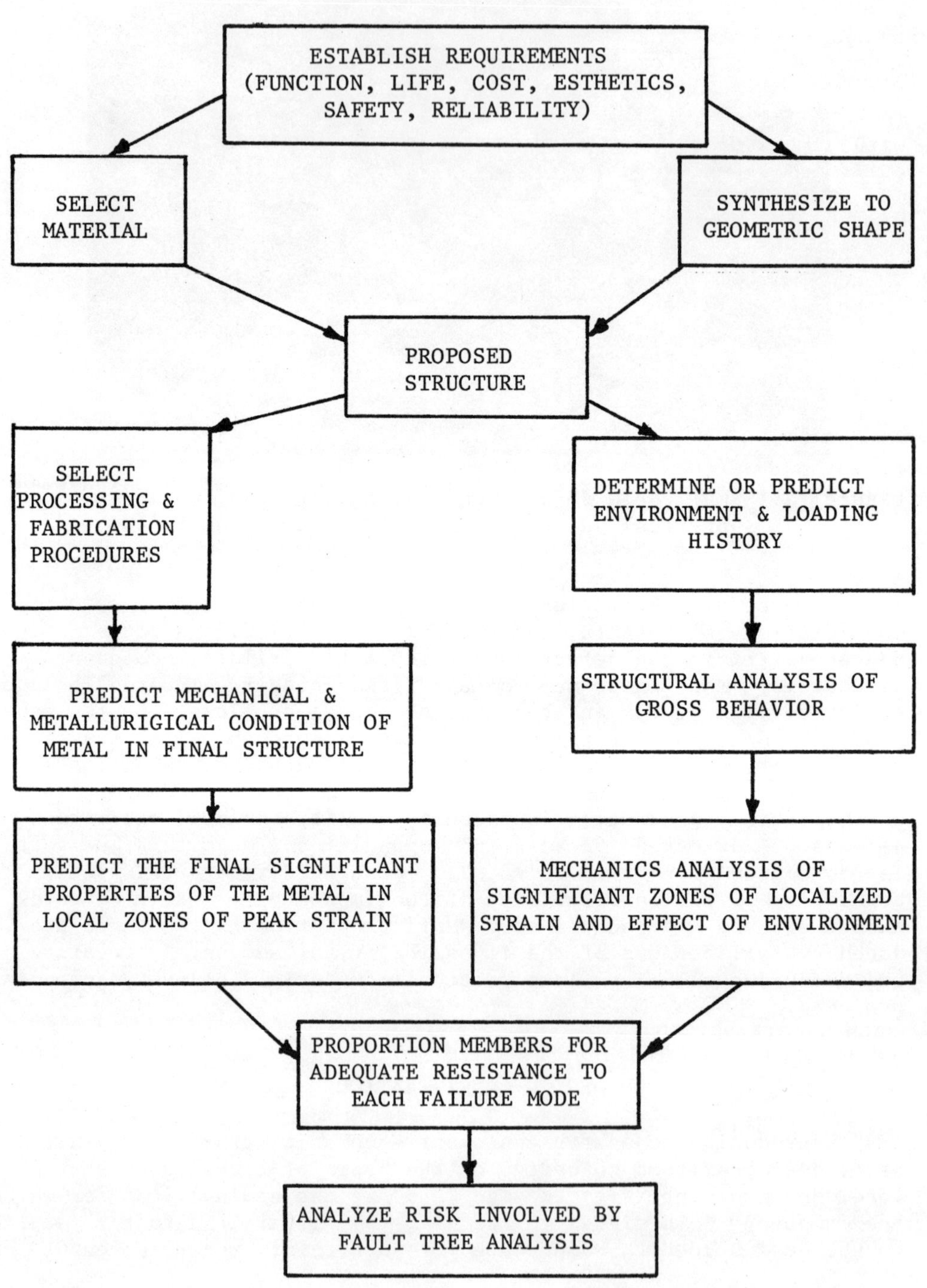

conditions of peak loading, and (c) "factor of safety" as a ratio of the ultimate load or yield load to the limit load. The "ultimate load" is specified as the load which will induce a stress equal to the "minimum guaranteed" ultimate strength of the material. In general, this policy is intimately related to static load-carrying capacity of a component that is loaded in tension. It is embarrassing to have fatigue cracks develop in an airplane wing that was designed with a "factor of safety" of, say, 1.5. Realistically, everything built has a small probability of failure.

For each mode of failure, a different mechanical resistance or "property" of the material is involved in measuring the nearness to structural damage. One observes in Table 1 that the tensile strength is not a significant or reasonable criterion to measure the nearness to failure (or the "ultimate load" in the failure mode), except in a few very limited cases. A so-called "factor of safety" based on tensile strength is meaningless in measuring the nearness to failure by brittle fracture, fatigue, buckling, stress-corrosion cracking, etc. Increased tensile strength is often detrimental where the potential mode of failure may be brittle fracture, low-cycle fatigue, stress-corrosion cracking, etc.

In Table 3 a general approach for mechanical design is outlined to indicate some of the considerations that must be handled by the designer. Each step in this diagram involves a number of factors which cover a broad field of interactions and modifications of the material parameters, the loading history and significant stresses. In many instances it will be necessary to do development and prototype testing to evaluate some of the unknowns and to complete a realistic analysis of the risks involved.

There are many man-machine interactions that should be studied from a psychological viewpoint to eliminate risks of failure or accidents because of the difficulty of communication between the instrumentation and operating facilities and the visual or aural observations of the man operating the equipment. While many operating and control instruments, switches, and levers may be subject to errors or malfunction, there is the other equally important problem of the reaction of an operator, and the interpretation of readings that he obtains from various pieces of equipment such as altimeters, temperature indicators, velocity measurements, fuel gauges, gun indicators, rates of fuel consumption, oil temperature indicators, warning lights and a host of other devices incorporated in modern, complex systems. Results of a number of psychological experiments should be consulted in the design and location of various instruments of these types so that the operator is not thoroughly confused in an emergency or misinformed during the course of his mission; minor mistakes can lead to disastrous results. A number of these failures have been documented in detailed studies

Figure 13(a). Failure of auto bumper brackets when trailer towing. Bumper with trailer hitch after the accident.

Figure 13(b). Brackets failed in fatigue from square holes.

of aircraft and missile accidents and an appraisal of the equipment must be made in terms of the risks involved as well as the potential modes of failure from such difficulties with man-machine interaction.

As a final precaution the designer must anticipate all possible dangers to personnel in the use of his product even for unplanned but foreseeable uses. Warnings about hazards and dangerous uses must be displayed so that every operator can see and understand them.

For example, Figure 13 shows a bumper that was pulled off an automobile while pulling a relatively small trailer. The trailer travelled across the median and hit a car coming in the opposite direction, resulting in a death and severe injuries. In this instance the stamped sheet metal brackets holding the bumper on the automobile developed fatigue fractures and were inadequate for this foreseeable use; no warnings were placed in the instruction manuals to prevent this usage.

NONLINEAR ASPECTS

Many nonlinearities and heterogeneous instabilities influence the progress of damage. It currently appears that no idealized "model" or physical representation of fatigue or brittle fracture is always readily transferable on a quantitative basis to the prototype in practical cases. Surface effects and environment may play important roles in the initiation and propagation of damage; scale effects of unknown magnitude are often inherent in the process of fatigue or brittle fracture. This leads to a nonlinearity in relationship between load and peak stress (the size and shape of the plastic zone may have no relationship to the size of the sample).

In carrying over concepts of fatigue mechanisms in material, a composite fabricated structure may have shorter life than estimated on the basis of tests of simple specimens for the following reasons:

1. Fatigue cracks can be expected to originate at rivets, bolts, seams and other discontinuities developed by fabrication and methods of fastening or joining. These develop severe stress raising effects that are difficult to appraise in simple tests. In many redundant structures yielding occurs in local zones and the significant stresses are not proportional to the external loads.
2. The multiplicity of adjacent stress raisers in composite structures may have the effect of multiplying together the strength reduction caused by two or more separate factors.
3. Fabrication techniques often develop patterns of residual stresses in redundant structures that are difficult to measure,

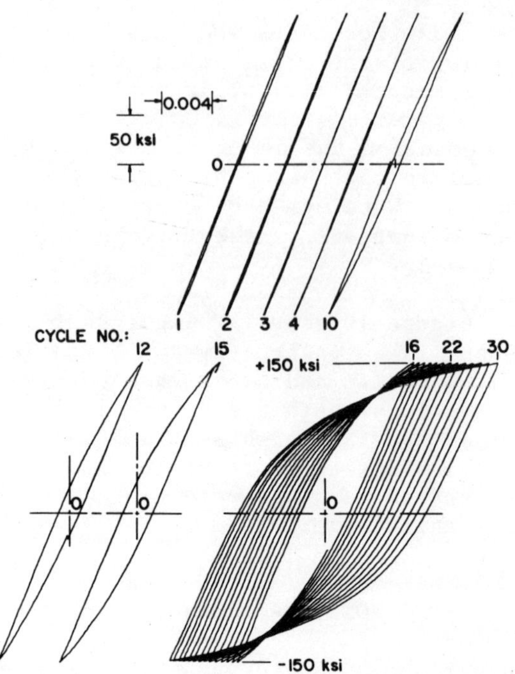

Figure 14. Cycle dependent hysteresis under reversed stress, SAE 4340 steel [10].

but which also may be altered by the first few cycles of imposed loading in service. In recent years the emphasis on low-cycle fatigure phenomena has added much new knowledge to the nonlinear behavior of metals at high cyclic strains in local zones of peak stress. For example, Figure 14 represents an actual recording of cyclic stress-strain behavior during the first few applications of a controlled stress amplitude. Note that the behavior was elastic and no hysteresis developed until about the tenth loading. Thereafter the amplitude of strain (and width of loop) increased every cycle in an unstable manner. Similarly Figure 15(a) shows a continuing extension (or cycle dependent flow) when stress cycled with a mean tensile stress. Following this test Figure 15(b) was conducted with completely reversed stress, and the sample remembered the prior extension and tried to recover during the cyclic stressing. When cycled under controlled zero to maximum strain, as in Figure 16, the stress response tends rather rapidly to approach a completely reversed stress cycle. Understanding of these nonlinearities has helped greatly in developing better methods of calculating cumulative fatigue damage.

4. It has become common practice to fatigue test critical components and entire structures in order to locate fatigue sensitive details and check the estimates on fatigue life. However, the fatigue sensitive areas in service may not coincide with those developed in constant amplitude fatigue tests. The occurrence of an overload or a load reversal in service may modify the stress situation in critical zones, and hence the evidence suggests the importance of placing primary concern on "peak stress modified structures" instead of the original structure for reliance on safety and reliability. If grossly accelerated loadings are used in a laboratory test, the peak cycles of stress influence the yielding and stress redistribution by a large amount. The

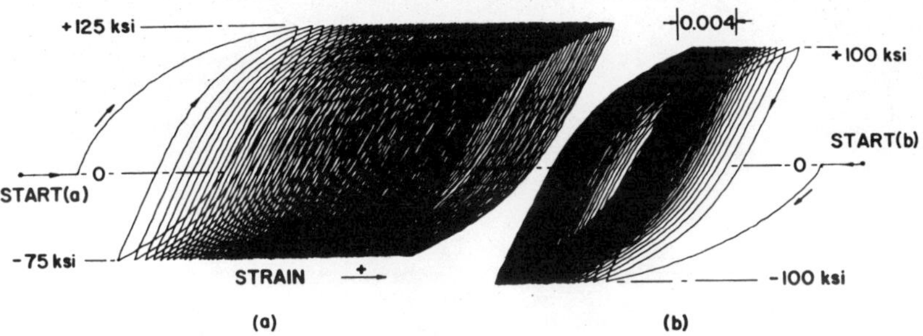

Figure 15. Cycle dependent plastic flow under controlled stress cycling (10].

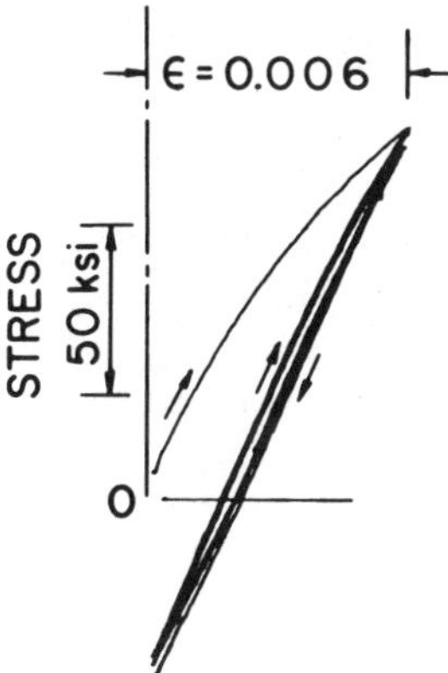

Figure 16. Cyclic stress relaxation under repetition of zero to maximum strain [10].

effect of localized yielding in redistributing the stresses dominates the results in aircraft wing specimen tests; thus service usage involving negative load cycles can not be replaced by tests involving only positive load cycles. With the stress redistribution among redundant load carrying components the significant stresses are not usually known.

LOAD HISTORY

The structural readjustments that occur upon changing load amplitudes are very complex and some of the paradoxical effects observed by Schijve [21] are shown in Figure 17 for riveted joints of aluminum alloy. Using a block loading spectrum to represent typical aircraft gust loadings, he found that the cumulative cycle ratio for failure was 2.90. However, when a reversed overload cycle was applied at the end of each block of loading as shown in line 2, this reduced the cumulative cycle ratio to 1.1, as might be expected.

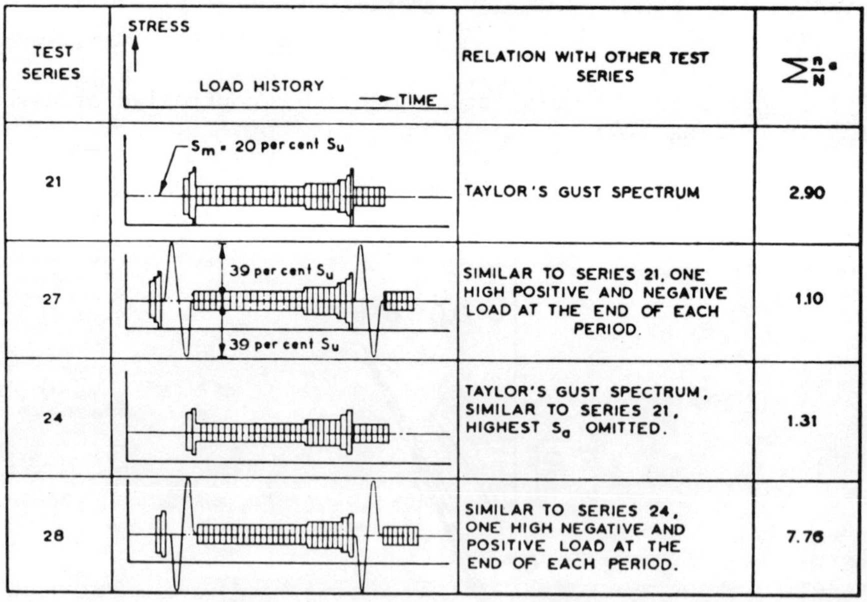

TEST SERIES	STRESS / LOAD HISTORY → TIME	RELATION WITH OTHER TEST SERIES	$\sum \frac{n}{N}$*
21	S_m = 20 per cent S_u	TAYLOR'S GUST SPECTRUM	2.90
27	39 per cent S_u; 39 per cent S_u	SIMILAR TO SERIES 21, ONE HIGH POSITIVE AND NEGATIVE LOAD AT THE END OF EACH PERIOD.	1.10
24		TAYLOR'S GUST SPECTRUM, SIMILAR TO SERIES 21, HIGHEST S_a OMITTED.	1.31
28		SIMILAR TO SERIES 24, ONE HIGH NEGATIVE AND POSITIVE LOAD AT THE END OF EACH PERIOD.	7.76

*Values of $\sum \frac{n}{N}$ are the average of seven test results.

Figure 17. Effect of high peak loads on the fatigue life of a 2024-T3 riveted lap joint under program loading [21].

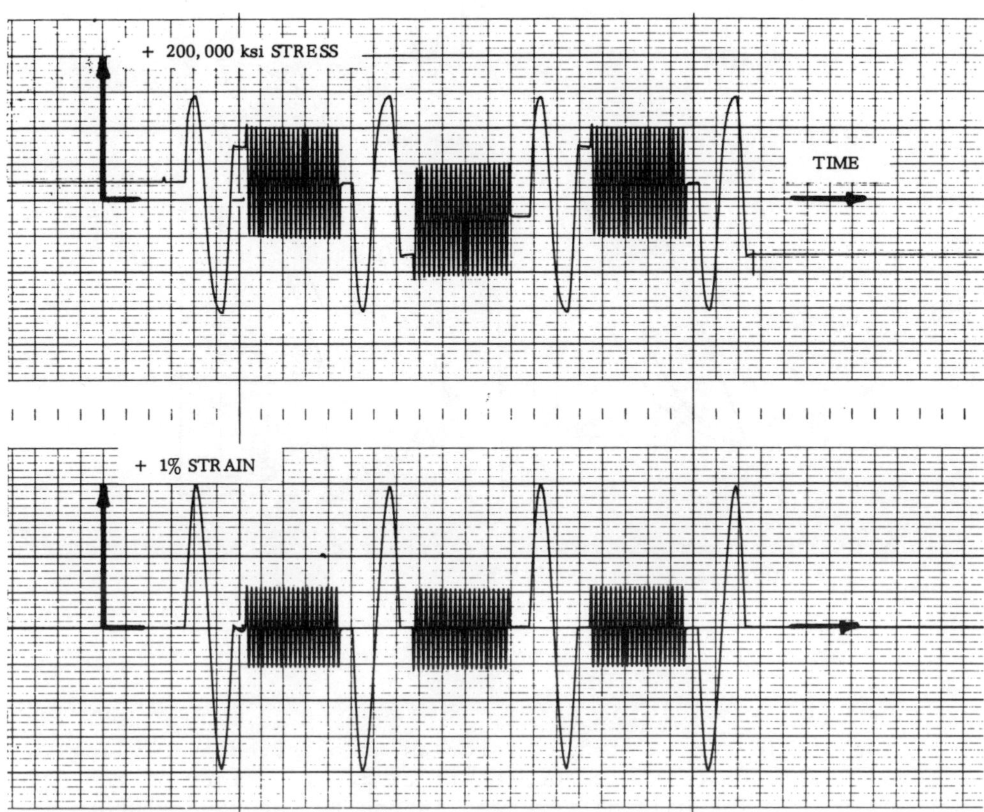

Figure 18. Two level completely reversed strain controlled fatigue test data; unnotched 4340 steel. Note nonlinear history dependent stress response.

What is surprising, however, is that if the same overload cycle is applied at the end of each block loading but in reverse order (that is, the tension part of the cycle applied after the compressive peak load) the life of the sample is greatly prolonged, resulting in a cumulative cycle ratio of 7.76. Thus the history effect imposed by the sequence in the overload cycles is of great importance. When the negative part of the overload just precedes the block diagram, the unfavorable residual stresses left by this unloading cause a more rapid accumulation of damage in the subsequent gust loading spectrum. Figure 18 illustrates the shift in mean stress when the peak strain is reversed in the sequence. New techniques have been developed by several investigators to account for the quantitative method of developing or predicting the cumulative damage from randomly applied loading histories with these sequence effects included.

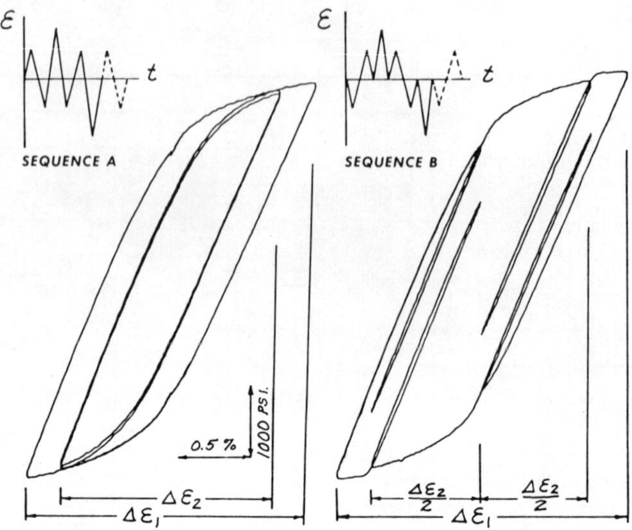

Figure 19. Strain history sequence effect due to material memory; 2024-T4 aluminum [19].

The stable stress-strain histories as observed by Dowling in Figure 19 illustrate the sequence effect due to memory of the prior strain cycle. The only differences in the two sequences shown is that the signs of the small strain cycles are reversed. In comparing fatigue lives of these two strain histories on 2024-T4 aluminum, ordinary range counting methods give unrealistic predictions of fatigue life. Note that for sequence B the small strain cycles developed narrow hysteresis loops accompanied by substantial mean stress, whereas for sequence A, broad hysteresis loops developed with stresses nearly completely reversed. These varying residual stresses (and the varied widths of hysteresis loops) inherent in a fluctuating load sequence can materially affect the fatigue life. In the analyses of complex strain histories the newly developed techniques involving low cycle fatigue, rainflow counting [19], and cumulative damage theory have been developed to handle many unusual cases with greatly improved reliability of the estimates.

NEW TECHNIQUES

As the result of widespread research during the past decade, several new methods of approach for better analysis of design and functional performance of mechanical products have been developed.

Fracture mechanics [17], low cycle fatigue analysis, rate of propagation of fatigue cracks, and cumulative fatigue damage theory and analysis [19,20] (as assisted by rainflow counting of complex stress histories) have all contributed markedly to the improved ability to predict fatigue life or resistance to rapid catastrophic fracture of a wide variety of components.

The idea of expecting the existence of an imperfection, flaw, or small crack has led to new concepts of the need for a quantitative quality control in the detection of flaws. The evolution of methods for the evaluation of toughness characteristics of metals now permit the calculation of a critical flaw size for the initiation of brittle fracture. The important contributions to knowledge of low cycle fatigue phenomena have reemphasized the nature of the damage encountered, and the need for ductility to resist the cyclic plastic deformations which occur [24]. In some instances a fatigue failure may constitute the cyclic growth of a crack to a critical size; the design may be goverened by the resistance to cyclic flaw growth. Therefore, methods of evaluating rates of crack growth in terms of da/dN are important in enabling the designer to predict growth to critical size and to determine when in the life of a product a major overhaul or maintenance becomes necessary [23].

For many years there has been controversy and lack of information about correct methods of counting or evaluating a stress history when it involves randomly applied loading. A method of assessing the important structural damage caused by cyclic excursions has been patterned after the "rainflow counting" method [19] developed in Japan and now expanded to the utility range in which it can be coded for computer to determine the quantitative amount of damage accumulated from a given stress history. Along with the counting of cyclic excursions the technique permits calculating cumulative damage occurring from each loading including the effects of mean stress, and the sequence of occurrence of high loads versus low loads during the course of the stress history. It is imperative in modern design that one make use of these methods and be assured that all of the latest technology has been utilized. The modifications of material from processing and fabrication plus any expected service induced defects need further consideration as a part of the overall risk analysis in the complex component.

RISK AND RELIABILITY

Because of the statistical variability and chance effects involved in material behavior, loadings and service environment, engineering judgment must be included in the final decision-making process and proportioning of members for a particular application. Formalized procedures for assessing the risk of failure of a product in service are useful design tools.

In the 1960's an aerospace recommended practice [14] was developed in the Society of Automotive Engineers with the designation ARP 926, "Design Analysis Procedure for Failure Mode Effects and Criticality Analysis". Application of this practice was made by NASA in evaluating the risk and reliability of missiles in the space program. In general, the method is a design evaluation procedure to document potential failures in individual components, and to determine by these analyses the effect of each failure on the system operation. It identifies all failures critical to operational success or safety, and ranks each potential failure according to the severity index and probability of occurrence. While this procedure becomes tedious and complex when applied to modern equipment involving a large number of components, it has proven to be invaluable in obtaining a quantitative answer for the probability of success in a particular mission by NASA.

At about the same time, a procedure entitled "Fault Tree Analysis" [12,13] was developed and was used as a new innovation in appraising the Minute Man safety program. This procedure placed responsibility on engineering for the traditional role of safety in all phases of the system - design, development and operation. The fault tree provides a concise and orderly description of the various combinations of possible occurrences within the system which can result in a predefined undesirable event. It gives the capability of defining potential problem areas, but also to evaluate their overall system impact. The original concepts were developed by Bell Telephone Laboratories in analyzing electronic switching circuits, but the procedure has proven to be ideally suited to the application of probability theory to numerically defined critical fault modes in a wide variety of components.

It is desirable to have reliability data from tests run on the specific equipment to be used and performed under the identical conditions of use. Frequently such data are not available. The analyst must collect and use information from experience or tests performed on past programs with components or equipment similar to that in the system under consideration. The reliability of a complex system comprising many elements, however, can be rather difficult to analyze because of the interactions of various independent components involved. In some instances the failure of one component may be a minor event, whereas in others it may lead to a sequence of catastrophic occurrences. The interaction of two or more different components may result in a failure rate which is the higher of two components failure rates, whereas in other instances the potential for failure from both of the two components must be added for a realistic evaluation.

One must consider the criticality of the failure of any particular component to failure of the system as a whole. These might be categorized into different levels such as: (a) failure which results in severe injury or potential loss of life; (b) failure which results in a potential mission failure or failure to operate the equipment involved; (c) failure which results in delay or loss of operational availability and excessive down time; (d) failures which result in excessive unscheduled maintenance in commercial operations of equipment. Failures in categories (c) and (d) may result in very high rates of excessive cost due to down time and replacement problems.

In summary, the risk analysis must include the failure mode effects of each component involved and the integrated effect on the system with a criticality analysis regarding the probability of failure or risk of the operation in each particular category of severity.

In their engineering education young engineers are subjected to a variety of mathematical and analytical procedures; the student devotes nearly all of his attention while in college to this type of approach. A proper perspective requires that he take a new view of the overall engineering process, and the synthesis and risk analysis that should be involved in every project. Engineers must be vitally interested in product reliability and safety, not only as an ethical responsibility but also because society will not tolerate accidental deaths and disabling injuries from products that can be modified to make them safe and reliable. In the future there may be criminal penalties imposed upon the designer where engineering negligence results in injury or death. The manufacturer must avoid the various pitfalls of drawing board errors, failure to install safety devices, failure to make a safety control check after a manufacture, and must develop a manufacturing process which does not lead to critical "defects" in the product. Important also is the failure to foresee the consequences of ordinary wear and tear or improper maintenance on the performance in future years. As a final caution, the recognition of any hazard in using a product must be documented and a warning displayed prominently that any user can see and understand.

PROTOTYPE TESTING

Simulated service testing has become an important facet of industrial development of all kinds of mechanical equipment. Investigations may be conducted on individual parts or assemblies, and frequently service-performance measurements are made in an actual field operation. In either event, tests should include all factors of service loadings: time, temperature, environment, prior processing operations, strain history in critical zones, interface effects at surfaces of the component, and other conditions (such as wear) that may affect its functional operation in actual service.

This is difficult, in most instances, and the results must be interpreted with caution. It may be necessary to simulate service defects such as the nicks in an aircraft propellor from thrown up stones to appraise the seriousness of these factors. When high temperatures or corrosive conditions are expected in service, the performance cannot quantitatively be simulated by simple accelerated testing in the laboratory. The influence of time, temperature, and dosage are difficult to simulate and evaluate; one is mixing a mode of failure from cyclic loading (fatigue) with the continuing damage that is time dependent from temperature or corrosion. Because of the inherent statistical variability in the fatigue life of a component it is not usually possible to draw significant conclusions from a limited number of tests of full size parts. The chance effects of sampling and localized irregularities in processing produce marked changes in fatigue life that make it difficult to obtain quantitative answers within a high degree of certainty.

Accelerated testing of a part is frequently found necessary to produce failure in a limited period of time. This usually requires the application of excessive loads or temperatures not usually expected to be encountered in service. Such tests are open to suspicion since the relative trends observed in comparing two materials (or two alternate designs) might be reversed if tests were repeated at lower load or temperature levels. Final check tests should always be made for conditions more nearly representative of actual service conditions.

The fatigue date in Figure 20 illustrate one of the difficulties that may arise from overload testing. In this case, the smooth specimens of quenched and drawn steel had higher fatigue strength than the annealed. However, when tested in a notched condition, the reverse was true. Trial tests at 25,000 to 30,000 psi would have been misleading if the part must withstand millions of cycles of loading in service (say, for a design stress of about 20,000 psi).

Even without the complicating effects of corrosion or high temperatures, overload testing is frequently misleading. Overloading may disperse or readjust residual stresses by yielding or redistributing the peak stresses in a complex redundant structure. The beneficial effect of shot peening on leaf springs is well known; but, in a reversed bending test, it would not show up as well because the compressive residual stresses would be reduced by yielding when subjected to high reversed stresses.

Since factors of strength and stress and both part of the problem in interpreting the tests of full size parts, it is useful to conduct experimental studies of the strain distribution in critical zones. Methods of experimental stress analysis are valuable in locating and evaluating regions of dangerous strain concentration as well as in obtaining supplementary measurements of acceleration,

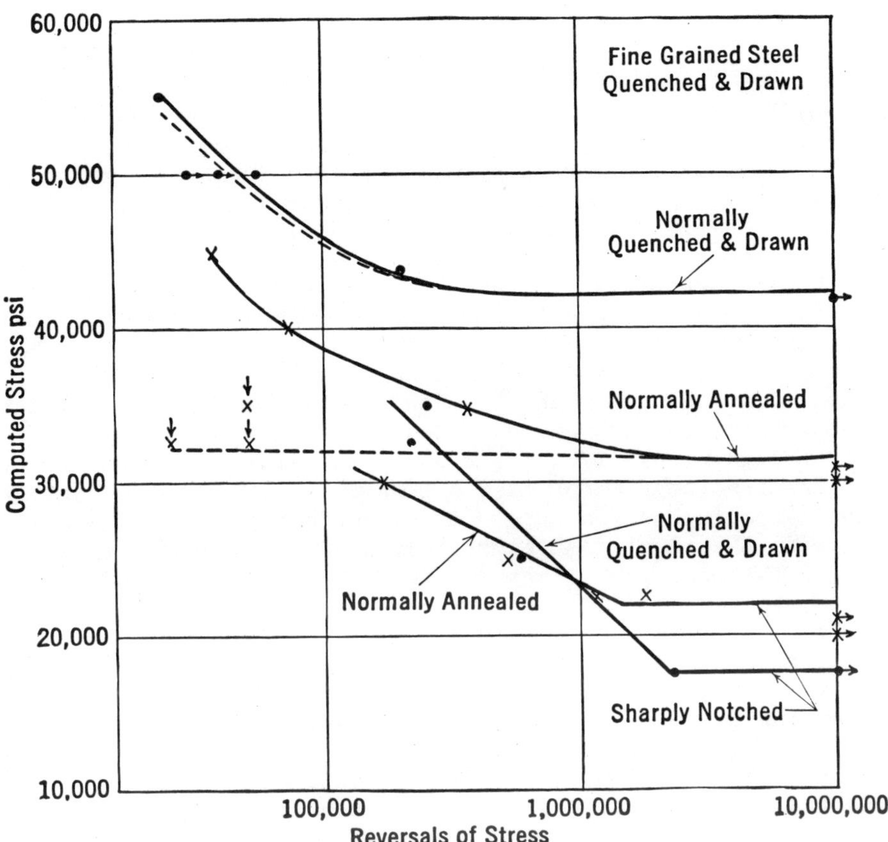

Figure 20. Flexural fatigue data for a steel in two conditions of heat treatment.

velocity, frequencies of vibration, temperatures, etc., that help to define the overall service condition.

Too frequently the failure of a part in service is taken as evidence of the necessity for a "test" without sufficient information regarding the immediate conditions which led to the failure. What measurements should be made to obtain realistic data which measure the significant structural action associated with the service condition, and which was responsible for the failure? If the failure observed in a test differs from that of a prototype in service, it is direct evidence that the service conditions have not been adequately reproduced. All experimental work should be supplemented by careful analysis of the important variables that are likely to contribute to the phenomena observed and each new problem presents specific features that must be understood.

"PROOF" TESTING

Various procedures of proof testing of pressure vessels, piping, etc., are widely used throughout industry, but the interpretation and significance of these "overloads" are little understood. A proof test is not really a quality-assurance test; rather it is an overall inspection test to determine whether gross defects exist that might cause failure on the first loading or which may cause flaws to be enlarged to permit detection. The fact that the component withstood an overload needs to be carefully interpreted and regarded with skepticism. The significance depends upon the type of intended service condition; whether the loads are to be static or repeated, etc.

When dealing with high-strength steels, catastrophic fracture will occur if the flaws are extended to a "critical" size, as illustrated in Figure 12. For tougher materials or higher temperatures, the tendency is for flaw extension and blunting of the flaw without catastrophic fracture. In either case, the lack of obvious fracture in the proof test gives only partial assurance that it would withstand the same load a second time.

Unfortunately, the plastic deformations that occur in localized zones of peak strain during proof testing use up some of the (local) available ductility that is necessary to resist failure. Any repetitions of the proof-testing cycle will reduce the normal cyclic operational life of the vessel because of this depletion of available local ductility. Proof testing more than once provides no additional useful information and, thus, has a deleterious effect on flaw growth, and on the impact strength or fatigue life. Since the critical flaw size will be smaller at low temperatures, proof testing at room temperature does not guarantee against catastrophic brittle fracture for parts that must operate at low temperatures.

Care must be taken in setting a proof pressure at a value which will not cause excessive plastic deformation (such as in a flexible expansion joint or bellows in a pipeline). The material ductility and toughness are reduced by large localized deformations, and the resulting distortions may change functional behavior of the component. The determination of what is excessive deformation depends upon the material ductility, the function of the component, and the types of loading cycles expected in normal service.

SUMMARY

In recent years industry has embarked upon a continuing search for structural materials with high static strength in order to achieve minimum weight. The premature adoption of some of these exotic materials, however, has resulted in a number of embarrassing failures because high static strength does not necessarily insure improved performance and, in fact, may make a part more susceptible to brittle fracture from small flaws, stress-corrosion cracking, and other modes of failure. For this reason, a philosophy of design based on avoiding failure modes needs to be emphasized with better analysis of performance in fatigue, adoption of fracture-mechanics methods of prescribing toughness characteristics, and determination of maximum tolerable sizes of flaws. This chapter reemphasizes the necessity of careful evaluation of changes in mechanical properties caused by fabrication processes and service components. Better methods are needed for appraising the resistance of materials to the environmental conditions of operation that lead to modes of failure such as fretting, hydrogen embrittlement, or slow flaw growth. A broader outlook of the designer and project director in the development of complex equipment is necessary to supplement standardized codes and specifications to provide sound utilization of materials. A comprehensive analysis based on minimizing the risk of all modes of failure under the foreseeable circumstances in future service must be the basis for prediction of the safe and satisfactory performance of the product with a high degree of confidence.

REFERENCES

1. Dolan, T.J., "Preclude Failure: A Philosophy for Materials Selection and Simulated Service Testing", Experimental Mechanics, January 1970, pp. 1-14.
2. Shank, M.E., "Brittle Failure in Carbon Plate Steel Structures Other Than Ships", Welding Research Council Bulletin Series, No. 17, January 1954.
3. Srawley, J.E. and Esgar, J.B., "investigation of Hydrotest Failure of Thiokol Chemical Corporation 260-in. Diameter Sl-1 Motor Case", NASA TMX-1194, January 1966.
4. "Failure Analysis of PVRC Vessel No. 5", Welding Research Council, Bulletin No. 98, August 1964.
5. Cottel, G.A., "Lessons to be Learned from Failures in Service", International Conference on Fatigue of Metals, ASME-IME, Session 7, Paper 1, 1956.
6. Wulpi, D.J., "How Components Fail", Metals Park, Ohio, ASM, 1966.
7. Samans, C.H., "Results of the Survey of the Study Group on Oil Storage Tank Failures", API Proc. Section III, 34, pp. 143-63, 1954.
8. "Civil Aircraft Accident. Report of the Court of Inquiry into the Accidents to the Comet G", Her Majesty's Stationery Office, London, 1955. See also T. Bishop, "Fatigue and the Comet Disasters", Metal Progress, 79, May 1955.
9. Dolan, T.J., "Simultaneous Effects of Corrosion and Abrupt Changes in Section on the Fatigue Strength of Steel", Appl. Mech., ASME, A-141, December 1938.
10. Dolan, T.J., "Nonlinear Response Under Cyclic Loading Conditions", Proc. 9th Midwest Mech. Conference, New York, J. Wiley & Sons, 1967.
11. Moisseiff, L.S., "Investigation of Cold Drawn Bridge Wire", ASTM Proc., 30, 1930, p. 313.
12. Haasl, D.F., "Advanced Concepts in Fault Tree Analysis", System Safety Symposium, Univ. of Washington & Boeing Co., Seattle, Juen 8-10, 1965.
13. Mears, A.B., "Fault Tree Analysis: The Study of Unlikely Events in Complex Systems", System Safety Symposium, Seattle, Wash., June 8-10, 1965.
14. "Design Analysis Procedure for Failure Mode, Effects and Criticality Analysis", Aerospace Recommended Practice, ARP 926, Soc. of Automotive Engineers, September 15, 1967.

15. Dolan, T.J., "Failure Analysis of Metal Components", Metals Engineering Quarterly, ASM, Vol. 12, No. 4, November 1972, pp. 32-40.

16. Highway Accident Report, "Collapse of U. S. 35 Highway Bridge, Point Pleasant, West Virginia, December 15, 1967", Report: NTSB-HAR-71-1, National Trans. Safety Board, 1970.

17. "Fracture Toughness Testing and Its Applications", ASTM STP 381, 1965. See also: ASTM STP 410, 1967.

18. Dolan, T.J., "Product Liability and Material Failures", Trans. SAE, Paper No. 710710, September 1971.

19. Dowling, N.E., "Fatigue Failure Predictions for Complicated Stress-Strain Histories", ASTM Journal of Materials, Vol. 7, No. 1, pp. 71-87, 1972.

20. Landgraf, R.W. and La Pointe, N.R., "Cyclic Stress-Strain Concepts Applied to Component Fatigue Life Prediction", SAE Paper No. 740280, Automotive Engineering Congress, Feb. 25-March 1, 1974.

21. Schijve, J., "Estimation of Fatigue Performance of Aircraft Structures", ASTM STP 338, pp. 193-215, 1962.

22. Swanger, W.H. and Wohlgemuth, G.F., "Failure of Heat Treated Steel Wire in Cables of the Mt. Hope, R.I., Suspension Bridge," Proc. ASTM, 1936, Part 2, pp. 21-84.

23. "Fatigue Crack Growth Under Spectrum Loads", ASTM STP 595, June 1976.

24. Morrow, J., et al., Journal of Materials, March 1969, pp. 159-209 (four papers on low-cycle fatigue behavior).

CHAPTER 2

NDT - AN AID TO FAILURE ANALYSIS

H. P. Hatch

Army Materials and Mechanics Research Center

Watertown, Massachusetts

ABSTRACT

Improved performance and reliability are dependent upon the material quality characteristics of individual components, and nondestructive testing (NDT) techniques aid in predicting premature failure by the detection of critical size defects or by the detection of material property gradients which can be equally detrimental in brittle materials. However, quantitative NDT results are dependent upon a number of variables. To illustrate the effectiveness of NDT to assist with the analysis of suspect material, results of metallurgical and NDT analyses of two transmission gears are presented. Inclusions much larger than specifications allow were detected ultrasonically and confirmed metallographically. The mechanical properties and structure of the gears seemed otherwise to be consistent with good metallurgical practice.

INTRODUCTION

To place the title of this chapter within the context of the subject of this book it should be pointed out that nondestructive testing is not a failure analysis technique, but rather a technique to assist in predicting possible failure by the detection of defects of near critical size or by the detection of material property gradients which can be equally detrimental in brittle materials [1]. Historically, designers have used published material property data such as strength, ductility, and fatigue life to match up materials with service performance requirements and the defects that were detected by NDT were accepted or rejected on the basis of fear rather than knowledge. But with the advent of fracture mechanics, a material's fracture characteristics are now also used as major design considerations. We can calculate critical flaw size or the largest flaw a material can sustain without fracture when subjected to design stresses and environmental conditions. Therefore, in order to produce hardware to fracture control design criteria, it is only necessary to assure that the hardware contains no flaw approaching critical size and it is at this point where nondestructive testing is called upon to contribute.

The preceding generalization is rather simply stated and sounds straightforward, but its implementation is somewhat more complex when one considers the variables which can affect quantitative NDT results. They include the material, its condition or processing history, the geometry of the component with respect to accessibility to probing energy of the critically stressed area where defect detection is essential, the types of defects anticipated and, probably most important, defect orientation. In order to be effective, NDT must be involved in the initial stages of design and materials selection and even materials development. However, it has also been of value in after-the-fact problem solving and the following case history is selected as an example to illustrate the benefit NDT can provide to a metallurgical analysis of material in unknown condition.

PROBLEM STATEMENT

The problem involved the catastrophic failure of a transmission gear from a twin engine helicopter [2]. The spiral-bevel combiner gear, almost a foot in diameter, is shown in Figure 1. This critical component couples the output from both engines to the power train and is bolted to a splined drive shaft through the bolt holes in the flange area. The initial failure analysis showed that the gear failed under fatigue conditions. A crack initiated in the flange area at a sizeable subsurface inclusion which had a major dimension of some 90 thousandths of an inch. Because of AMMRC's participation in a subsequent Critical Parts Review, we were asked to conduct a materials analysis and obtained two additional gears produced from the same mill heat of steel. The two gears were traceable through serial numbers, not only to the mill heat, but also to the heat treat lot. The material specification for this component calls for a carburizing grade 9310 steel produced by the consumable electrode vacuum arc remelt process. The gear teeth are carburized whereas the flange area containing the bolt holes is not.

NONDESTRUCTIVE INSPECTION

Before any metallurgical sectioning was performed, the two gears were inspected by radiographic, magnetic particle, and

Figure 1. View of spiral-bevel combiner gear from the gear side.

ultrasonic techniques. Both the X-ray and magnetic particle results were negative; however, one significant flaw indication was observed in the flange area of each gear by an ultrasonic scanning technique. The ultrasonic C-scans were recorded using a 10 MHz damped transducer by rotating a gear on an underwater turntable as the transducer indexed in a radial direction. With this arrangement, a polar plot of the flange area can be easily produced without traversing the complex geometry of the remaining portion of the gear. The resulting scan of one of the gears (SN M819) is illustrated in Figure 2 and will serve as reference for the following comments meant to assist in interpreting the ultrasonic record.

An electronic signal gate is adjusted in both time and duration to occur between the reflected top and bottom surface echoes of the ultrasonic pulse which propagates through the thickness of the flange. Consequently, any signal reflected from a discontinuity within the material will appear in the gate and be recorded as intensity marks. The bolt holes, however, also produce intensity indications because the ultrasonic energy passes through the holes and is reflected from and propagated into the supporting stainless steel turntable. Inasmuch as the signal gate position and width are fixed with respect to the first water/steel interface signal, the back-echo from the turntable falls within the gate and is therefore recorded (turntable thickness is less than gear flange thickness).

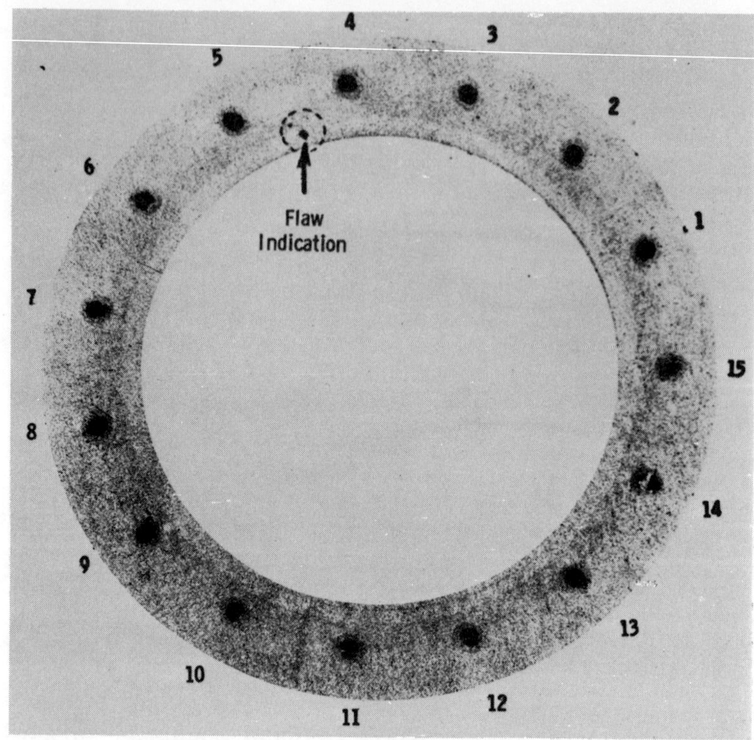

Figure 2. Ultrasonic C-scan recording of gear M819 flange area.

The most significant feature of Figure 2 with respect to this investigation is the flaw indication encircled by a dotted line between bolt holes 4 and 5. The presence of an actual ultrasonic reflection within the signal gate is confirmed by Figure 3, which shows the oscilloscope record of the ultrasonic signal obtained at the indicated flaw location. The flaw indication can be seen to occur at 5/6 the distance between the two large off-scale top and bottom (from left to right) surface echo signals. The C-scan (Figure 2) provides the X-Y coordinates for a given signal and the A-scan (Figure 3) gives us the Z axis or depth dimension as well as quantitative amplitude information. In Figure 3, the 0.36 inch dimension is the flange thickness.

The ultrasonic scan record of the second gear (SN M826) was almost identical to that of the first except the flaw indication appeared adjacent to bolt hole 9. The corresponding A-scan is shown in Figure 4. In this case, the flaw signal is located at 2/3 the distance between the top and bottom surfaces and is somewhat smaller in amplitude than the signal detected in gear M819

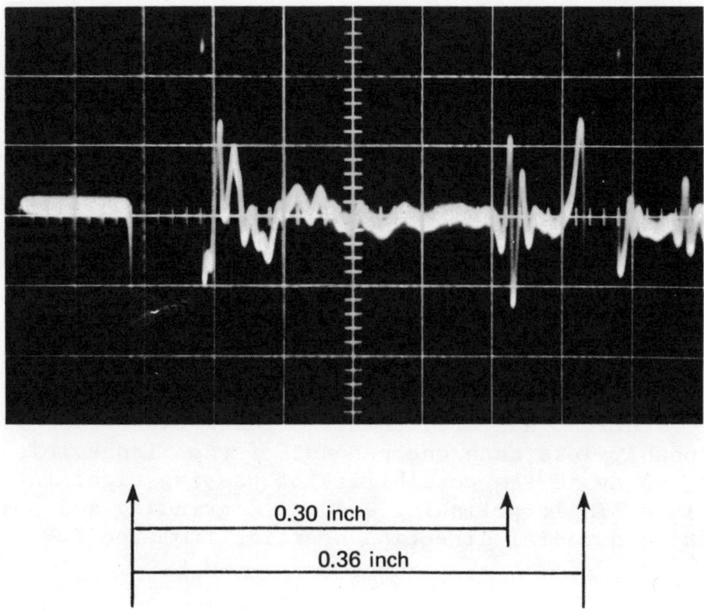

Figure 3. Oscilloscope presentation of ultrasonic signal from defect between holes 4 and 5, gear M819. Vertical deflection factor = 0.05 volt/cm; horizontal sweep = 0.5 μsec/cm.

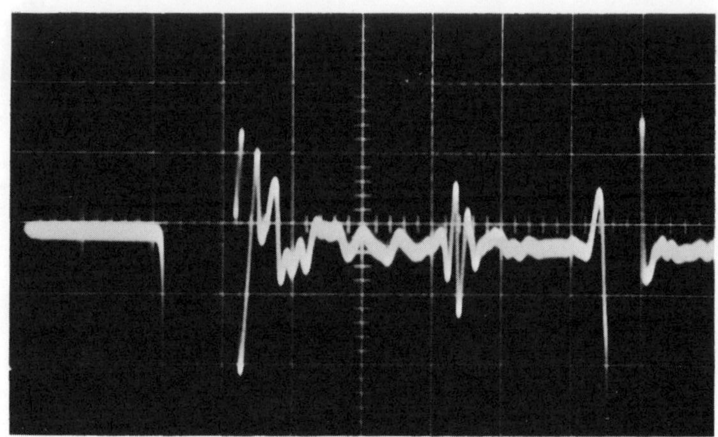

Figure 4. Oscilloscope presentation of ultrasonic signal from defect in gear M826. Vertical deflection factor - 0.05 volt/cm; horizontal sweep = 0.5 μsec/cm.

METALLOGRAPHIC EXAMINATION

In order to evaluate the nature of the inclusions tentatively located by ultrasonics in the flange areas of both gears, specimens were cut from each flange and mounted metallographically perpendicular to the anticipated long dimensions of the inclusion. While the exact dimensions of the inclusion could not be determined by ultrasonics, the most probable orientation of the long dimensions could be specified based on mechanical flow patterns observed in macroetched cross sections. It is perhaps worthwhile to point out that we at first considered approaching each flaw from either the top or bottom surface because we did have a fairly close estimate of the flaw locations relative to those surfaces based on the oscilloscope records. However, the fact that no radiographic image was detected at either location indicated that the thickness of the flaws was small, probably less than one percent of the flange thickness. Therefore, to avoid the possibility of passing right through the evidence in a single polishing sequence, grinding and polishing progressed in a radial direction starting from the I.D. of the flange.

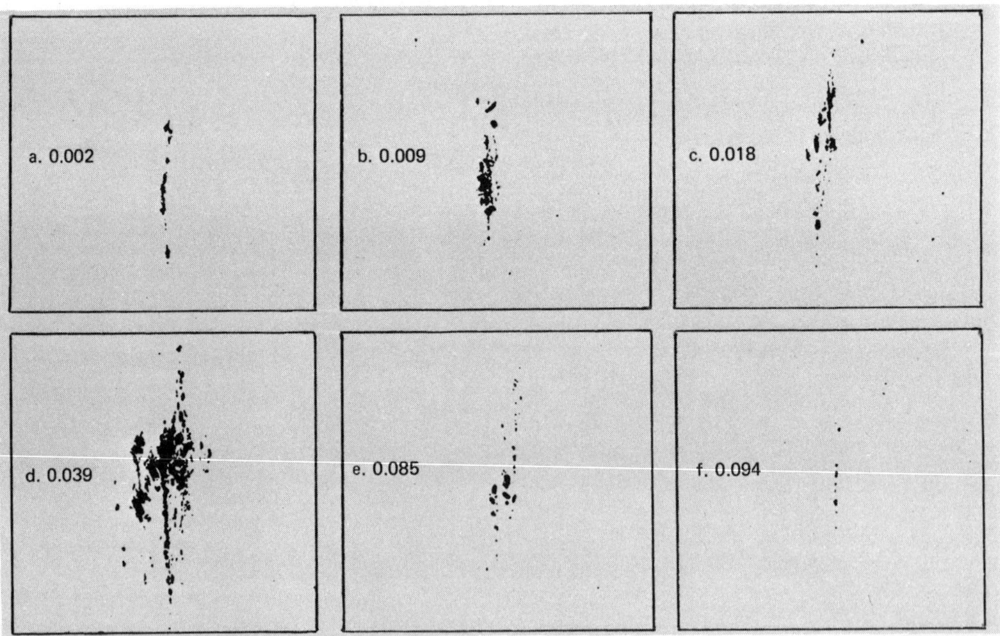

Figure 5. Serial section views of inclusion in M826. Depths in inches from I.D. of flange. (80X, reduced 25% for purposes of reproduction.)

Figure 5 shows selected photomicrographs of the inclusion in gear M826 taken at various locations during the subsequent serial sectioning. Magnification is 80X and the top surface of the flange would be to the left of each view. The numbers on each view represent the radial distance in inches from the inner diameter of the flange. Similarly, the inclusion in gear M819 is illustrated in Figure 6 and has a somewhat different appearance. It is more spread out as indicated by the 50X magnification and, in this case, the top of the flange would be above each view. The maximum depths at which the inclusions were observed were 0.108 inch for the M826 gear and 0.115 inch for the M819 gear. The cross sections of the inclusions for the M826 gear are 0.010 x 0.108 inch and for the M819 gear 0.080 x 0.115 inch where in each case the thickness is a few ten thousandths of an inch. These observations, however, represent dimensions of inclusions which have been flattened out by the forging operation. If we assume that the original inclusion in the forging billet was of cylindrical shape with an aspect ratio of 2 to 1 and, if we further assume that Figure 5b represents an average cross section, we are able to calculate that the dimensions of the original inclusion to be about 0.010 inch in diameter by 0.020 inch in length. These dimensions are far greater than the relevant ASTM specification allows for VAR steel. Since we could not metallographically polish the very surface fo the inner diameter of the flange (our first metallographic observation occurred at a maximum of 0.002 inch below this surface), it is not possible to unequivocally say that either of these inclusions intersected the surface of the flange.

While the morphology and orientation of these inclusions were consistent with the forging operation, there is no way that they could have been isolated metallographically without first having been located by the foregoing ultrasonic scanning procedure. Aside from the one flaw in each gear, the mechanical properties and structure of the examined gears represented good standard metallurgical practice. If both gears had been sectioned at any other location without benefit of prior nondestructive testing, one would have assumed the gears to be free of flaws and of good quality, suitable for the intended application.

This one case history also points out the advantage of using more than one NDT method when examining material of unknown condition. In this case, the flaws were preferentially oriented for ultrasonic detection, but definitely unfavorably oriented for radiography. If, on the other hand, the thin dimension of the inclusions was normal to the flange surfaces, the converse would be true. It is for this reason that quantitative defect size information is difficult to obtain from many NDT methods when defect orientation plays such an important role.

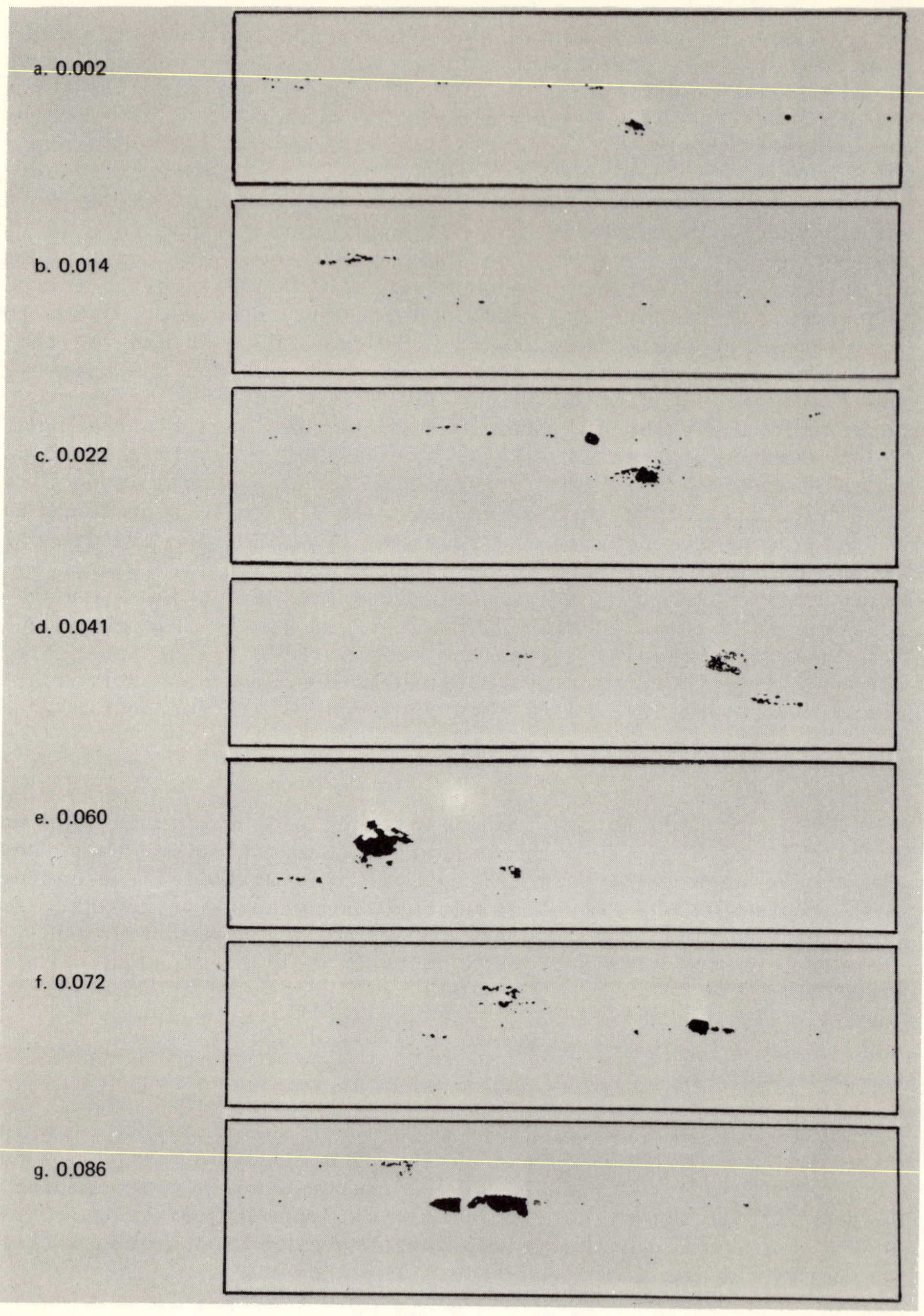

Figure 6. Serial section views of inclusion in M819. Depths in inches from I.D. of flange. 50X

OTHER USEFUL TECHNIQUES

For those who are unfamiliar with nondestructive testing techniques and who are involved with failure analyses or materials characterization, there exists a number of conventional NDT methods which are generally available in most institutions which can be effective for isolating either surface or internal defects [3]. For surface defects, these include magnetic particle, penetrant, eddy current, and ultrasonic surface wave techniques. For internal defects, radiography and ultrasonic pulse-echo, through transmission, and angle beam shear wave techniques are applicable. In addition to defect detection, bulk properties and local variations in properties can be monitored by ultrasonic velocity and attenuation measurements as well as by a number of electromagnetic measurements. There are, of course, other more sophisticated and advanced techniques and a few will be presented in other chapters of this book. All NDT methods have limitations as well as advantages and, therefore, must be used judiciously with a thorough understanding of the material to be inspected.

METHODOLOGY FOR FAILURE PREVENTION

In keeping with the subject of this book, it should be stated that it is the task of NDT to devise techniques for measuring material properties and for detecting flaws. However, as C. H. Hastings has professed on a number of occasions, the task of improving performance, extending life, or guaranteeing reliability has to be a joint undertaking involving the designer and fracture mechanics engineer, the materials engineer concerned with materials characterization and processes, and the NDT engineer who must be involved at the start before specimens are tested or destructive sectioning is performed. To assume that fatigue specimens or fracture toughness specimens are ideal, homogeneous, flaw-free material may be misleading. As Professor Dolan pointed out in Chapter 1, the real materials we make into hardware are variable in properties and all contain defects or imperfections of some size and distribution. Although some defects have no influence on service behavior, they should be considered suspect until proven irrelevant. Knowledge of test specimen condition (variability) or defect location gained through NDT prior to testing may very well account for data scatter experienced in many cases. If an interdisciplinary group effort is able to identify the significant defects which do indeed occur for a new material and design, then the use of NDT which can find these defects is paramount, not only during production fabrication, but for pre-service inspections as well as later in-service inspections for critical components which undergo cyclic loading in service.

ACKNOWLEDGMENT

The case history described in this chapter is the result of a cooperative effort between AMMRC's Metals Research Division and Nondestructive Evaluation Branch which was reported in Reference [2]. Dr. Fopiano, as principal author, conducted the metallurgical analysis which represented the bulk of the effort. Although this chapter highlighted only the NDT phase of the work, details of the metallurgical investigation are described in the referenced report.

REFERENCES

1. Hastings, C.H., "Nondestructive Tests as an Aid to Fracture Prevention Mechanics", Journal of the Franklin Institute, Vol. 290, No. 6, December 1970, pp. 589-98.

2. Fopiano, P.J., Hatch, H.P. and Brockelman, R.H., "Materials Analysis of Two Combiner Gears from a CH-47 Helicopter", Army Materials and Mechanics Research Center, Watertown, Mass., Report No. AMMRC TN 76-7, October 1976.

3. Nondestructive Testing Handbook, Robert C. McMaster, ed., New York: The Ronald Press Company, 1959.

CHAPTER 3

X-RAY DIFFRACTION TECHNIQUES IN ANALYSIS AND PREDICTION OF FAILURE

H. K. Herglotz

Engineering Department, E. I. Du Pont de Nemours and Company
Wilmington, Delaware

ABSTRACT

When it was found that X-ray diffraction (XRD) was capable of unraveling the atomic arrangement in solids, it aroused an enthusiasm which has never dwindled. Crystallographers use the method with success to define the idealized structure, but knowledge about the deviations from this perfection is equally important, particularly to the materials scientist.

What the engineer calls "stress" is know to the materials scientist as a change of lattice dimensions; "failure" means a separation of atoms in the lattice or in the intercrystalline material. The structural status of a solid, as revealed by X-ray diffraction, is therefore capable of measuring stress, testing strength, predicting the time and location of failure, and diagnosing its cause.

In this chapter the underlying concepts of XRD techniques, the commonly used instrumentation, applications, limitations, and pitfalls will be reviewed and illustrated by a few representative examples.

BACKGROUND AND INTRODUCTION

The subject of this chapter interests the materials scientist as well as the engineer, and also the X-ray physicist. Since an X-ray physicist was chosen to write it, the engineer and materials scientist are asked to bear with his approach. It starts from the capability of X-ray diffraction to determine the position of atoms

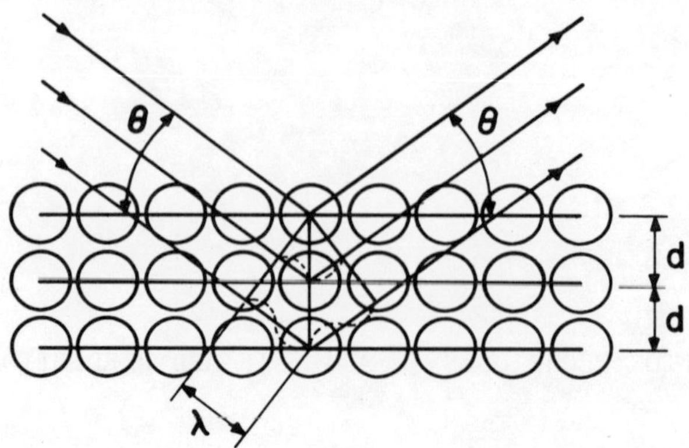

Figure 1. Bragg equation $n\lambda = 2d \sin\Theta$ with n = integer number, d = distance between consecutive "planes", and Θ = Bragg angle.

in crystalline matter, the common state of most technically important materials (metals, ceramics, polymers). The first and still a flourishing application of this capability is structure analysis, the description of the periodic arrangement of atoms, molecules and ions in solids by their coordinates in the unit cell. The first determination of the crystalline structure, a simple one, namely NaCl, we owe to W. H. Bragg, shortly after M. von Laue et al. had demonstrated diffraction of X-rays on crystalline matter. The same Bragg also deserves credit for description of X-ray diffraction effects by the ubiquitous Bragg equation:

$$n\lambda = 2d \sin \Theta \tag{1}$$

(See Figure 1) It describes in a simple and easily derived way the relationship between X-ray wavelength λ, the distance d between two consecutive lattice planes, and the "Bragg angle" Θ. It will accompany us throughout this chapter, since it adequately meets our objectives. It is, however, insufficient for the structure analyst, because it does not reveal the intensities of these "Bragg-reflections".

Since Bragg's first structure analysis of NaCl in 1913, the structures of innumerable compounds with ever-increasing complexity have been unraveled.

Bragg's equation also holds the key which the engineer uses for the analysis and prediction of failure. Any force applied to a solid crystalline material on an atomic scale, as represented by Figure 1, will alter the distance d in this figure. A compressing force will diminish it, a tensive force increase it. The precision

TABLE 1

Polycrystalline Aggregate
Triple Scale Discontinuum

Entities	Observation
o Grains	Visible → Microscopic
o Subgrains Crystallites Coherent Domains	Microscopic → Electron Microscopic
o Atomic Arrangement Molecular	X-Ray Diffraction (not Imaging)

obtained with X-ray methods is adequate the measure the strains in materials due to stresses. But, what appears simple in concept can become rather complicated when more detail is involved. The volume of materials shown in Figure 1 for the derivation of Bragg's equation is only a few hundred $(Å)^3$. A practical analyzing X-ray beam is about one millimeter in diameter and penetrates fractions of a millimeter, representing $\sim 10^{21}$ $Å^3$ of material. But the objects to be analyzed are many orders of magnitude larger! What was true for the small volume of Figure 1 would remain true for our large objects if they were homogeneous, perfect single crystals. However, we work with polycrystalline aggregates as typified somewhat drastically in Figure 2. These aggregates are a triple scale discontinuum described and defined in Table 1.

Any stress applied to such a pile of pebbles will affect the members of this aggregate in a different way. Some crystallites will get the brunt of the force while others will be spared. It becomes worse if one realizes that Bragg's equation is highly selective and that diffraction occurs only from those few crystallites whose proper plane lies at the correct angle Θ. It should be remembered that an infinite number of these fictitious "planes" of the Bragg equation can be drawn in a crystal (Figure 3). Figure 4, together with Figure 3, intends to impress upon the reader the rarity of a diffraction event in the aggregate.

The problem at hand looked simple after Figure 1 and hopeless after Figures 2-4, but it is neither. The next two sections will show that X-ray methods can provide very valuable information.

STRESS MEASUREMENT

Pulling on a wire of cross-sectional area A with a force F is synonymous with applying a stress $\sigma = F/A$ which increases the

Figure 2. Micrographic picture of the etched surface of a metal to demonstrate a "polycrystalline" aggregate.

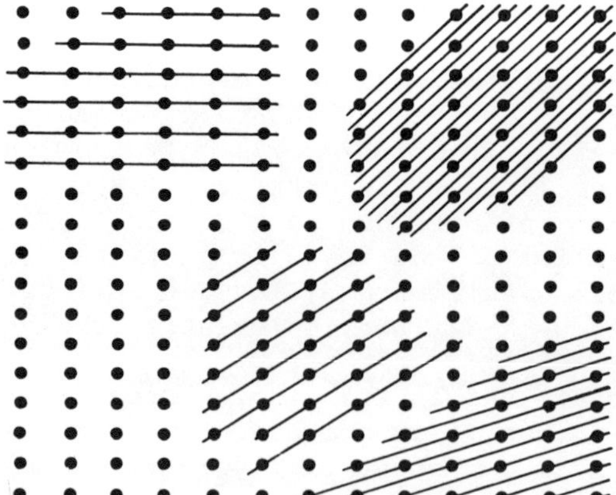

Figure 3. Various "planes" drawn through a planar array of points. The third dimension can be easily imagined.

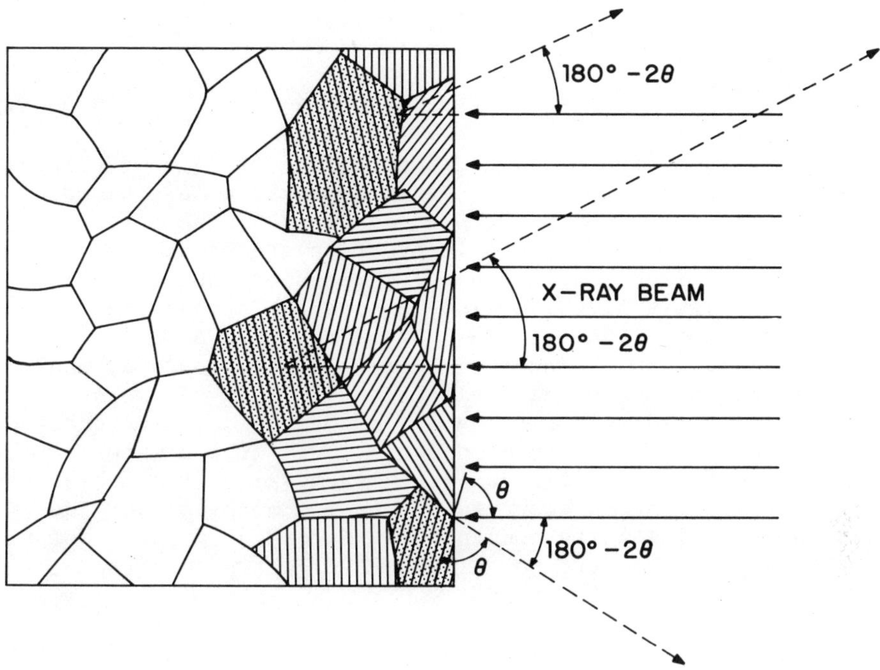

Figure 4. Selectivity of XRD: The plane hkl (symbolized here by the parallel lines within the crystallites) reflects at Bragg angle Θ. Only if a crystallite has the very unlikely position of providing this condition to the X-ray beam does reflection occur (Grey identifies these "privileged" crystallites in this drawing).

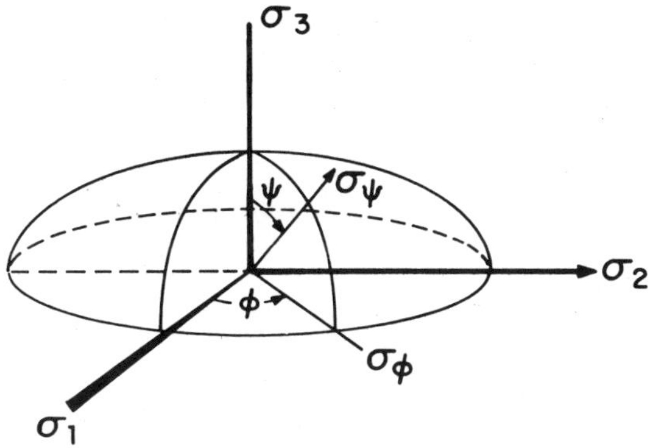

Figure 5. Stress-elipsoid and definition of symbols.

wire's length ℓ by $\Delta\ell$ or in other words, produces a strain $\varepsilon = \Delta\ell/\ell$. Stress and strain are related by Hook's law:

$$\varepsilon = \sigma/E \quad (2)$$

where E is a proportionality constant called Young's modulus. The increase in length causes a contraction of the wire's diameter, and this effect is described by Poisson's number ν, the ratio of this contraction over ε. Few cases in practice are simple and linear as that of the wire. The extended, three-dimensional case is described in Figure 5, which also defines the symbols in the following sections. Any strain $\Delta\ell/\ell$ is equal to $\Delta d/d$ [d of Equation (1)] and in the case of cubic symmetry to $\Delta a/a$, where a is the dimension of the unit cell. Therefore, X-ray methods can measure strains and we can relate them to stresses via Equation (2) which requires the validity of Hooke's law. If the material returns to its stress-free condition after the removal of the stress, an elastic deformation has occurred. [Young's modulus E of Equation (2) is called "elasticity constant" in German.] Plastic deformation in contrast leaves a different, more lasting type of strain, attributed to "residual" stress.

Application of X-ray methods for the measurement of stresses was tried first in the 1930's by R. Glocker and his school [1,2]. There was success in some cases, while others yielded only spurious results. We are already aware of the reason for the on-off success: the composition of the aggregate would be intractable to measurement if it looked like that of Figure 2, while a fine-crystalline, quasi-homogeneous aggregate would be amenable to analysis.

The 1960's brough a kind of renaissance to the field. There was better instrumentation, but equally important were more refined data extraction and interpretation techniques, such as those described in references [3,4] and summarized in Figure 6 (using the symbols defined in Figure 5). Measuring $\varepsilon_{\phi\psi}$ for various angles of ϕ and ψ, then plotting ε (at constant ϕ) vs. $\sin^2 \psi$ results in a linear function for each constant value of ϕ. The slope m* is related to σ_ϕ by

$$m^* = (S_2/2)\, \sigma_\phi \quad (3)$$

where $S_2/2$, the so-called Voight-abbreviation, equals

$$S_2/2 = (\nu + 1)/E \quad (4)$$

The large number of single measurements, made possible by faster experimental methods and the avilability of rapid computational tools for data acquisition and reduction (e.g., smoothing, least square refinement) have made the method applicable where earlier, more primitive attempts failed. We will have to say a

$$\varepsilon_\psi = s_2/2\ \sigma \sin^2\psi + s_1\sigma$$
$$= \left(\frac{\Delta d}{d}\right)_\psi = \frac{d_\psi - d_o}{d_o}$$

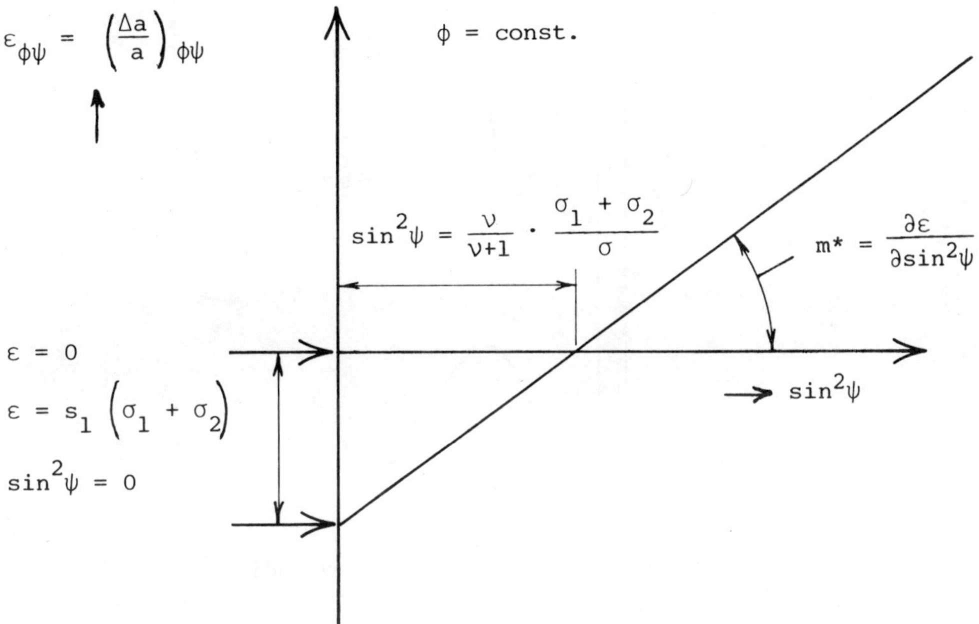

Figure 6. ε vs. $\sin^2\psi$ - Diagram [3].

little more about instruments under 'Experimental Methods'.

In spite of instrumental advances and computational refinement, results of the type presented in Figure 7b, c are encountered. They are not a consequence of the method's shortcoming but are inherent in the sample. After plastic deformation of metals, residual strains are distributed nonuniformly and the X-ray measurement with its selectivity cannot be expected to arrive at a smooth function $\varepsilon = f(\sin^2\psi)$.

Let us look at this previously mentioned selectivity with a more-than-casual eye. The spherical projection was chosen for this glance because it is very illuminating and needs less abstraction than the stereographic projection which is derived from it and is frequently used by metallographers. Figure 8 illustrates the concept of the spherical projection on the octahedral planes (111) in the cubic system. Each plane is represented by the point on the

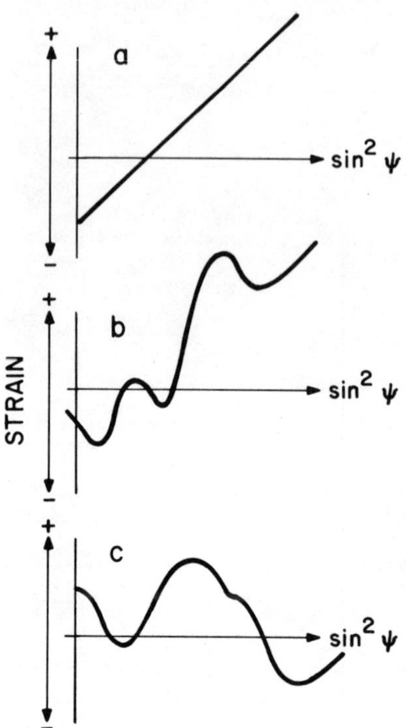

Figure 7. "Healthy" (ε vs. $\sin^2\psi$) - Diagram indicating elastic deformation (a); "crooked" diagrams b, c, indicate stresses of second kind, varying over short distances, as usually happens with plastic deformation [3].

sphere's surface generated by the plane's normal vector. The spherical projection of any crystallographic plane in a random polycrystalline aggregate (without preferred orientation which can occur in rolling or drawing operations) covers the sphere's surface homogeneously as indicated in Figure 9. For reasons of clarity, the points of Figure 8 are replaced by small circles.

To be in diffracting position for an X-ray beam x_o, the vector of any of the innumerable planes (h k l) with Bragg angle Θ_{hkl} has to form an angle (90° - Θ_{hkl}) with x_o. All vectors fulfilling this condition form a cone with an apex angle (180°-2Θ) and leave a circular trace on the sphere's surface, as evident from Figure 9. Changing the angle ψ of Figures 5 and 6 means moving x_o with the (180°-2Θ) and the x_d-cones coupled to it, through various angles along one of the meridians of the sphere (because ϕ = const.). The more angles ψ one chooses, the more reliable will be the plot of Figure 6.

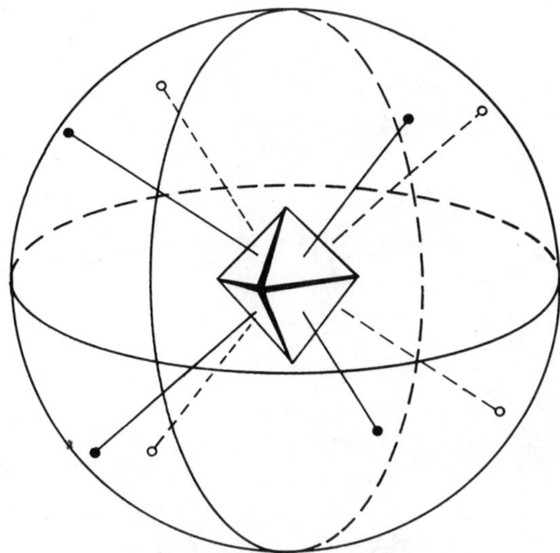

Figure 8. Spherical projection of the octahedral planes of a crystal of the cubic system.

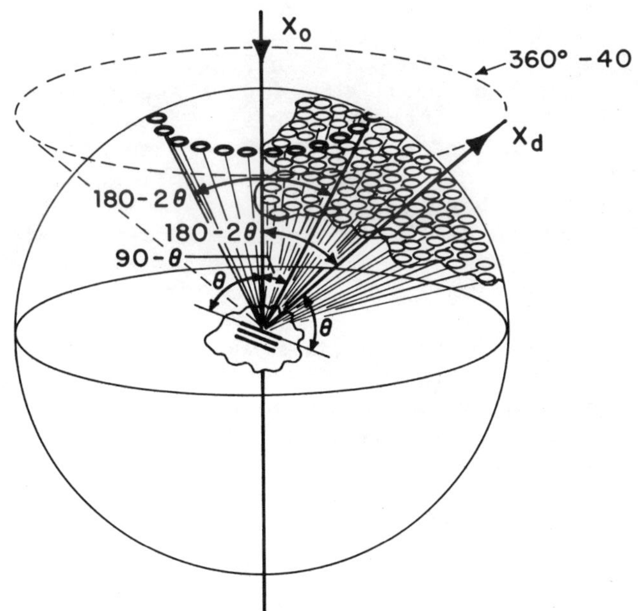

Figure 9. Spherical projection of a polycrystalline aggregate with random arrangement of the plane which reflects at Θ; x_o x = incident and reflected X-ray directions.

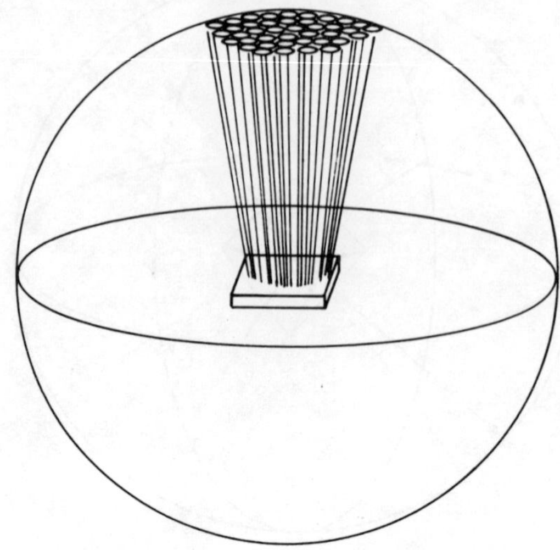

Figure 10. Spherical projection of a plane with preferred orientation nearly parallel to the surface of the sample.

The results of Figure 7 b, c are now better understandable if one applies the thoughts condensed in Figure 9 to materials consisting of coarse grain, where sometimes these grains are themselves aggregates of much smaller crystallites. The situation can become even worse if materials carry a preferred orientation. Figure 10 gives an example for a plane that has been oriented, e.g., by a rolling operation so that it is nearly parallel to the surface of the rolled sheet, such as the (001)-plane in α-Fe. Stepping through varying ψ-angles in such a case is rather restricted, because of the paucity of planes away from the pole.

In spite of its limitations, X-ray stress measurement can be a very important diagnostic tool if applied discriminately and with full awareness of what it can and cannot do. It is, however, by no means the only X-ray method that can contribute to the analysis and prediction of failure.

OTHER X-RAY MEASUREMENTS

Table 2 presents a summary of all observables by X-ray diffraction (XRD) that are useful to our subject.

1. Line Shift: We have dealt at length with the first, the shift of the Bragg angle $\Delta\Theta$ due to elastic stress which is homogeneous over the observed volume (often called "of the first kind")

TABLE 2

Effects Measurable by XRD

Observed Effect	Cause	Measured Variable
1. Line Shift	Elastic Stress (of 1st kind)	$\partial\Theta = f(d)$
2. Line Broadening (Radial Breadth)	Plastic Deformation Stresses of 2nd kind	$FWHM = f(\partial d)$
3. Peripheral Broadening	Crystallite Displacement	$\partial\delta = f(\partial\rho)$
4. Crystallite Fragmentation	-	-

2. Line Broadening: If stresses of the 2nd kind, varying over short distances, are present, the multitude of Bragg angles from the sample leads to a broadening of the XRD-line.

   Figure 11 presents an example [5]. To ensure a statistically significant line width FWHM (Full Width at Half Maximum, defined in Figure 12), it might be necessary to homogenize the line in samples with coarse grain by scanning during exposure. The fact that the FWHM is functionally related to the degree of deformation is shown in Figure 13. A line width close to the maximum value measured in a workpiece would indicate an impending failure by fracture.

3. Peripheral Broadening of Diffraction Spots: For interpretation and utilization of the peripheral width we have to abandon the much-quoted Bragg equation and resort to the Polanyi formula, which connects the angles ρ, δ and Θ by spherical geometry:

$$\cos\rho = \cos\delta\cos\Theta \tag{5}$$

   Figure 14 defines the symbols. Any displacement, rotation, etc., of crystallites in samples which fall into the range of crystallite sizes that produce spots will broaden these spots in peripheral direction (Figure 15) according to the derivation of (5) at constant Θ:

$$(\partial\delta/\partial\rho)_{\Theta = \text{const.}} = \sin\rho/\cos\Theta\sin\delta \tag{6}$$

   Measuring the peripheral width of a statistically significant number of spots on a diffraction line and knowing the original width before deformation, wear, etc., can become a sensitive indicator and predict failure [5, 6 p. 332]. Recent application of these concepts is found in the work of S. Taira and collaborators [7].

4. Crystallite Fragmentation: The last item on Table 2, cyrstallite fragmentation, is experienced in the case of heavy plastic deformation or after many cycles of pulsating load in the

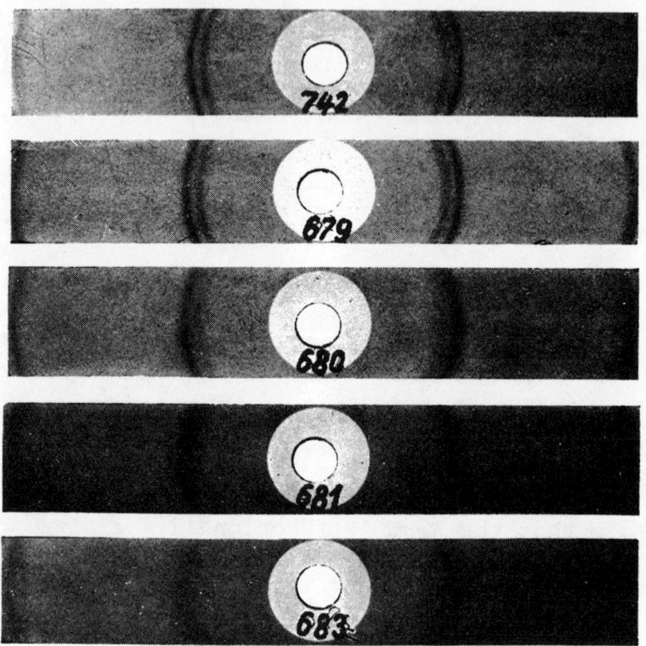

Figure 11. Changing line width (FWHM, Figure 12) of aluminum due to increasing plastic deformation (Cu-radiation).

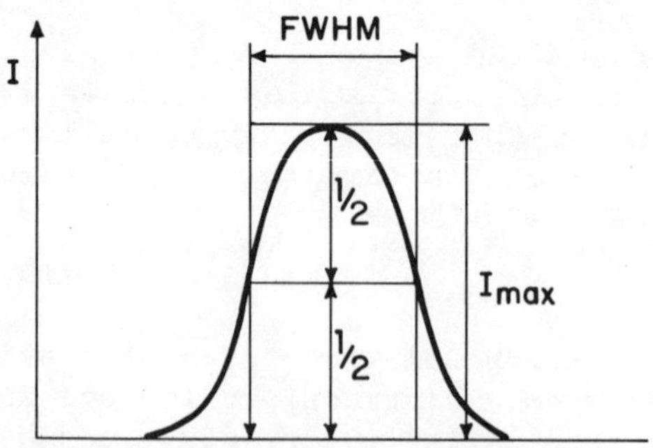

Figure 12. Definition of FWHM (Full Width at Half Maximum).

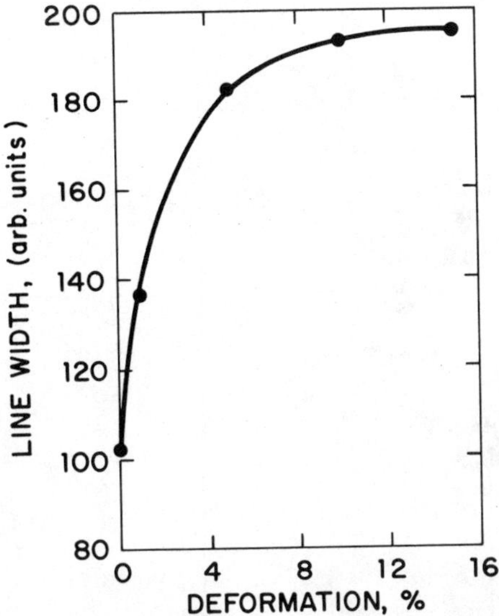

Figure 13. FWHM (in arbitrary units) vs. degree of deformation.

elastic regime. The X-ray pattern reveals the disappearance of individual diffraction spots in the diffractogram of the original material and their replacement by continuous diffraction rings. Since this effect has been observed in samples that were never stressed beyond their elastic limit, it testifies to the uneven distribution of effects in the aggregate [8].

EXPERIMENTAL METHODS

Because XRD methods should be nondestructive to extended objects, back-reflection methods with $\Theta > 45°$ are a necessity. They are also a virtue since the sensitivity to lattice dimensions increases with Θ according to

$$(\partial\Theta/\partial d)_{\lambda\,=\,\mathrm{const.}} = -\tan\Theta\ /d \tag{7}$$

This formula is the partial derivative at constant λ of the Bragg equation. To find a suitable plane hkl and a λ^* (* = by choice of the X-ray tube) to produce a Bragg angle close to or larger than $\Theta = 80°$ is therefore the first step to any application. A wavelength that excites strong X-ray fluorescence in the sample, as, e.g., copper radiation in iron, it to be avoided. Figure 16 shows the distorted back reflection cone from a sample under stress,

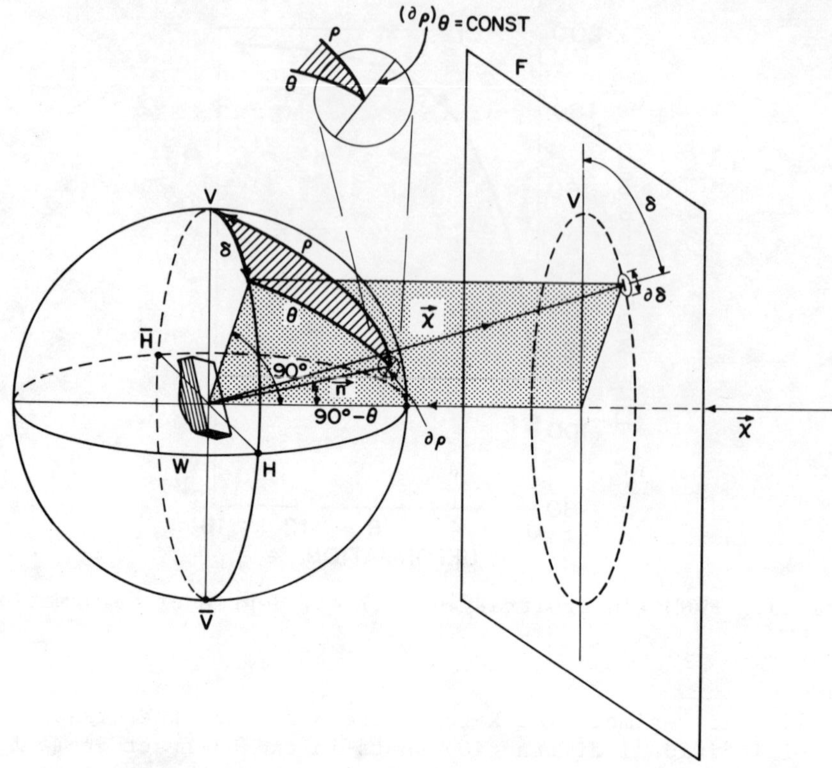

Figure 14. Definition of the Polanyi formula for back-reflection and for the spread $\partial\rho$ with consequent spread $\partial\delta$.

V-$\bar{V}$, H-$\bar{H}$	vertical and horizontal direction (references only; no crystallographic significance)
$\vec{x}^{\circ}$, $\vec{x}$	incident and diffracted X-ray directions
W	workpiece (diffraction planes indicated)
$\vec{n}$	plane vector of diffracting plane
Θ	Bragg angle
ρ	angle between plane vector n and vertical direction V-$\bar{V}$
$\partial\rho$	spread of plane vector due to deformation
δ	angle between vertical on X-ray pattern and diffraction spot; appears also in rectangular spherical triangle $\Theta\rho\delta$
$\partial\delta$	spread of δ due to $\partial\rho$
F	film
	The plane in which $\vec{x}_o$ $\vec{n}$ and $\vec{x}$ lie is marked by graytone.

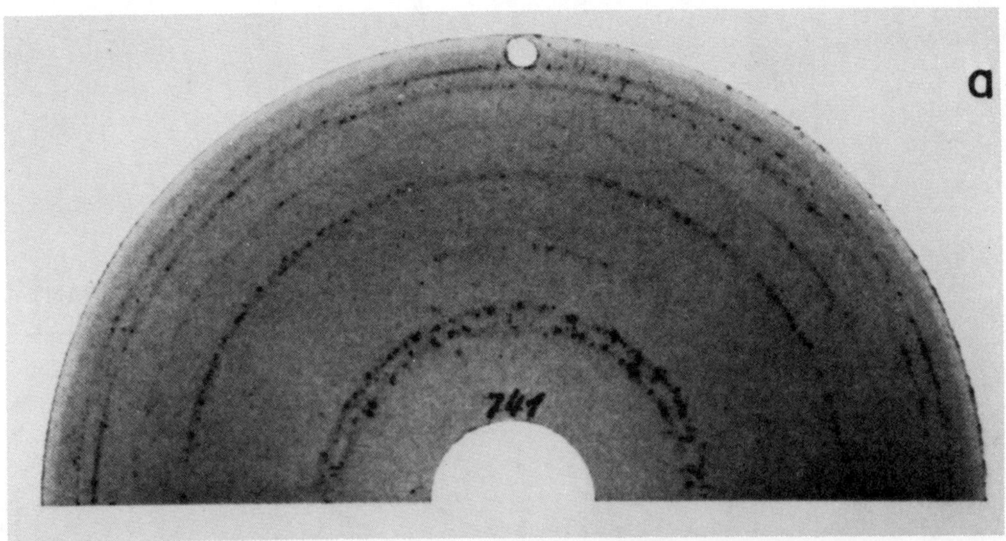

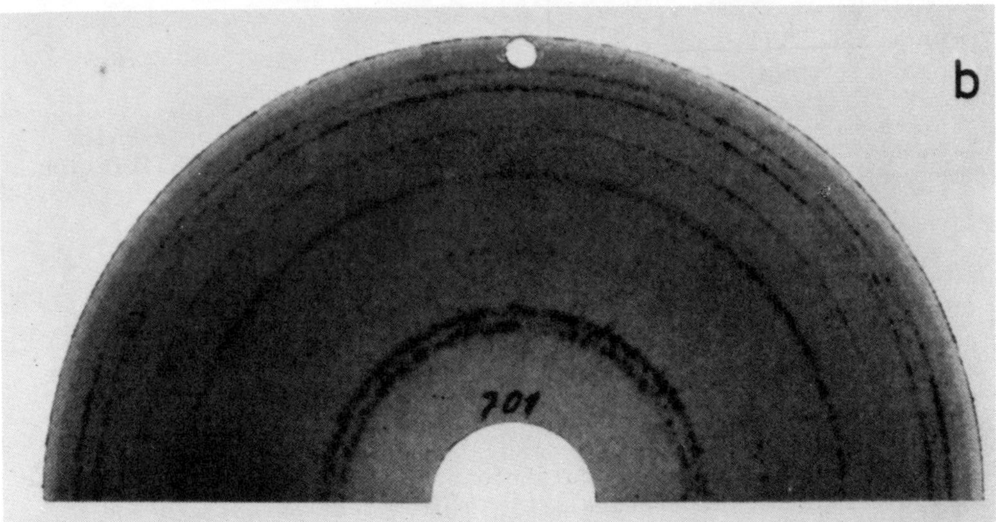

Figure 15. Demonstration of the peripheral broadening (along the reflection circle at constant Θ) due to deformation in aluminum. (a) undeformed sample; (b) after 1% permanent stretching.

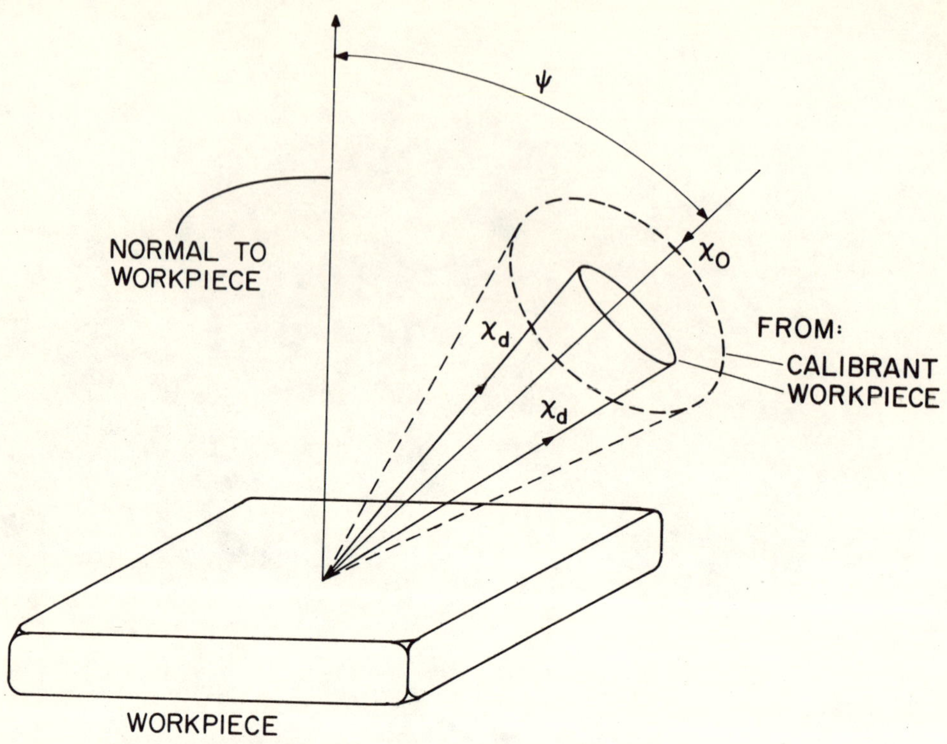

Figure 16. Distorted back-reflection cone from a sample under stress; undistorted cone (dashed) from a calibration material (e.g., gold-powder).

together with the undistorted cone from a calibration material which is applied to the surface. The calibrant has to have a Θ close to the tested material, be stress-free, and inert. Gold powder is frequently used.

Recording and data acquisition can be accomplished by photographic film or electronically. The inherent advantages and disadvantages of the two modes are shared with other test methods. To record the diffraction cones of Figure 16 photographically, many forms of film and camera arrangements are used, and some of them are shown in Figure 17 [6, p. 328]. All of these methods are focusing: a divergent X-ray beam probes larger sample volumes than a collimated beam and shortens exposure times without loss of definition of the recorded X-ray line. A goniometer recording with electronic readout is exhibited in Figure 18 (see also Ref. [9]). An example of a modern, mobile stress measuring instrument is

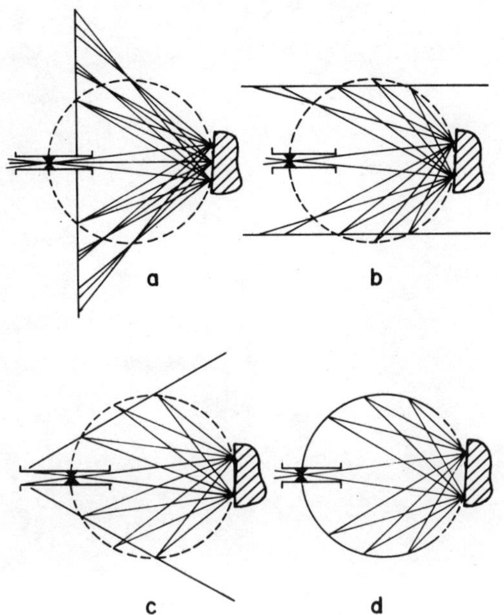

Figure 17. Four examples of film shape for nondestructive recording of focusing back-reflection patterns. (a) plane, (b) cylinder, (c) cone, (d) ring (circular)

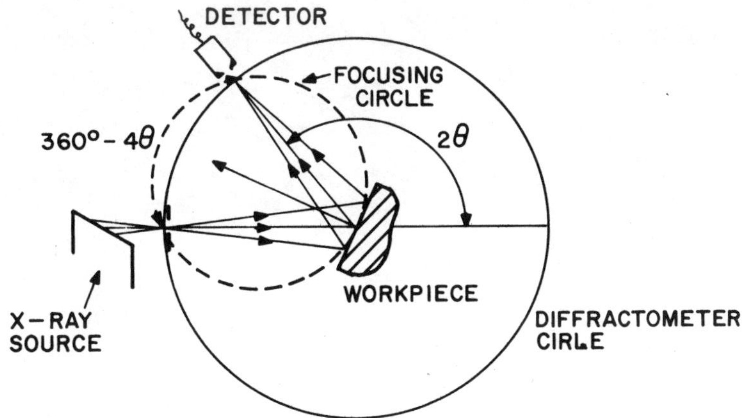

Figure 18. Scheme of goniometer with electronic detector serving the same purpose as the film devices of Figure 17.

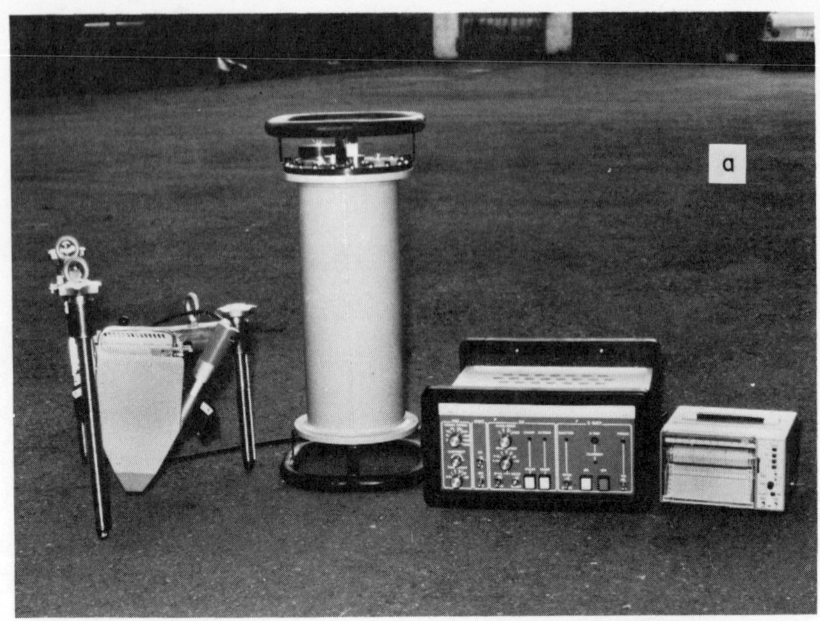

Figure 19. Modern commercial equipment of "In-Situ" stress measurements. (a) Components: tripod, X-ray generator, electronics, X-Y recorder. (b) In use in a pipeline.

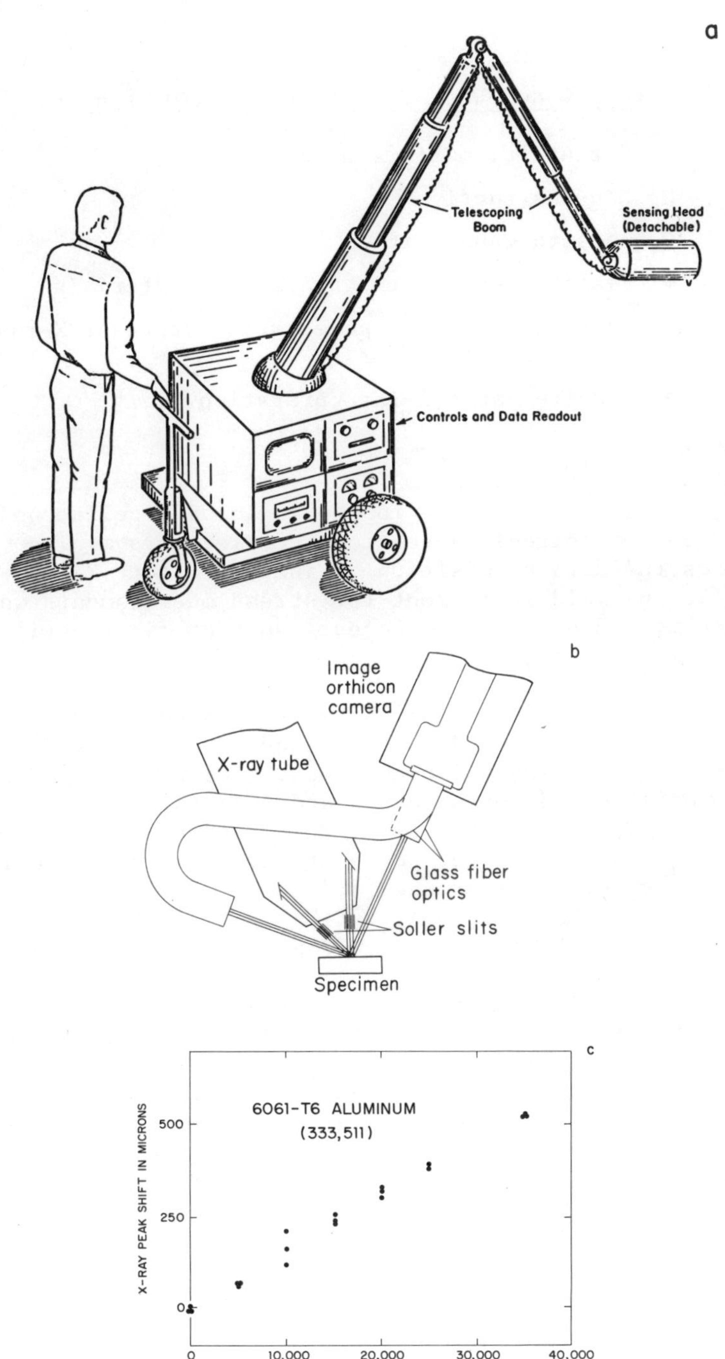

Figure 20. Advanced instrumentation for residual stress measurement [10]. (a) Overall view. (b) Sensing head. (c) Example of readout.

TABLE 3

Strong O and Weak ● Features of XRD Methods

- O Versatile, multifaceted
- O Non-destructive
- O In situ capabilities
- ● Feasible only with crystalline materials
- ● Valid only within penetration depth of X-rays (λ-dependent)
- ● Require careful interpretation

presented in Figure 19, while in use inside a pipeline under construction. Electronic detection, coupled to a mini-computer, streamlines the data acquisition. Finally, Figure 20 is an example of a mobile in-field instrument for stress measurements which combines some unconventional concepts such as fiber optics and image intensifier [10].

For the sake of thoroughness, another group of diffraction methods should be mentioned which are not used in the field, but in research: topographical techniques. Their imaging capabilities make them well suited for stresses and defects in semi-conductor chips in the electronic industry. To elaborate on these methods, however, would be beyond the scope of this chapter. The interested reader is referred to Reference [11].

CONCLUSIONS

It is hoped that previous sections have left the reader with the impression that XRD methods play a definite role in the analysis and prediction of failure. They are a powerful tool for gaining unique insight into structural changes caused by stress, fatigue and wear, but they are neither simple nor unambiguous and require expertise for proper application and interpretation. Table 3 summarizes advantages and weaknesses.

XRD methods have more often supplemented than replaced other methods of nondestructive testing. In spite of the shortcomings, XRD techniques are indispensable to materials science and to the engineer where critical decisions have to be made.

REFERENCES

1. Glocker, R., Hess, B. and Schaber, O., Zeitschrift f. techn. Phys., 19, 1938, p. 194.

2. Glocker, R., Materialprüfung Mit Röntgenstrahlen, 4th edition, 1958.

3. Macherauch, E. and Müller, P., Zeitschrift für angewandte Physik, 13, 1961, pp. 305-12.

4. Macherauch, E., Proc. Soc. Exp. Mechanics, 23, 1966, p. 140.

5. Nachtigall, E., Regler, F. and Herglotz, H., Verformung von Aluminumwerkstoffen. Aluminum Ranshofen Mitteilungen, 2, January 1955.

6. Regler, F., Einführung in Die Physik der Röntgen-und Gammastrahlen. Verlag Karl Thiemig, Munich, 1967.

7. Taira, S. and Kamachi, K., Detection of Fatigue Damage by X-rays, Proceedings of the 23rd Sagamore Conference.

8. Herglotz, H.K., A Historical Example of Fatigue Damage, Proceedings of the 23rd Sagamore Conference.

9. Azaroff, L.V., Elements of X-Ray Crystallography, McGraw-Hill, New York, 1968

10. Ruud, C.O. and Farmer, G.D., An Advanced X-Ray Instrument for Residual Stress Measurements, Proceedings of 23rd Sagamore Conference.

11. Weissmann, S., Topography, Proceedings of 23rd Sagamore Conference.

CHAPTER 4

HOLOGRAPHY FOR DEFECT DETECTION ON ARTILLERY PROJECTILES

Paul J. Kisatsky

US Army Armament Research and Development Command

Dover, New Jersey

INTRODUCTION

A matter of grave concern in the production of artillery projectiles is the presence of cracks in the steel projectile body. During rough handling and active usage by the troops, small sub-critical cracks can grow to critical dimensions. Ultimately, the high stresses induced in the shell by the set back forces of gun launch can cause the cracks to completely fracture the shell while still in-bore. An in-bore premature is extremely serious, resulting in destruction of the weapon and possible loss of life.

Inspecting for cracked projectiles is now accomplished by two conventional methods; magnetic particle penetrant or ultrasonic pulse echo testing. The first of these has the advantage of visual defect observation but it does not distinguish between critical cracks and cosmetic blemishes. Further, being visually inspected, the process is subject to operator fatigue and error resulting in critically defective shells often being accepted. In comparison, ultrasonic echo testing lends itself to automation but is highly dependent on proper use of "standards" to adjust the accept-reject pulse threshold margins. It requires water immersion of the part, is fallible to false echos, and is intrinsically a nonimaging method requiring elaborate scanning and multiple channel transducer configurations.

As part of the R & D effort in holographic nondestructive inspection (NDI) in our laboratory, we initiated a series of experiments to ascertain the feasibility of employing holographic interferometry for cracks in projectiles. When feasible and practical, holographic NDI offers certain intrinsic advantages while doing

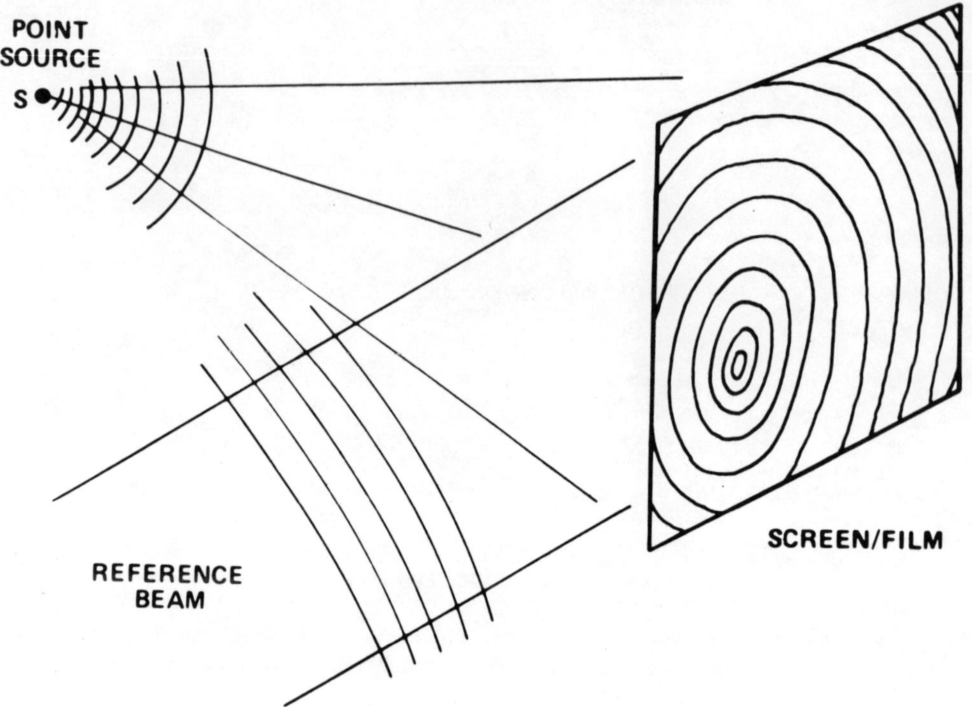

Figure 1. Constructing an elementary hologram.

away with some of the main disadvantages of magnetic particle or ultrasonic NDI. The artillery shells worked with and currently under investigation are the 105mm HEP M393, the 155MM RAP, M549 and the 155MM, GPC, M483.

PRINCIPLES OF HOLOGRAPHIC INTERFEROMETRY

To better understand the experiments described in this chapter, the basic principles of nondestructive inspection by holographic interferometry are reviewed here. The construction of an elementary hologram is shown in Figure 1. If we simultaneously illuminate a screen with coherent radiation from a point source and that from a columated beam, later known as the reference wave, it will be noticed that a series of fringes will form on the screen due to the interference of the two waves. These fringe patterns are shown in a highly exaggerated form in Figure 1. If the screen is replaced with photographic film which is exposed and developed into a transparency, the fringe pattern formed by the source and the reference beam will be recorded. If the developed film is then placed back into its original position and illuminated with the same reference beam in the absence of the original point source, upon looking through the

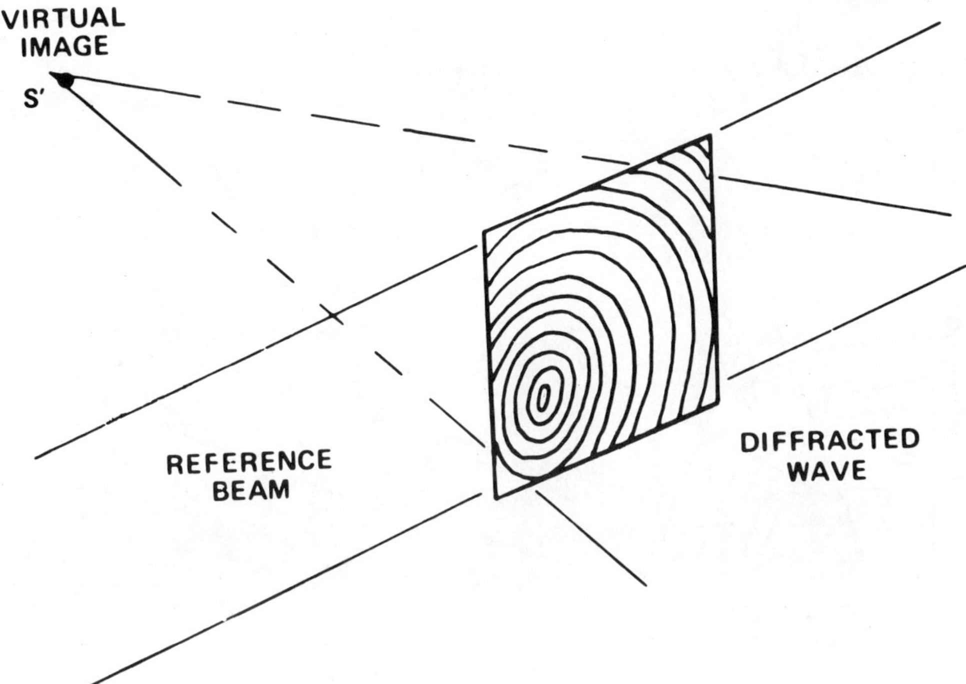

Figure 2. Reconstruction of elementary wave from a hologram.

hologram one would see an apparent virtual image of the point. This is called reconstruction and is shown in Figure 2. It is well known that the fringe pattern formed by the interference of two coherent waves contains all the information of both of the original waves in terms of recording their amplitudes and phases. It is a property of the laws of diffraction that when such a hologram is illuminated with one of the two waves, the light will pass through the transparency and will be diffracted in such a way that a replica of the second wave will automatically be created. The eye therefore sees an image of the point source from its reconstructed wave. With the above explanation of holograming a point source, one can easily extend the concept to making a hologram of a full, three-dimensional object. Since an object can be considered to be a distribution of point sources, then if all the points are simultaneously recorded on the hologram, upon illuminating the hologram with the reference beam, a real three-dimensional virtual image of the full object is re-created (Figure 3). In all holographic arrangements, a laser generated coherent light source is split, one portion illuminating an object and a second portion serving as a reference beam. The light scattered from the object falls onto the holographic plate simultaneously with the reference beam creating the fringe pattern, or hologram

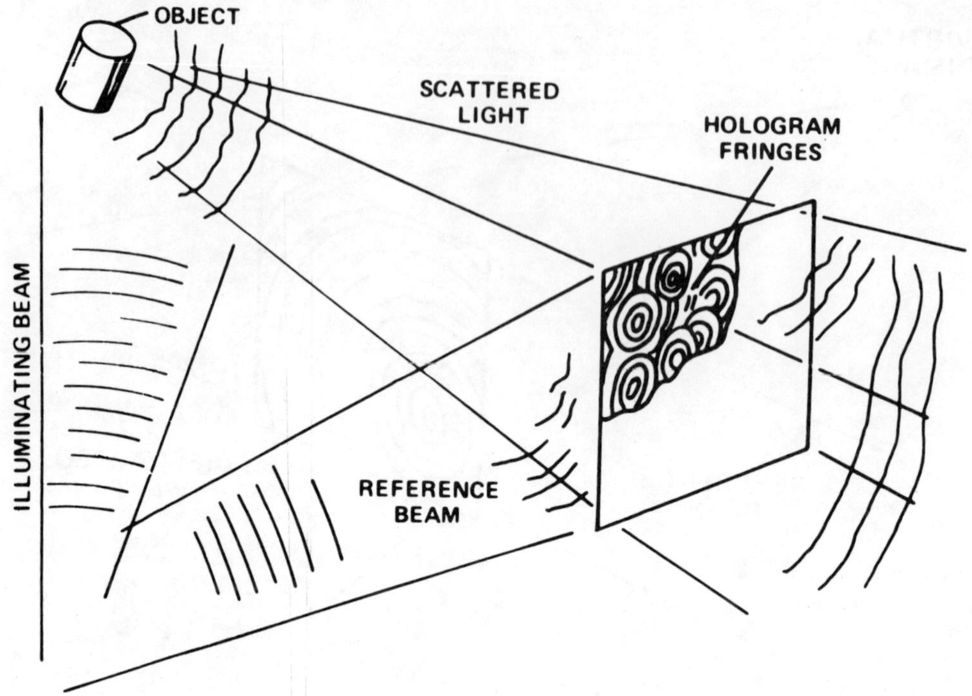

Figure 3. Hologram of a three-dimensional object.

Figure 4 demonstrates how a hologram of an object can be used in nondestructive testing. Since the hologram "remembers" the exact form of the scattered light from an object, it is possible to observe small differential changes due to displacement of the object surface. When small stresses result in surface movement such as in heat expansion, pressure distension, elastic deformation, etc., a slight misregistration occurs between the real object and its image. The object wave then interfers with its own "remembered" image wave, resulting in another set of fringes easily visible to the eye. This method is often known as time-lapsed interferometry. This can be done in several ways but the method employed here is that of double exposure holography, in which a hologram is made of the projectile in the undeformed state, and a second hologram of a deformed projectile is then made on the same plate by double exposure. The fringes seen are the result of interference between the two projectile states. In holographic interferometry surface displacements on the order of a wavelength of light (approximately 12 microinches) are easily seen anywhere over the object surface.

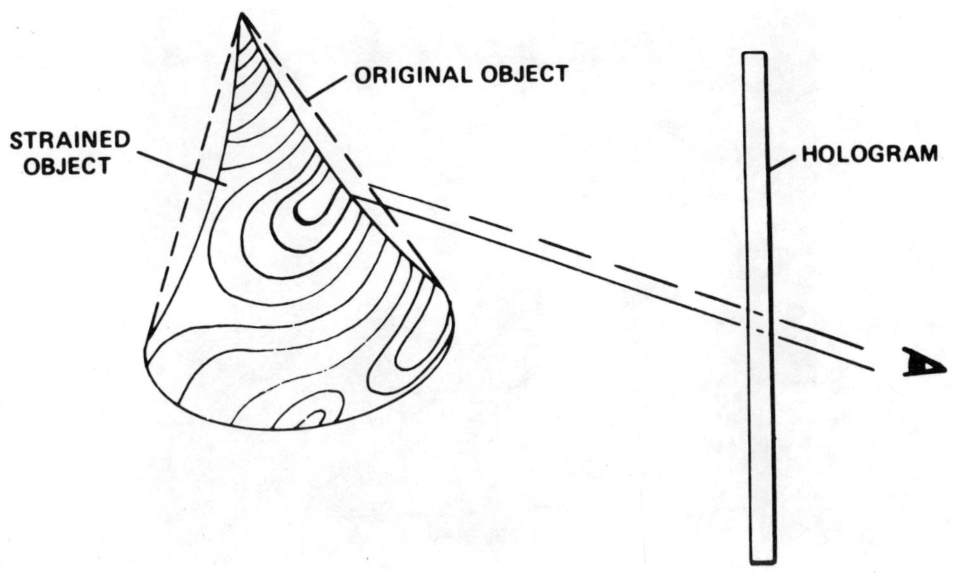

Figure 4. Measuring distortion by holographic interferometry.

EXPERIMENTAL RESULTS

As described above, defects are often seen in holographic interferometry by virtue of the fact that they sometimes cause anomalously large or small surface displacement in their vicinity when the item is stressed. Finding the appropriate stressing mechanism is part of the holographic investigation. The ideal stress maximizes the unusual behavior of the defect region but minimizes gross and rigid body motions (translations and rotations) which can produce trivial fringes which in turn can mask or be confused with the defect induced fringes.

Since a projectile is a hollow cavity, a natural form of stress is applying small internal air pressure. Pressure, being balanced in nature, tends not to shift the overall body center of mass. Yet even mild pressures will cause an outward ballooning of the relatively heavy steel projectiles. The experiments described here display the holographic fringe characteristics of both normal and cracked projectile bodies when stressed with mild internal air pressure, from ten to two hundred PSI.

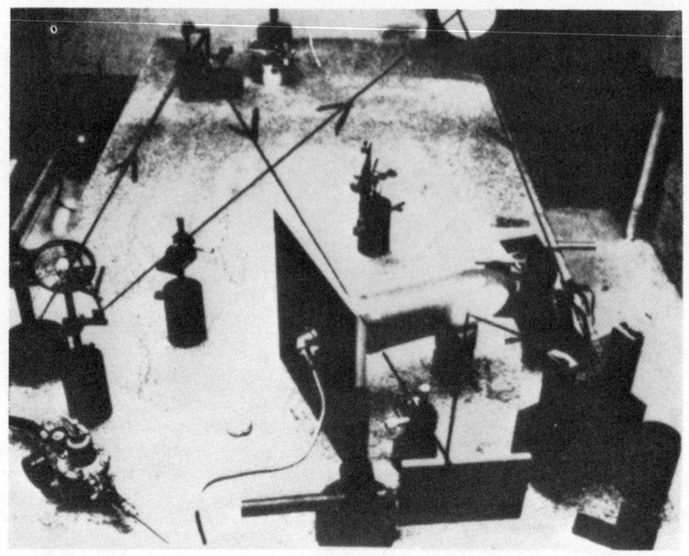

Figure 5. Laboratory arrangement for holographic interferometry with pressure stressing.

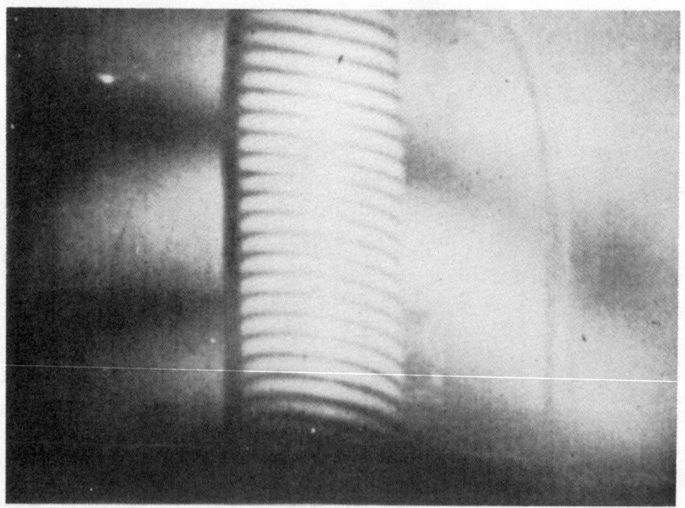

Figure 6. Expansion fringes with 55 PSI pressure.

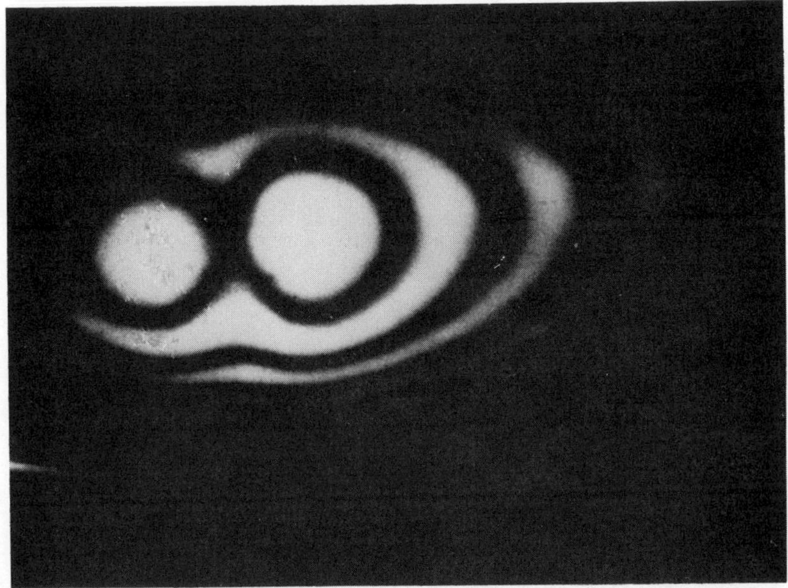

Figure 7. Expansion fringes in ogive area.

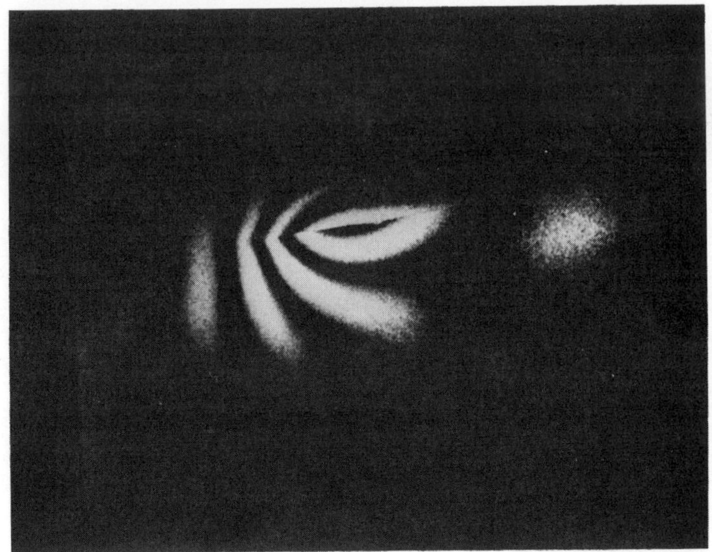

Figure 8. Anomalous fringes around crack at 10 PSI.

Figure 9. Anomalous fringes around crack at 20 PSI.

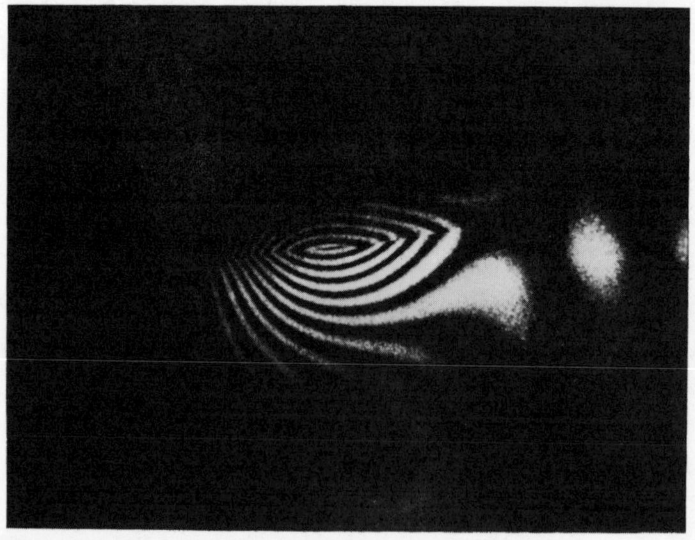

Figure 10. Anomalous fringes around crack at 40 PSI.

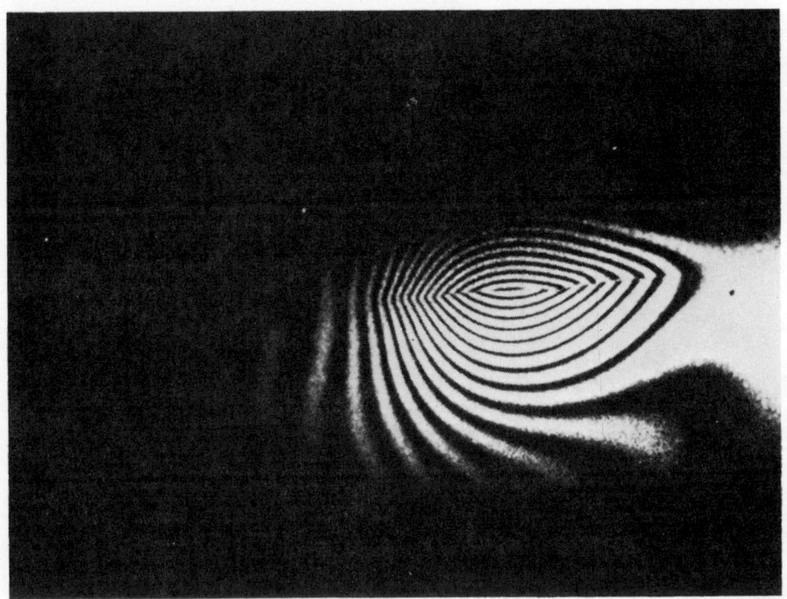

Figure 11. Anomalous fringes around crack at 70 PSI.

The experimental apparatus with projectile visible, is shown in Figure 5. The typical fringe pattern resulting from applying 55 PSI air pressure to a normal shell is shown in Figure 6. The fringes seen are relatively smooth, approximately periodic, and lie on the cylindrical portion of the shell body. In Figure 7 the fringes seen are those that fall around the nose of ogive area of the shell. Two figures are used here to display the entire fringe pattern but in the actual holographic arrangement the entire shell can be seen in one view. The nature of the fringes in the ogive are due to the change in shape and structural properties of the shell in its nose area.

A shell containing a very tight natural crack was then hologramed. Figure 8 indicates the fringe pattern that is seen in the vicinity of the crack. This pattern was produced with a pressure differential of only 10 PSI. Figures 9, 10 and 11 display the fringe pattern about the crack as the pressure in the shell was increased to 20 PSI, 40 PSI, and 70 PSI, respectively. The anomalous fringe patterns formed in the vicinity of the crack are due to anomalous displacement of the adjacent surface area where the structural integrity of the steel was weakened due to the presence of the defect.

Encouraged by these results, a series of artificial defects was introduced by "elox" cutting. A typical fringe pattern for

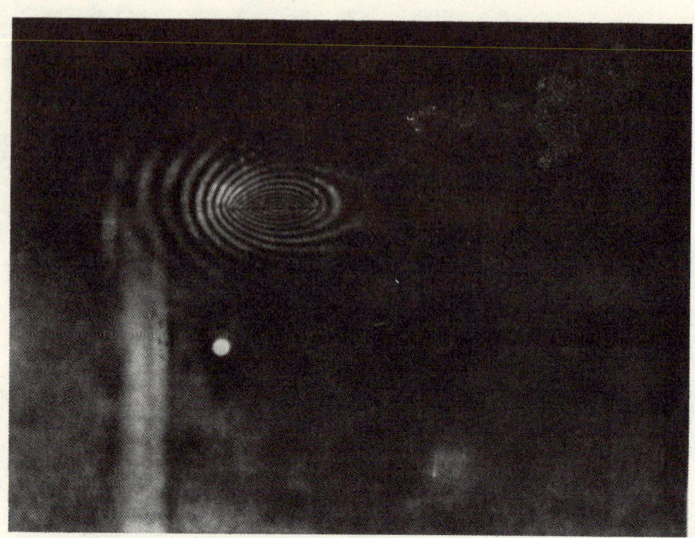

Figure 12. Fringes around artificial crack.

such an artificial crack is shown in Figure 12. In every case where the defect extended a significant portion of the way through the shell body, the crack was easily observable by the type of holographic fringe display shown here. The feasibility of detecting cracked shell bodies by means of holographic interferometry has been clearly demonstrated in this series of experiments.

ADVANTAGES OF HOLOGRAPHIC NDI

When compared with other means of NDI, holographic pressure inspecting of projectiles offers certain advantages and disadvantages as reviewed in Figure 13 which summarizes some of the major advantages.

Holography is a nonscanning, noncontact optical technique and, therefore, does not require water immersion of the part or the complexity of scanning transduces as in ultrasonics. It enables one or possibly many shells to be seen simultaneously while displaying defects anywhere over the shell surface.

Holography, if done on film, can yield a permanent record of the test. This may be of value when keeping test records of large quantities of produced shells in order to track field performance with shell quality. If malfunction occurs from a known lot it is sometimes possible to correlate the failure with the test records from its production.

HOLOGRAPHIC NONDESTRUCTIVE INSPECTION

* Non Contact Optical Technique
* Large Area Inspection (Zero Scan Time)
* Permanent Record
* Picture Readout - Common Sense Interpretation
* Independent of Defect Orientation
* Ignores Cosmetic Surface Blemishes
* Sensitivity of Technique is Proportional to Defect Severity

Figure 13. Advantages of holographic NDI.

An advantage of holographic interferometry not to be overlooked is that it yields a pictorial readout of a shell which is more easily interpreted than transducer signals. In this sense holography is somewhat like radiography since a picture of the item appeals to that which the human is familiar. Pictorial display of the data can be considered to be both an advantage and a disadvantage in some respects. While there is no argument that an image conveys much more information and lends itself to common sense interpretation compared to analog channels of transducers, there are those who consider that visual inspection requires a person and is therefore already at a disadvantage. The desires of the automated NDI community to eliminate the human from the inspection loop are very great. Except for the simplest cases, automated recognition of spatial imagery is not yet available and replacing the powerful interpretive powers of the human eye-brain will not be done by machine for some time. Despite this, intensive research is being conducted in the area of automated pattern recognition and image interpretation. To this end automated holographic interpretation is no exception and will benefit from these advances. Methods are currently being considered for the automated readout of holographic fringe patterns in order to eventually eliminate the human inspector from the loop.

An extremely important advantage of holographic inspection in this case is that the sensitivity of the technique is proportional to the defect severity. Unless the defect is of sufficient magnitude to cause weakening of the shell structure, the anomalous displacement due to the air pressure will be inconsequential and will not produce unusual fringes. It is of interest to make a comparison. In the cases shown so far the crack was long and deep, seriously impairing the basic sheel structure, but was so tight that it was invisible to the naked eye even when the paint was scratched from the shell

Figure 14. Easily visible surface blemish on shell.

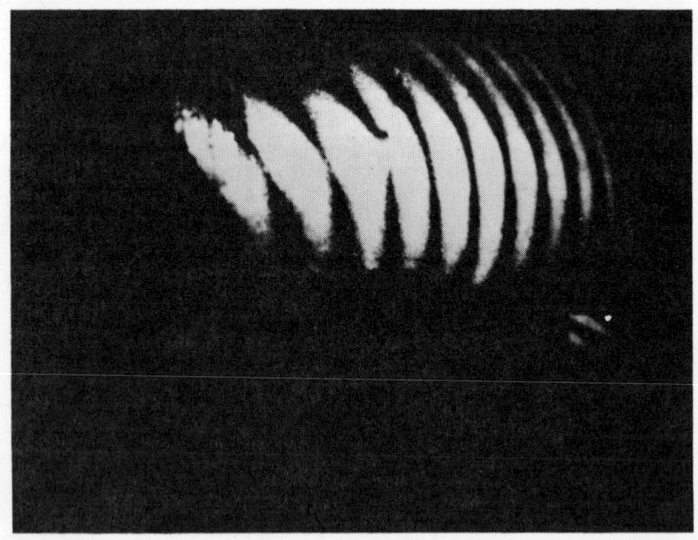

Figure 15. Expansion fringes over surface blemish.

metal. In fact, this particular shell was passed as acceptable by a magnetic particle inspection. In contrast, Figure 14 displays a rather easily visible defect by virtue of its size. The "large" defect is easily visible under magnetic particle inspection and can cause the shell to be rejected. The effect shown in Figure 14 is, however, a mere surface mark, which can be classed as a cosmetic blemish. It is of no consequence to the structural property of the shell. The holographic fringe pattern for this shell is shown in Figure 15. Since no unusual fringes are seen, and none should be expected on a structurally sound shell, the shell would pass a holographic inspection. Thus we are led to an important observation. Anomalous holographic fringes are formed only if the defect resulted in significant weakening of the shell wall but cosmetic defects are otherwise ignored. In nondestructive inspection we are often more interested in what a defect will do rather than the defect itself. Holography does just that. It sees the shell motion due to weakening by the crack, not the crack itself. This extremely valuable feature results in a test method whose sensitivity is proportional to the severity of the defect in causing shell wall distension under pressure. Few NDI techniques can claim this feature.

INDUSTRIALIZING HOLOGRAPHY

The principle advantages of holographic pressure inspection were reviewed above. The disadvantages have always been primarily in the area of practical industrialization of the test equipment and cost per item inspection. Although holography was introduced as a useful nondestructive test method in the early sixties, until recently it has found very little practical application outside the diagnostic research laboratory. During this period holography has been under continued development and has been maturing as a new NDI technology. Most of the limiting disadvantages of the past have now been removed as a result of intensive efforts to bring holography into the industrial plant. Figure 16 summarizes some of the pacing problems in industrializing holographic NDI. Since holography is fundamentally an interferometric method, it is necessary to utilize vibration isolation systems. It should be remembered that the fringes being recorded on the hologram are submicroscopic and therefore vibrations or movements of the apparatus can result in fringe washout and loss of the holographic data. For this reason it is always necessary to employ vibration isolation systems to reduce system movements to less than a wavelength of light when employing continuous wave (CW) holography.

Traditionally, holography had to be done with film as a result of the stringent spatial resolution requirements of the fringe recording media. Only film has been able to achieve this requirement concurrent with the appropriate light sensitivity needed to

INDUSTRIALIZED HOLOGRAPHIC NDI
PACING PROBLEMS

* Vibration Isolation
* Rugged Optics
* Film versus Real Time Recording
* Operator versus Automated Readout
* Laser Stability in Rough Environment

Figure 16. Industrializing holography.

get holographic exposures in short time periods.

As indicated earlier, the pros and cons of operator versus automated readout must be considered since the holographic data is in the form of fringes superimposed on an object image.

Holographic systems require lasers and one must consider the stability and operational life of a laser in the rough environment of the production factory. Lenses, morrors and other optical components must remain clean in dirty environments. The past problems of industrializing holography have been sufficiently formidable to keep the method primarily in the laboratory.

FUTURE OBJECTIVES

The lead in industrialzing practical holography systems has been taken by the tire industry. Here, the data rendered by holographic frainges has been so valuable in determiningdefects in tires that the drive has persisted to introduce practical holographic machinery into the tire testing environment. As a result, the problems mentioned above (with the exception of automated readout) have now been removed. Today, holography is done in tire factories and retread shops where vibrations abound. The optics are now sufficiently rugged to survive the environment and, as a result of the maturing of laser technology, long hours of maintenance-free laser life are now achieved. Although data is still presented in the visual form (usually on a television screen) the method of inspection has been reduced to simple signaturization of the defect patterns. Unskilled personnel with a minimum amount of training soon become experienced defect locators with the holographic system. The continual drive to replace film with real-time media has resulted in research and development efforts and now real-time, non-photographic recording media has been entering the market. Without photographic holographic methods, the cost and complexity of automated film developing equipment is eliminated and it is

Figure 17. Industrialized holograhic tire tester.

expected that virtually all industrial holography in the coming years will be done without film.

Because of the successful methods employed by the tire industry in bringing holography to the production plant environment, it has been decided to take maximum advantage of these developments in order to approach the practical holographic inspection of projectile bodies. Figure 17 is a picture of a typical holographic inspection system utilized for tires. The dome shown in the figure is lowered over the tire, automated equipment adjusts pressure in the dome, and a double exposed hologram is automatically made. This approach is not significantly different from that which can be employed in inspecting projectiles. With a dome of the proper design, a ring of projectiles can be inserted and the pressure and/or vacuum differential stress

INDUSTRIALIZED HOLOGRAPHIC TIRE TESTER

* Modify Tire Tester to Projectiles
* View Entire Shell
* View Multiple Shells
* Address Plant Interface
* Nonfilm, Video Holographic Readout
* Fully Automated Fringe Interpretation

Figure 18. Planned objectives.

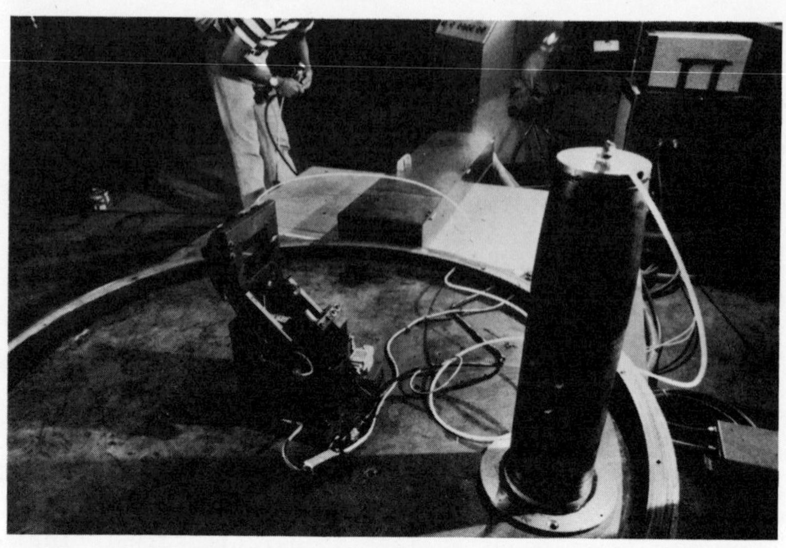

Figure 19. Adapting holographic tire tester to projectiles.

applied, while viewing the holographic data in real-time. The current thrust of our effort is to address the problem of industrializing holographic inspection of shells by modifying to the maximum extent possible the machinery already employed in the tire testing industry. Figure 18 summarizes the primary objectives of our current efforts. In the on-going project it is planned to bring multiple shells under a modified dome of a tire tester. Figure 19 demonstrates an M483 projectile being experimentally fixtured into a tire tester base as part of the on-going work. The holographic "camera" arrangement is seen in the center. Eventually, an entire ring of projectiles will be inserted. Attempts will be made to demonstrate the handling of multiple shells in the mass production environment. An envisioned schematic of testing shells in full flow production is shown in Figure 20. Real-time holographic methods will be merged into this technology in order to use non-film, video system readout on the shells.

As part of our current objectives, we wish to expand the data base of observable defects in the shells by producing as many shells as possible with a wide variety of defects in order to signaturize them holographically. It is our intent to perform as much of the data base as possible in an actual, industrialized rig in order to compare results with those obtained in the holographic laboratory. The actual fringe pattern seen in a given holographic setup is a function of the optical geometry employed. It is, therefore, expected that the fringe patterns seen both for normal and defective shells in an industrialized device will be somewhat different from

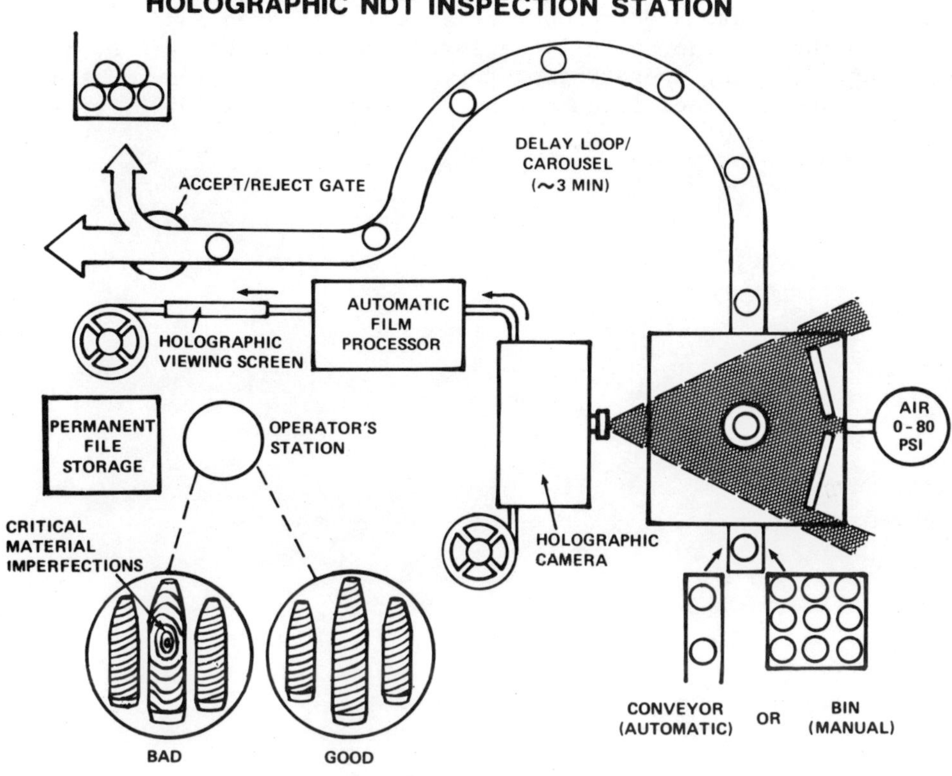

Figure 20. In plant inspection scheme.

those seen when taking fundamental data in the laboratory.

Efforts are currently under way to address the feasibility of automated readout of the fringe pattern. This is perhaps the most formidable remaining problem in completely automating holographic systems. If this is accomplished, the readout system will probably be dependent on a computer interpretation of a video pattern. The algorithms to determine when a defect is present would have to be written about the signature pattern as seen in an actual machine which is as close as possible to the system eventually designed.

CONCLUSIONS

It has been shown that holographic interferometry is feasible to detect serious defects in artillery projectile bodies. Efforts to industrialize practical holographic systems have now paved the way to developing practical machinery for the projectile production plant. Fully automated readout remains a risk area and the answer as to whether it can be fully deployed awaits further R & D, now underway. A plant operating prototype is planned for automated shell production facilities now being built for the 1980's time frame.

CHAPTER 5

DUCTILE FRACTURE ANALYSIS AND SAFETY OF NUCLEAR PRESSURE VESSELS

F. J. Loss

Naval Research Laboratory

Washington, D. C.

INTRODUCTION

The past two decades have witnessed the development of commercial nuclear power into a mature industry. Because of the potential for release of radioactive materials there has been a parallel growth in procedures for the prevention of fracture in nuclear pressure vessles and associated components. A catalyst to this development has been the discovery of radiation embrittlement in the ferritic steels used in pressure vessel construction. This chapter reviews current fracture analysis methodology for nuclear vessels from the viewpoint of linear elastic fracture mechanics and code procedures which are applied to assure structural integrity.

The continuing necessity to quantify the margin of safety against fracture and to qualify existing conservatisms in the operation of these critical structures has spurred new research relating to the characterization of elastic-plastic fracture. Developments in this area are summarized with emphasis on the J-integral approach. Elastic-plastic fracture mechanics is placed in perspective with respect to the likely benefits and implications of current research relating to light water reactor systems.

CURRENT FRACTURE ANALYSIS PROCEDURES

A typical pressurized-water reactor pressure vessel, shown in Figure 1, contains several critical locations that must be analyzed with respect to the potential fro fracture. Prior to 1972, the USA approach to provide assurance against fracture of the vessel was based largely upon analysis procedures developed at the Naval

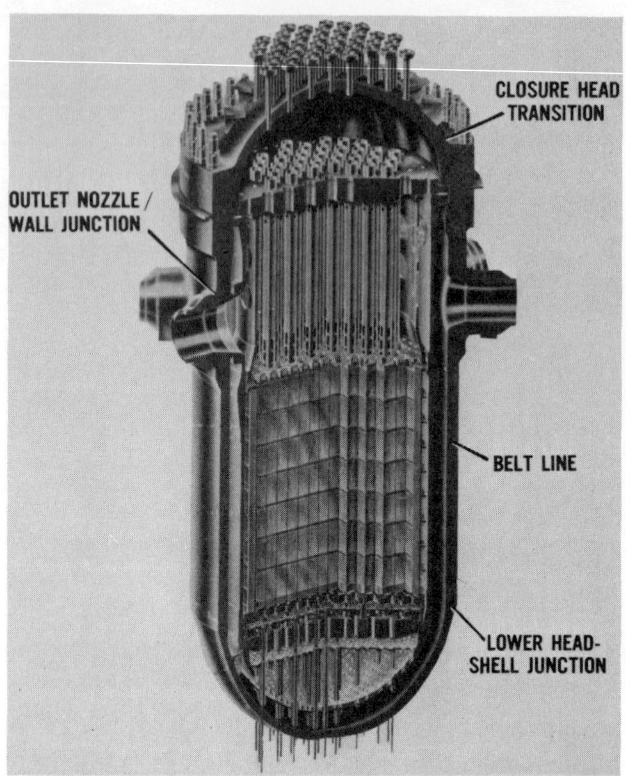

Figure 1. Typical pressurized water reactor pressure vessel illustrating critical locations with respect to vessel integrity.

Research Laboratory (NRL) [1,2]. This approach relied upon the characteristically sharp, brittle-to-ductile transition in fracture toughness which evolved with increasing temperature above the Nil Ductility Transition (NDT) temperature [3]. Full power operation was restricted to temperatures in excess of 33°C above the NDT temperature where the metal was presumed to exhibit sufficient ductility to preclude brittle fracture.

The past decade has produced significant changes in the analysis methods for fracture safety of nuclear structures. By far the most significant development has been in the area of linear elastic fracture mechanics (LEFM). This analysis tool was first introduced into the ASME Boiler and Pressure Vessel Code [4] in 1972 and has also been adopted by the U. S. Nuclear Regulatory Commission (NRC) as specified in the Code of Federal Regulations,

10CFR50 [5]. The reason for the introduction of LEFM was basically the need for the safety analysis to specify the degree of conservatism in a quantitative manner; this assurance could not be easily provided by the single requirement of a minimum operating temperature as embodied in the transition temperature approach.

Radiation Embrittlement

The beltline region of the pressure vessel (Figure 1) requires special attention in view of the degradation in fracture toughness which can be caused by neutron bombardment emanating from the reactor core. This change in fracture toughness is reflected by a temperature elevation, ΔT, of the NDT temperature as well as a drop in the ductile or upper shelf toughness (temperature independent) as illustrated in Figure 2. Because of its small size the Charpy

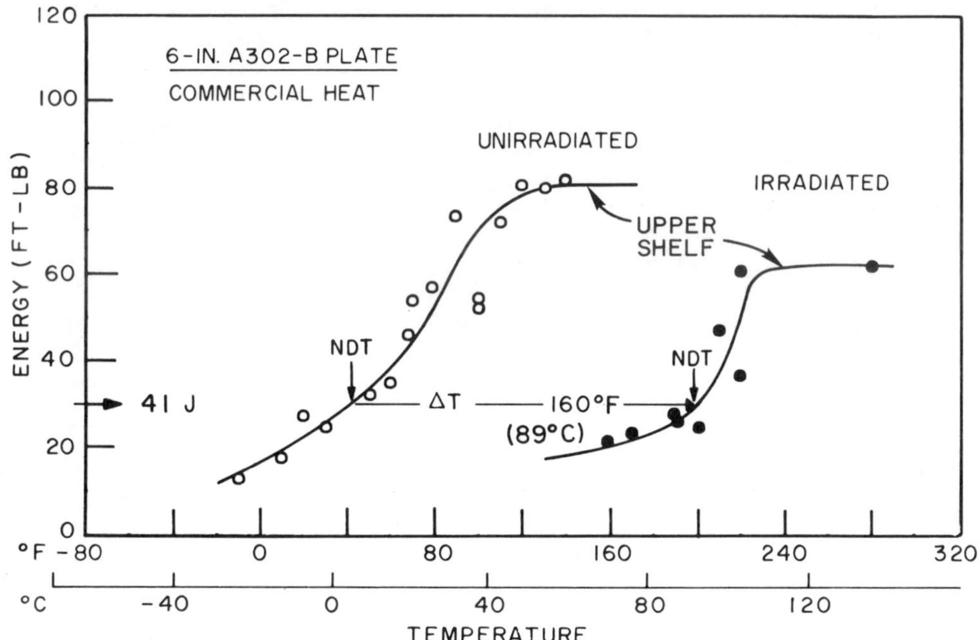

Figure 2. Illustration of the type of data developed with the C_V energy to predict embrittlement resulting from neutron bombardment. The ΔT elevation of the irradiated curve corresponds to the elevation in the NDT temperature. This same ΔT is used in ASME code, Section III, to predict the position of the irradiated K_{IR}.

V-notch (C_V) specimen is currently used to define the NDT temperature elevation in terms of the temperature difference at which a fixed energy level is achieved in both the unirradiated and irradiated conditions. Correlations developed by NRL [6] have shown that the NDT temperature elevation is essentially identical to the elevation of the C_V curve at the 41-J energy level.

While the C_V specimen is still in widespread use, this test has several disadvantages. The major drawback is that the energy absorbed in breaking the specimen cannot be correlated with the fracture toughness in a consistent manner. Because of its small size, however, the C_V specimen continues to be used in surveillance programs of irradiation embrittlement in service. More recently, LEFM specimens have been introduced to supplement the C_V data and to provide a more definitive procedure to measure changes in toughness.

Linear Elastic Fracture Mechanics Approach

The most fully developed quantitative concept for fracture characterization is the linear elastic or plane strain fracture toughness approach. The development of the theoretical and practical bases covers the period which approximates that of commercial nuclear power in the USA. LEFM provides a quantitative means with which to relate a property of the material, i.e., the plane strain fracture toughness, K_{Ic}, to the critical flaw size and nominal stress level in the structure. For example, with an idealized crack length of 2a in an infinite plate subject to nominal tensile stress σ, the stress intensity K_I is

$$K_I = \sigma\sqrt{\pi a} \tag{1}$$

This LEFM approach has been readily accepted by the designer who normally thinks in quantitative terms. A standard method for K_{Ic} measurement is given in an ASTM Standard [7].

One major advantage of the LEFM approach is that the toughness may be characterized in terms of the stress singularity at the crack tip. This singularity is common to any crack regardless of whether it is in the test piece of in the structure, provided, of course, both bodies are subject to the maximum (plane strain) constraint. This highly significant fact is what makes LEFM "work". In other words, it provides a means for direct translation of measurements from simple test specimens in the laboratory to predict the behavior of structures of radically different proportions.

However, LEFM is not without its disadvantages. Chief among these is the necessity to perform an accurate stress analysis of the structure. Also, a crack size must be established from

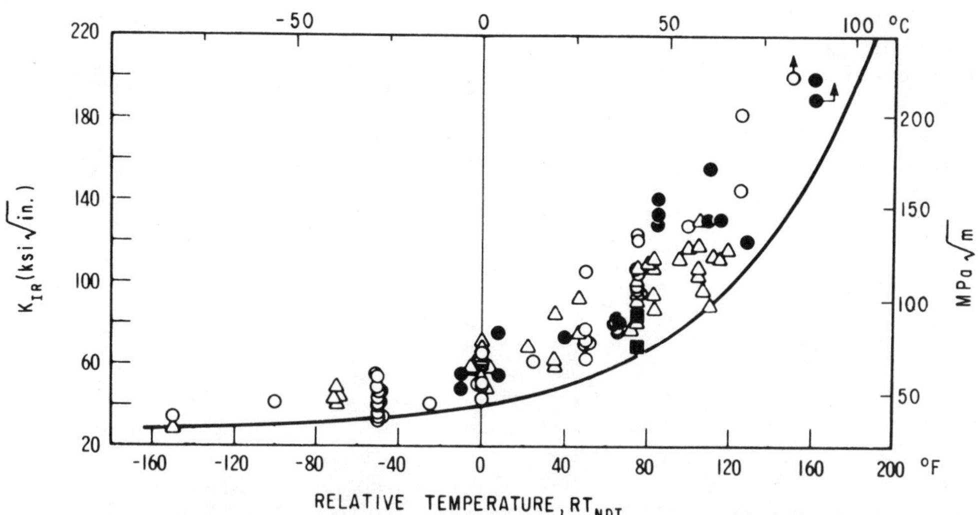

Figure 3. Reference stress intensity factor (K_{IR}) curve showing the type of basic data. Includes data from Types A533-B plate steel and A508 CI 2 forging steels and includes both K_{Id} and K_{Ia} (crack arrest) data [4].

measurements or as an assumed flaw. Thus, successful applications are limited to relatively simple structures amenable to rigorous analysis and inspection. Another problem is that LEFM applies only to linear elastic behavior whereas nuclear pressure vessel steels generally exhibit considerable pre-fracture plastic flow under operating conditions. Because of this apparent dichotomy, the use of LEFM is rationalized on the basis of assuring that fracture will not occur even if the material were to exhibit plane strain behavior. If, as expected, a higher level of toughness is realized (e.g., elastic-plastic behavior) then this only adds to the margin of safety.

Likewise, the dynamic toughness (K_{Id}) values are often used to characterize the vessel material. This results not because the structure necessarily will be dynamically loaded but because the dynamic toughness, within the transition region, is less than that exhibited statically. The use of K_{Id} thus can provide an added degree of conservatism.

LEFM procedures are embodied in ASME Boiler and Pressure Vessel Code rules in terms of a reference (K_{IR}) curve (Figure 3) [4] of fracture toughness that defines a lower bound to existing data from unirradiated material. The shape of the K_{IR} vs temperature curve is considered to be identical for all materials to which it applies; this includes postirradiation toughness as well. To account for a temperature change of the brittle-ductile transition region, due to heat-to-heat variation or irradiation, an index or reference temperature, termed RT_{NDT}, was invented [4]. This temperature is based upon the NDT temperature as well as C_v properties (energy and lateral expansion). Because of the difficulty in irradiating the large specimens required to measure K_{Ic} directly, the toughness of irradiated material is defined by a simple temperature translation of the K_{IR} curve by an amount equal to the temperature elevation of the C_v curve with irradiation as shown in Figure 2.

The K_{IR} curve, as adjusted to account for irradiation, is used to establish operating limits for the start-up and shutdown of the reactor vessel. These operations result in a large change in the temperature of the vessel wall; this, in turn, results in a commensurate change in fracture toughness. Briefly, the vessel analysis proceeds as follows. An applied K_I level is computed on the basis of the vessel wall stress and the assumption of a flaw in the wall equal to one-quarter its thickness and oriented axially. This stress intensity level (multiplied by a safety factor of two for primary or pressure-induced stresses) must be less than the value of the K_{IR} curve at all temperatures of operation. Should this not be the case, then the reactor operation could be curtailed or the plant could even be required to shut down.

Flaws discovered during service are treated by Section XI of the ASME Code [8]. This Section attempts to provide a basis for acceptance standards for evaluating flaw indications detected during inservice inspection of the vessel and to describe quantitatively the way flaws may grow during service. The flaw evaluation procedures are based on curves similar to the K_{IR} curve but a further description is beyond the scope of this chapter.

ASSESSMENT OF CURRENT STATUS

The current procedures based upon ASME Sections III and XI [4,8], in conjunction with Federal rules [5], provide a conservative analytical framework for the avoidance of non-ductile failure. Nevertheless, there are areas for more rigorous or less assumptive bases for improving fracture characterization. A synopsis is provided here highlighting two ares of improvement that are receiving current emphasis.

Low Upper Shelf Toughness

Fracture is generally not believed to be a serious factor for structures whose operating temperatures are in the regime of the ductile upper shelf. If, however, a relatively low level of upper shelf toughness can be exhibited, then brittle fracture again becomes a possibility albeit associated with a ductile failure mode on a microscopic scale. Such a potential problem faces some reactor vessels. While the problem of irradiation embrittlement for new plants has been essentially solved through the specification of steels having a low residual impurity level [9], the early plants in the USA were not able to benefit from the advances in steel specifications that now limit the embrittlement. A current problem stems from the use of welds of high copper impurity content in the beltline region coupled with a relatively low pre-service upper shelf toughness, in terms of C_V energy.

Estimates of the toughness degradation caused by irradiation are provided by Regulatory Guide 1.99 [10] in terms of a temperature elevation of RT_{NDT} as well as the expected drop in C_V upper shelf energy as a function of fluence. Using this Guide, it has been predicted that a number of USA plants will exhibit a C_V upper shelf energy of less than 68 J (50 ft-lb) within the next several full power years. A toughness level of less than 68 J is especially significant in that it precludes the specification of the RT_{NDT} and thereby prevents the use of the K_{IR} curve analysis of ASME Section III.

The low shelf energy phenomenon has highlighted another area where additional research is required. This deals with an understanding of the structural significance of the C_V upper shelf energy. Since the C_V energy by itself is only of relative value in assuring adequate toughness, it is necessary to express this energy in fracture mechanics terms (K_{Ic},K_{Id}). It is not easy to evolve a correlation between C_V energy and K_{Ic} because of the large size specimen requirements needed to achieve the necessary mechanical constraint for valid K_{Ic} tests. In addition to this, it is difficult to simulate the degradation in shelf energy except by irradiation. The problem with the latter is that it is not feasible to irradiate test specimens having a thickness over 100 mm (4 in.) in a test reactor because of a non-uniformity in the irradiated properties that would occur. However, a C_V - K_{Ic} correlation has been devised by Rolfe and Novak [11] for other steels of higher strength which may prove useful in the interim period while current research programs become productive.

The need for quantitative assessment of upper shelf toughness has triggered a significant research effort on the part of NRC, the Electric Power Research Institute (EPRI), and industry to characterize the elastic-plastic toughness that can be exhibited by low shelf steels. This effort includes the development of new test methods such as the J integral. Progress in this area is discussed later. Over the near term, it is conceivable that plant safety requirements can be met provided that the irradiated material exhibits a toughness level of approximately 165 MPa$\sqrt{m}$ (150 ksi$\sqrt{in.}$). Thus, near-term objectives will be met if it can be shown that a particular low shelf steel is capable of exhibiting at least this level of toughness.

If, after a period of irradiation, the toughness is believed to be inadequate, then it is possible to recover a portion of the virgin toughness through periodic heat treatment (annealing) of the reactor vessel. Current research programs sponsored by NRC and EPRI are investigating this option. The objectives are geared to predict the rate of embrittlement by reirradiation following an anneal as well as to predict the toughness recovery by this procedure. Basically, the degree of embrittlement relief is a function of time and temperature with the larger values of both the variables producing the greater recovery [12].

Predictions From Surveillance Specimens

The embrittlement of the early USA pressure vessels must be assessed through surveillance specimens that consist almost exclusively of C_v specimen. Two approaches are being explored to produce more quantitative toughness information from this specimen. First, sophisticated analytical analyses of the C_v specimen have been undertaken under EPRI sponsorship to investigate the possibility of obtaining different, but more meaningful indications of toughness from this specimen. This effort is expected to bear fruit over the longer term and may define a better type of surveillance specimen. Second, interpretations of C_v energy are being investigated through correlation with K_{Ic} and J integral-type tests. This research will produce results over the next several years. The objectives in the latter program are (a) to provide a correlation between C_v energy and K_{Ic}, (b) to demonstrate the correlation between the temperature elevation in the K_{IR} curve as predicted from the shift in the C_v energy at the 68 J (50 ft-lb) level and that actually measured with fracture mechanics specimens, and (c) to investigate the correspondence between the static (K_{Ic}) and dynamic (K_{Id}) toughness of irradiated materials.

The NRC is also sponsoring a large irradiation program to aid in the better definition of the K_{IR} curve for irradiated material by direct measurement instead of by inference from the temperature shift of the RT_{NDT}. This effort involves compact specimens (CS) of up to 100 mm thickness. A critical objective in this program is to define the shape of the K_{IR} curve for irradiated, low upper shelf weld metal. It should be noted that the K_{IR} curve is based primarily upon the behavior of steel that exhibits a high level of upper shelf energy. It is also known that the shape of the post-irradiated C_v curve can exhibit a much less steeply rising trend in energy vs temperature in comparison to the preirradiated behavior (Figure 4) [13]. Therefore, it appears reasonable to infer that the K_{IR} curve for irradiated steel of low shelf toughness also may not exhibit the same sharp increase with temperature as is defined in ASME Section III. If this hypothesis is correct, then current assessments of irradiated toughness in the brittle-ductile transition regime based on the shape of the unirradiated K_{IR} curve may not always be conservative.

ELASTIC PLASTIC FRACTURE MECHANICS

A significant research effort is currently being directed toward the characterization of elastic-plastic fracture. This research stems from a need to provide greater confidence in current assessments of the margin against fracture as well as to permit relaxation of conservatisms inherent to plant operating limits that are based on the assumption of linear elastic behavior. One of the most fruitful products of this research has been to enable the prediction of thick section (LEFM) fracture performance based upon small specimens which exhibit considerable plasticity.

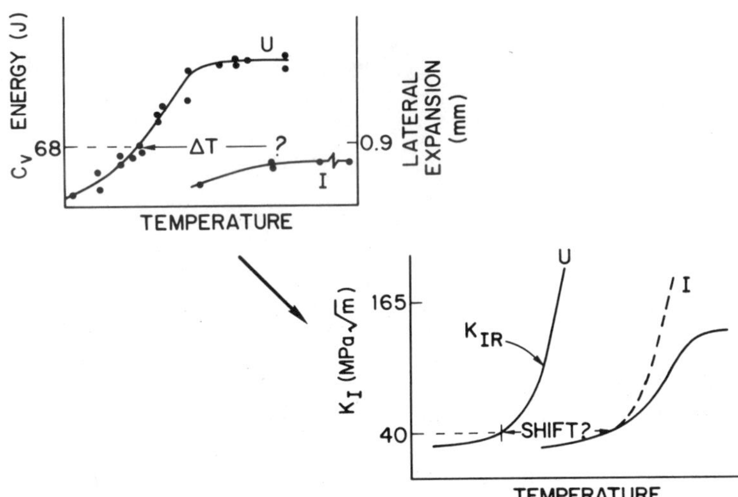

Figure 4. Schematic representation of the problem that can result in defining the K_{IR} curve for irradiated steels having upper shelf toughness less than 68 J (50 ft-lb). The unirradiated (U) C_V curve represents actual data from the A533-B1 plate that provided K_{Id} data for the K_{IR} curve (Figure 3). The irradiated (1) C_V curve represents data from an irradiated A533-B submerged arc weld of high copper level [13].

Much of the research in the field of elastic plastic fracture involves the J-integral concept. While the COD concept, used extensively in Europe, is directly relatable to the J integral, the former approach has received relatively little emphasis in the USA. Therefore, this discussion considers primarily the J-integral approach. Also, because of space limitations, only the experimental aspects can be highlighted.

J Integral - R Curve

The J integral as proposed by Rice [14] has been shown to characterize the crack tip stress and strain field under both elastic and plastic conditions in two dimensions. This parameter offers a logical extension of LEFM concepts to include the case of large-scale plastic behavior. Basically the J integral is the potential energy change, dU, caused by a small increase in crack length, da, in a notched body at a given deflection, δ_o, of the load point:

$$J = -\frac{I}{B}\frac{dU}{da} \tag{2}$$

where B is the specimen thickness.

The J integral has a firm mathematical basis, and for linear elastic behavior, J reduces to the strain energy release rate, G, for LEFM conditions [14]. Furthermore, the J integral can also be related to the stress intensity K [15] as follows:

$$J = G = \frac{K^2}{E} \tag{3}$$

where E is the elastic modulus. Under elastic-plastic conditions, J loses its physical interpretation in terms of the potential energy available for crack extension, but retains its physical significance as a measure of the intensity of the characteristic crack tip strain field [16].

Because of the relationship to K, the J integral was proposed as a fracture parameter for crack initiation, J_{Ic} [16]. Later, Landes and Begley [17] presented experimental evidence that J_{Ic} could be used to predict the K_{Ic} values obtained from much larger specimens that were tested under conditions of brittle (cleavage) crack initiation.

Rice and others [18] have developed a simple formula for J_I that applies to deep crack in pure bending:

$$J_I = \frac{2A}{Bb} \tag{4}$$

where A is the area under the load vs load-point-development curve and b is the length of the unbroken ligament. This equation can be successfully applied to CS (this application to a compact specimen also requires the use of correction factors which will not be discussed here) and three-point bend specimens for a crack depth-to-specimen width ratio (a/W) greater than 0.5 according to ASTM Committee E-24 on Fracture Testing.

Procedures for J_{Ic} measurement are being developed by the E-24 Subcommittee on Elastic-Plastic Fracture. At the present time, the value of J_I at initiation is defined by means of an R curve (Figure 5). The latter is simply a plot of applied J_I vs average crack extension (Δa) from the tip of the fatigue precrack. This crack extension is of a stable nature and requires a continuous application of load for the crack to advance. An extrapolation (in actual practice, the extrapolation is not to Δa of zero but to a computed crack extension indicative of the crack blunting that occurs prior to the actual extension) of the R curve to the point where crack extension begins defines J_{Ic}.

Because J is defined as a two-dimensional parameter, certain minimum thickness requirements must be met in order to measure a value that is thickness independent, J_{Ic}. The thickness limit currently suggested by ASTM E-24 is

$$B > 25 \frac{J_{Ic}}{\sigma_f} \tag{5}$$

where σ_f is the flow stress measured at the midpoint between the yield and ultimate stress.

J-Integral Applicability to Nuclear Structures

One of the major advantages of the J integral for nuclear applications is its ability to predict the behavior of the very large specimens required in K_{Ic} testing (i.e., 300 mm thickness) on the basis of a much smaller specimen (25 mm thickness). The resultant saving in material and manpower can be quite significant. A similar advantage is projected in the area of surveillance specimen testing. For example, it would be extremely difficult to irradiate the large K_{Ic} specimens required to characterize the toughness degradation during service. By current estimates, small specimens ($\approx$ 25 mm thick) are believed to be the largest size necessary to provide the J_{Ic} measurement capacity to demonstrate the level of toughness needed for accident analyses. However, J-integral applications are in the formative stages and there are still problems to overcome. Some of the areas that require further research are described in the following paragraphs.

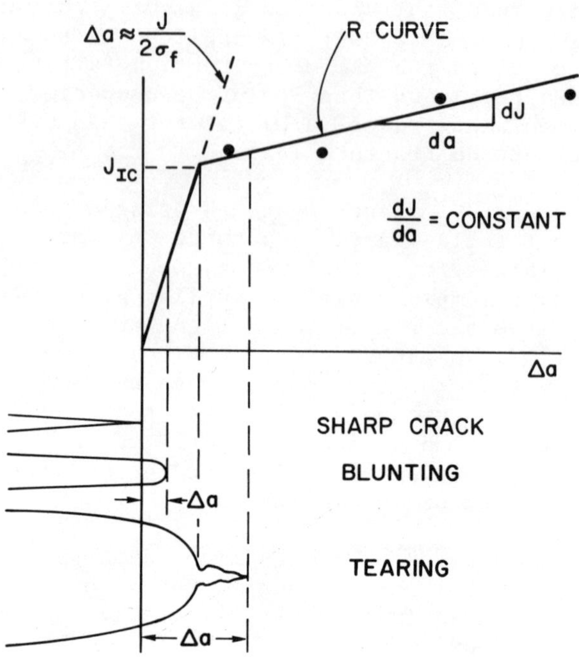

Figure 5. Representation of the R curve method for J_{Ic} measurement. The J_{Ic} value is defined by the interaction of the R curve with the crack tip blunting line ($\Delta a = J/2\sigma_f$). The R curve is a least-squares fit through experimentally determined values of crack extension.

Currently, the most reliable way to determine J_{Ic} is through an R curve that is generated with several specimens (Figure 5). Each one is loaded to a different level, heat tinted, and then broken apart to reveal the crack extension. This method is time consuming and wasteful of material, especially if it involves irradiated specimens. Developments are proceeding with a single specimen J-R curve test. The most promising of the various procedures is the unloading compliance method. With this method that specimen is unloaded slightly (< 10%) at various deflections and the elastic slope of the unloading trace, P/δ, is used to define the crack length via a prior compliance calibration (Figure 6). This method has been demonstrated [19] for non-nuclear steels and appears promising. However, great precision is required in applying the technique and it has not yet been proven for hot cell application.

A related problem involves the shape of the specimen crack extension. This extension creates a tunneled shape starting from

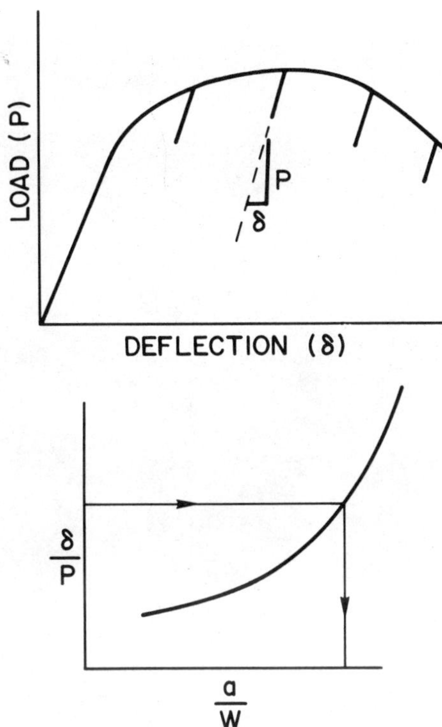

Figure 6. Illustration of the unloading compliance technique for determining the J-R curve of Figure 5 with a single specimen.

the fatigue precrack as illustrated in Figure 7. Obviously, the latter is not representative of what would be expected in the center of a very thick structure such as the reactor vessel. Ongoing research [20] has indicated that a relatively straight crack front can be developed with modest face grooves machined in the specimen. It appears that face-grooving can reduce the slope of the R-curve but there is disagreement among investigators on this point. While a change in R curve slope need not result in a different value for crack initiation it could influence the margin against instability as discussed next. However, a more urgent topic for investigation is the demonstration of the invariance of the R-curve with respect to specimen size. This topic is under current investigation at several laboratories. A specimen and size independence must be shown in order for J to find widespread application for structural integrity analysis.

Another developmental area with respect to the J integral relates to the problem of low upper shelf energy previously

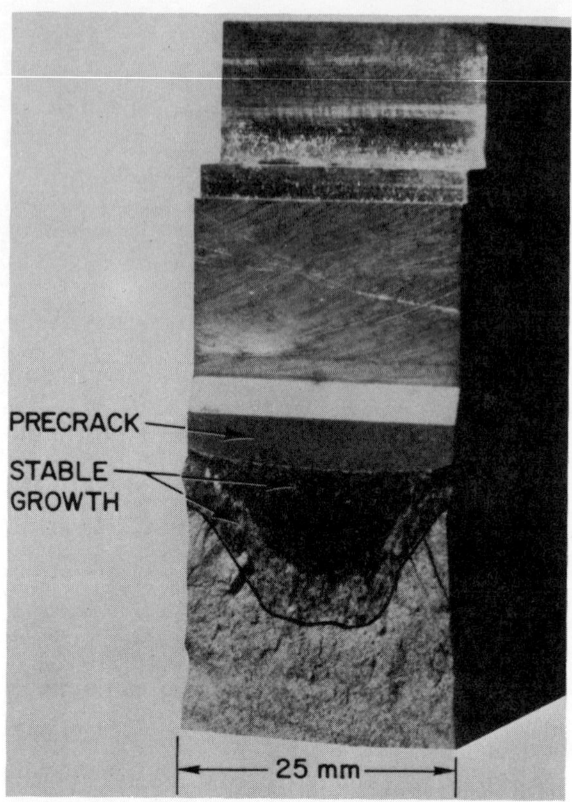

Figure 7. Illustration of stable crack extension in a compact specimen. The tunneled crack shape results from the relaxation of mechanical constraint at the surfaces, thereby permitting plasticity as opposed to crack growth.

described. Preliminary accident analyses for nuclear vessels reportedly suggest that an irradiated material with a K_{Ic} toughness of at least 165 MPa$\sqrt{m}$ (150 ksi$\sqrt{in.}$) can satisfy safety requirements. Because of the low C_v upper shelf energies that can be exhibited [13], it can be projected [11] that this requirement will not always be met for irradiated weld metal containing a high impurity copper level. The result is unfortunate and it implies that a toughness criterion based on crack initiation, e.g., J_{Ic}, cannot always be used to demonstrate the required toughness. Consequently, short of curtailing the operation of some plants with this potential problem, it is necessary to explore other aspects of the problem, such as the question of crack stability considered next.

Tearing Instability

In terms of a different viewpoint of low upper shelf energy discussed above, Paris and other [21] have developed a new approach based on crack stability. In this method the toughness level in terms of J_{Ic} or K_{Ic} is not of primary concern as long as significant extension crack cannot take place. Considering the R curve of Figure 5, it is easy to visualize a small crack extension associated with higher values of J_I. Provided the slope of the R curve (dj/da) is steep, it can be shown that a material of low J_{Ic} can exhibit a very high level of toughness of J_I after a certain crack extension.

Considering the above behavior, Paris and others [21] have defined a material parameter called the tearing modulus T (dimensionless), defined as follows:

$$T = \frac{dJ}{da} \cdot \frac{E}{\sigma_o^2} \tag{6}$$

where σ_o is the flow stress in simple tension. Space limitations preclude a detailed description of this method. However, certain essential points are discussed. Basically, a unique value of T will result for a given material so long as the R curve is thickness independent. Knowing the value of T, it is possible to predict an instability condition for various geometries. This is done by assuming the structure is loaded to the plastic limit load and that any further stretching of the specimen is contributed directly by the slip lines and will result in a lower limit load; this, in turn, releases energy from the unloading of the elastic portion of the specimen that is removed from the slip field. The latter provides the crack driving force. Instability will result if the elastic shortening exceeds the plastic lengthening required for crack extension. Applying this logic to a center-crack plate with fixed-grip loading yields the following stability condition:

$$T = \frac{dJ}{da} \cdot \frac{E}{\sigma_o^2} < 2\,\frac{L}{W} \tag{7}$$

where L and W are the length and width of the plate, respectively.

In the preceding example, the plate will be stable so long as T exceeds 2L/W. Note that the stability condition is independent of crack length. Similar instability conditions can be evolved for a deep surface crack in a reactor vessel. For this condition Paris has defined the stability condition for the "snap-through" of the crack to become a through thickness flaw as

$$T < \frac{2\ell}{t} \tag{8}$$

where ℓ and t are the crack length and vessel wall thickness, respectively. From this result it is clear that even a long crack in the vessel will be stable even with very low values of T. The above results, while still under study, offer significant potential benefit to reactor safety analysis and should be developed further.

SUMMARY

The current status of fracture characterization procedures has been reviewed for light water reactor structures. These methods are contained in the ASME Boiler and Pressure Vessel Code, Sections III and XI, as well as in Federal Regulations. The Code rules are based upon linear elastic fracture methods as a rational tool to provide a quantitative assessment of the margin of safety against fracture. In combination with associated ASTM Standards and Nuclear Regulatory Guides, the current procedures provide rules for materials testing, test and acceptance standards, standards for radiation damage assessment, and a means for relating material toughness to the requirements of the application.

While these procedures provide a relatively conservative analytical approach for the avoidance of non-ductile failure, there are areas where improvement must be sought to provide more rigorous or less assumptive bases. An assessment of our current status has shown that more information is required in the areas of low upper shelf energy and interpretation of surveillance specimen results. The potential problems associated with low shelf energy can affect several USA plants and this results primarily from the use of belt-line weld metals that are unusually sensitive to irradiation. Currently available analyses are inadequate to deal with this problem. Also, the C_V upper shelf energy expected in these plants in the near future has not been rigorously related to the flaw size tolerance of the pressure vessel nor to the margin or safety available. If this problem is not successfully resolved by analysis, it still appears possible to extend plant lifetime by producing embrittlement relief through periodic heat treatment for thermal correction (annealing) of radiation embrittlement.

Much of the future progress in fracture characterization rests in the advancement of elastic-plastic fracture. Of the various approaches to this area, a major research effort exists to investigate applications of the J integral. This method projects a significant advantage in the use of small specimens to characterize the toughness of large structures, such as reactor vessels, which are subject to a higher mechanical constraint. Current investigations with the J integral are promising for use with irradiated material, including surveillance specimens. With respect to structural integrity, a new approach is described that deals with the topic of tearing instability. This method offers potential for the analysis of vessel integrity when the material exhibits a low upper shelf toughness as well as high toughness.

In conclusion, it is believed that the past decade has produced significant research information and related codes and rules which help assure the safety and reliability of nuclear structures. Future research in the areas outlined will enhance our abilities to quantify the margin of safety, to assure the continued reliability of nuclear power, and may even allow some relaxation of current conservatisms.

REFERENCES

1. Pellini, W.S. and Puzak, P.O., "Fracture Analysis Diagram Procedures for the Fracture-Safe Engineering Design of Steel Structures", Welding Res. Council Bull. No. 88, 1963.
2. Pellini, W.S., Principles of Structural Integrity Technology, Office of Naval Research, Arlington, VA, 1976.
3. ASTM, "Conducting Drop Weight Test to Determine Nil-ductility Transition Temperature of Ferritic Steels", ASTM Designation E208-69, Part 10, 1976, pp. 330-49.
4. ASME Boiler and Pressure Vessel Code, Section III, Nuclear Power Plant Components, Sub-Section N, Appendix G, Amer. Soc. Mech. Eng., New York, NY, July 1, 1974, p. 487.
5. Fracture Toughness and Surveillance Program Requirements, Appendices G and H, Title 10, Code of Federal Regulations, Part 50 (10 CFR Part 50), U. S. Atomic Energy Commission, Federal Register, 38, July 17, 1973, p. 136, 19012.
6. Steele, L.E., Neutron Irradiation Embrittlement of Reactor Pressure Vessel Steels, Atomic Energy Rev. 7, 1969, p. 1
7. ASTM, Plane Strain Fracture Toughness of Metallic Materials, ASTM Designation: E399-72, Part 31, July 1973, p. 960.
8. ASME Boiler and Pressure Vessel Code, Section XI, July 1974, with addenda, Winter 1976, ASME, New York, NY.
9. Hawthorne, J.R., Koziol, J.J. and Groeschel, R.D., "Evaluation of Commercial Production A533-B Plates and Weld Deposits Tailored for Improved Radiation Embrittlement Resistance", ASTM STP 570, Amer. Soc. for Test. and Matls., Philadelphia, PA, 1975, pp. 83-102.
10. USNRC, "Effects of Residual Elements on Predicted Radiation Damage to Reactor Vessel Materials", Regulatory Guide 1.99, Revision 1, Office of Standards Development, U.S. Nuclear Regulatory Commission, April 1977.
11. Rolfe, S.T. and Noval, S.R., "Slow-bend K_{Ic} Testing of Medium Strength-High Toughness Steels", ASTM STP 463, Amer. Soc. for Test. and Matls., Philadelphia, PA, 1970.
12. Hawthorne, J.R., Watson, H.E. and Loss, F.J., "Exploratory Investigation of Cyclic Irradiation and Annealing Effects on Notch Ductility of A533B Weld Deposits", in Irradiation Effects on the Microstructure and Properties of Metals, ASTM STP (pending publication), Amer. Soc. for Test. and Matls., Philadelphia, PA, 1978.

13. Loss, J.F. (ed.), "Structural Integrity of Water Reactor Pressure Boundary Components, Progress Report Ending 31 May 1976", NRL Memorandum Report 3353, Naval Research Laboratory, Washington, DC, Sept. 1976.

14. Rice, J.R., Fracture - An advanced Treatise, Vol. II, Mathematical Fundamentals, Academic Press, New York, NY, 1968, pp 191-311.

15. Paris, P.C. and Sih, G.C., in Fracture Toughness Testing and Its Applications, ASTM STP 381, Amer. Soc. for Test. and Matls., Philadelphia, PA, 1965, pp 30-83.

16. Begley, J.A. and Landes, J.D., in Fracture Toughness, Part II, ASTM STP 514, Amer. Soc. for Test. and Matls., Philadelphia, PA, 1972, pp 1-20.

17. Landes, J.D. and Begley, J.A., "Recent Developments in J_{Ic} Testing", ASTM STP 632, Amer. Soc. for Test. and Matls., Philadelphia, PA, 1977, pp 57-81.

18. Rice, J.R., Paris, P.C. and Merkle, J.G., in Progress in Flaw Growth and Fracture Toughness Testing, ASTM STP 536, Amer. Soc. for Test. and Matls., Philadelphia, PA, 1973, pp. 231-45.

19. Clarke, G.A., Andrews, W.R., Paris, P.C. and Schmidt, D.W., "Single Specimen Tests for J_{Ic} Determination", ASTM STP 590, Amer. Soc. for Test. and Matls., Philadelphia, PA, 1974, pp 27-42.

20. Loss, F.J. (ed.), "Structual Integrity of Water Reactor Pressure Boundary Components, Progress Report Ending 30 Nov 1977", NRL Memorandum Report 3782, Naval Research Laboratory, Washington, DC, May 1978.

21. Paris, P.C., Tada, H., Zahoor, A. and Ernst, H., "A Treatment of the Subject of Tearing Instability", NUREG Report 0311, U.S. Nuclear Regulatory Commission, Washington, DC, August 1977.

CHAPTER 6

ENVIRONMENTALLY ASSISTED FRACTURING UNDER SUSTAINED LOADING

B. F. Brown

Department of Chemistry, The American University

Washington, D. C.

INTRODUCTION

One of the risks involved in the structural integrity of military equipment is environmentally assisted fracture under sustained (or, for that matter, under cyclic) loading. This risk has increased with increasing levels of technology, as illustrated by three examples: the brass cartridge case was all too prone to "season cracking", although its paper predecessor was immune; the aluminum airframe has had a long history of stress corrosion cracking (SCC), and indeed the titanium alloy counterpart has not been immune, whereas the fabric-covered fuselage was, of course, immune to SCC; and the modern high strength steel rocket motor case is susceptible to SCC and hydrogen embrittlement cracking whereas cast iron and bronze canon were not. No reason is evident for expecting this pattern of increased threat of SCC to subside, and it may even spread.

The form of environmentally assisted fracture under sustained load which is of most concern in current technology is SCC, and that will be the subject of this chapter. Two other modes of sustained-load environmental fracture are hydrogen embrittlement and liquid metal embrittlement. Many of the principles involved in SCC also apply to hydrogen and liquid metal cracking. Indeed, much if not all SCC in high strength steels is really hydrogen embrittlement, and some investigators would also include SCC cracking in aluminum alloys, titanium alloys, some magnesium alloys, and even austenitic stainless steels in the hydrogen embrittlement category.

CHARACTERISTICS OF STRESS CORROSION CRACKING

It is useful both in the development of structural integrity plans for the design, fabrication and maintenance of structures and in the performance of failure analyses to bear in mind the many special characteristics of SCC. These are as follows:

1. SCC produces macroscopically brittle fractures even though the alloys may be highly ductile in a simple tensile test. Thus SCC typically occurs without signs of general yielding. This statement does not mean that SCC cannot occur in plastically deformed metal - it can and does. But since most structures are designed to operate at less than yield, most service SCC failures are lacking general yielding.

2. Although general plastic yielding is not needed for SCC and is usually missing, a small amount of localized plastic flow is often seen associated with SCC in studies with high resolution fractography. This local plasticity often takes the form of "tear ridges" and may ultimately be found to be important to the theory of sustaining SCC.

3. Only a tensile stress will produce SCC. All SCC is of the "opening mode", and there is no SCC analog to the shear lip, though purely mechanical shear lips may form as borders of a stress corrosion crack.

4. There is a wide range of degree of susceptibility to SCC among various combinations of alloys and environments, even in a single alloy family.

5. There is an approximation of specificity about those combinations of alloy and environment which leads to the most severe cracking problems in service. In SCC, in other words, generally only a few chemical species cause serious cracking problems in a given alloy. These species may be awfully ubiquitous in some cases, such as water (or water vapor) with aluminum alloys, or chlorides with austenitic stainless steel. The degree of susceptibility of a given alloy to SCC by one damaging species may not be the same as that by another damaging species.

6. Theory has not yet advanced to the stage of being able to predict the species which may cause SCC in a given alloy. As a corollary to this statement, to know the response of a given alloy to a new environment, there is no safe alternative to a test.

7. The chemical species in the environment essential for SCC need not be present in large quantities or in high concentration (in the bulk environment). Where it has been measured, the concentration of aggressive species has been found to be greater in growing cracks than in the bulk environment.

8. The cominbations of alloy and environment which lead to the most severe cracking problems are usually such that the alloy is

almost, but not quite, inert to the environment.

9. SCC may occur in a given alloy in a given environment at one potential but not at another. As a consequence of this behavior, cathodic protection is a practical solution to the SCC problem in some circumstances.

10. SCC occurs in alloys but not in pure metals, though it should be noted that phosphor copper, which is nearly pure, has experienced SCC. (There are laboratory cases of SCC in high purity metals which may be important to the theory but which are not likely to be of importance to engineering applications.)

11. Stress corrosion cracks may be either branched or not branched. If they are branched, it is usually an indication of either high stress intensity or of a complex stress field as may be left over from forming operations. The cracks may be either intergranular or transgranular, and very subtle changes in governing parameters may cause the crack path to switch from one to the other morphology.

12. It is worth emphasizing that SCC is a nucleation-and-growth phenomenon, and the factors which affect nucleation may affect growth in a different manner. In a large proportion of service failures the necleus of SCC is a corrosion pit, and in some other failures the nucleus is a crack-like flaw, such as a quench crack.

SCC PROPERTIES OF VARIOUS ALLOYS

The theory of SCC has not been fully developed. In its present state of development the theory is useful in guiding our thinking about the problem and in helping organize what would otherwise be a jumble of assorted observations. But theory has not yet developed to the stage of being capable of giving the needed guidance for engineering applications. Therefore, we must rely upon the behavior of macroscopic specimens to indicate the SCC properties of various alloys. A large number of different types of specimens and test methods have been used to make the evaluations upon which our present-day SCC control measures are based. These specimens and test methods are described in Reference [1].

There is no single standard environment which has been established as providing characterization of universal significance even for one alloy family. There has been a practice of evaluating aluminum alloys and high strength steels in salt water, and copper alloys in ammoniacal environments, with the presumption that the properties in these environments will provide useful guidance in the behavior of such alloy families in many common service environments. It should be emphasized, however, that our understanding of the phenomenon of SCC is not sufficiently developed to enable us to interpret the behavior of an alloy in one environment in terms of the behaivor to be expected in a greatly different environment.

The SCC properties of the major structural alloy families are summarized in Reference [2]. A much more comprehensive compilation of SCC properties is in the final stages of preparation and should be available in the near future [3].

Since stress corrosion cracks commonly occur in elastic bodies, it is appropriate to characterize the stress factor by linear elastic (Irwin-Kies) fracture mechanics. Such application of fracture mechanics is difficult to do except there be only a single crack, and that one not a braching crack. This simplification is often realizable by using a fatigue precrack to localize the SCC and using low enough stress intensity values K to avoid branching. There appears to be a threshold value of K designated K_{Iscc} below which SCC is not detected, at least in some important systems. The magnitude of K_{Iscc} is dependent upon the alloy, the heat treatment, the environment, and if there is a strong texture, upon the orientation.

The K_{Iscc} values determined for a number of titanium alloys are shown in Figure 1 as a function of yield strength. This figure is taken from a relatively old publication [4], but there has not been a comparable more recent compilation.

There is nothing fundamental to be derived from Figure 1. It is simply a "shopper's guide" and shows the values of K_{Iscc} available in various alloys having yield strengths as shown. It can be shown that the load-carrying capacity of a fixed-size member in the pertinent environment and containing a standard size surface flaw is linearly proportional to K_{Iscc}.

The lines in Figure 1 are drawn from the Irwin equation [5] simplified to the conditions of a long surface crack and yield-point stresses, under which conditions the equation becomes

$$a_{cr} = 0.2 \left(\frac{K_{Iscc}}{\sigma_y}\right)^2$$

where a_{cr} is the depth of the long surface flaw, σ_y is the yield strength and K_{Iscc} is as defined above.

The value of the fracture mechanics characterization is that, having determined empirically the K_{Iscc} of an alloy/environment system, and if one knows the <u>total</u> operative stresses, one can set an inspection standard for the maximum allowable flaw size.

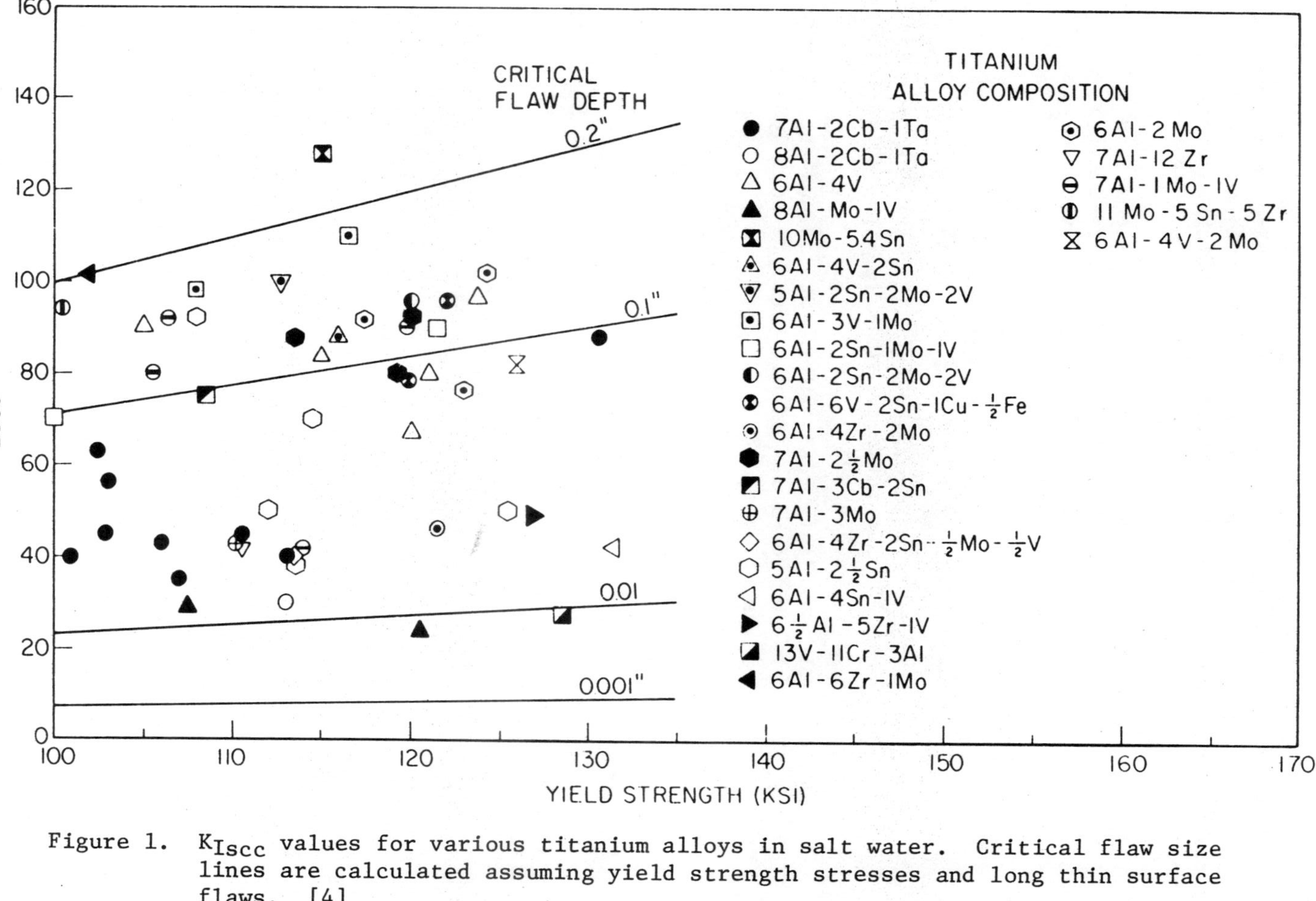

Figure 1. K_{Iscc} values for various titanium alloys in salt water. Critical flaw size lines are calculated assuming yield strength stresses and long thin surface flaws. [4]

It should be emphasized that much SCC in practice is a nucleation-and-growth phenomenon, that data such as in Figure 1 represent a truncation of much of the necleation process, and that Figure 1 is therefore an incomplete representation of SCC characteristics even of those titanium alloys shown thereupon in salt water.

The SCC characterizations of other alloy systems tend to be even less quantitative than the titanium alloys of Figure 1, so much so that pricatical ratings for some alloys may be simply categorized along the following lines:

Category A - Very high resistance. No record of service problems and SCC not anticipated in general applications.

Category B - High resistance. No record of service problems and SCC not anticipated if well recognized precautions are taken.

Category C - Intermediate resistance. SCC is not anticipated if the total sustained stress is kept less than a specified value.

Category D - SCC failures have been observed in service or would be anticipated if there are sustained tensile stresses in vulnerable directions with respect to the texture of wrought products.

These schematic categories will serve to illustrate the largely qualitative characterizations upon which much of our present day control technology is based.

CONTROL MEASURES

All control measures fall into three categores: (1) Reducing stress. Most of the stresses responsible for service SCC failures are fabrication stresses caused by forming, heat treating, welding, grinding, or misalignment, and predicting these stresses is not always easy. (2) Using alloys of high resistance to SCC. There are compilations of relative resistance, such as References [2] and [3]. (3) Ensuring where possible a benign environment. This involves, for example, low chloride levels in stainless steel steam systems. It may alternatively involve changing the environment by adding inhibitors, which have been of great practical utility in several engineering problems involving SCC. One many change the electrochemical environment by cathodic protection, and this move has proved to be of practical importance in, for example, cladding high strength aluminum alloys with sacrificial cladding material.

The practical control measures in current use for major alloy systems are summarized in Reference [2]. It should be noted that frequently the practical approach is a combination of several measures, such as modifying alloy choice somewhat (without going to a totally inert alloy), reducing residual stresses somewhat (without totally eliminating them, which may be difficult), and modifying the environment somewhat, as may be practicable without going to a totally benign environment.

FAILURE ANALYSES

There is one simple anion about SCC crack morphology: All stress corrosion cracks are "opening mode" (as in the opening of a book) - if a fracture is a <u>slant</u> fracture, it is <u>not</u> a stress corrosion crack. But an opening mode crack can be generated not only by SCC but also by liquid metal cracking, hydrogen embrittlement, "brittle fracture", fatigue, and corrosion fatigue.

The failure analyst has a large and increasing number of procedures at his disposal to distinguish between these opening mode fracture processes in many instances. These procedures include first of all a look with the unaided eye, then metallographic sectioning, scanning electron microscopy, high resolution electron microscopy of replicas, Auger electron microscopy, etc.

There are also failure analysis aids in the form of recent publications [6-8]. Reference [9] carries some especially useful suggestions and cautions as well as much sage advice in fracture identification.

It was once popular to say that stress corrosion cracks are typically branching. This is true of some, but not all, stress corrosion cracks. A branching morphology should alert the failure analyst to the possibility of SCC, but certainly the absence of branching does not exclude SCC.

There is a regime of corrosion fatigue which merges into cyclic SCC and then into simple SCC as the stress factor is changed, and in such circumstances it is not easy, and perhaps not always possible to draw a sharp line of demarcation between the fracture modes.

It has not been found possible to distinguish between SCC and hydrogen embrittlement of high strength heat treated steels, even by the most skilled high resolution electron fractography, perhaps because there may not be any difference between these failure modes.

Studying pedigreed laboratory fractures of various modes in the failure material or similar material is of great help to the failure analysis.

ACKNOWLEDGEMENTS

The author should like to record his appreciation to Dr. Volker Weiss, Syracuse University, for the opportunity to participate in the 24th Annual Sagamore Army Materials Research Conference. The final typescript was prepared by M. H. Mason.

REFERENCES

1. Ailor, W.H. (ed.), Handbook on Corrosion Testing and Evaluation, John Wiley, New York, 1971

2. Brown, B.F., Stress Corrosion Cracking Control Measures, National Bureau of Standards, Washington, D.C., 1977.

3. Staehle, R.W., et al., ARPA Handbook on Stress Corrosion Cracking (and Corrosion Fatigue), in preparation, 1978.

4. Judy, R.W., Jr. and Goode, R.J., NRL Report 6564, 1967.

5. Irwin, G.R., Trans. A.S.M.E., Vol. 29, Series E, 1962, pp 651-4.

6. Metals Handbook, 8th Ed., Vol. 10, Failure Analysis and Prevention, A.S.M., Metals Park, OH, 1975.

7. STP 600, C.D. Beachem and W.R. Warke, eds., A.S.T.M., Philadelphia, PA, 1976.

8. STP 645, B.M. Strauss and W.H. Cullen, Jr., eds., A.S.T.M., Philadelphia, PA, 1978.

9. Meyn, D.A., Reference 8, p 49ff.

CHAPTER 7

FRACTURE MECHANICS APPLICATIONS FOR SHORT FATIGUE CRACKS

M. H. El Haddad and T. H. Topper

University of Waterloo

Waterloo, Ontario, Canada

ABSTRACT

Elastic and elastic plastic fracture mechanics solutions are modified to predict the growth of short fatigue cracks by introducing an effective crack length, which is equal to the actual crack length increased by an amount ℓ_o, into the solutions for intensity factors and the J integral method of analysis. Since the threshold stress approaches the fatigue limit of the material as the crack length tends to zero the value of ℓ_o can be obtained from the threshold stress intensity factor and the fatigue limit. Crack growth results for short cracks, in both elastic and plastic strain fields of smooth and notched specimens, when interpreted in terms of the modified solutions, show excellent agreement with elastic long crack data. The accuracy of the term ℓ_o in predicting crack growth rates for short cracks is found to be independent of the applied strain level. It varies linearly with grain size for low carbon steels and can be considered at the surface as a measure of the reduced flow resistance of surface grains due to their lack of constraint. It also accounts for the effect of the free boundary on the distribution of stresses around the crack tip. Non propagating cracks lengths and the minimum stress levels required for failure are correctly predicted based on the fracture mechanics solutions.

INTRODUCTION

Fatigue cracks in structural elements usually initiate at a geometric stress raiser such as holes, changes of member cross section or weld toes. Cracks thus formed advance first through the highly stressed region close to the notch and then continue through the body of the structural element until final fracture occurs. Analysis of the early stage of crack growth is not, however, possible in terms of linear elastic fracture mechanics which is neither applicable to the plasticity generated in notches at high load levels nor accurate for very short cracks. The importance of small cracks, from both scientific and practical viewpoints, has recently been recognized [1-3] and there is substantial evidence that in some cases small cracks do not follow the crack growth laws derived from larger cracks [1-5]. Therefore, the description of fatigue crack propagation rate by the Paris law [6] which is based on the singular part (ΔK) of the stress field becomes questionable for extremely small cracks, unless ΔK solutions are modified. Otherwise, a higher crack propagation rate would be expected for smaller cracks [1-4].

The authors [1-2] have developed an elastic plastic fracture mechanics solutions which appear to be applicable to both short and long cracks in smooth and notched specimens. These solutions are employed in this chapter to predict the growth of fatigue cracks in smooth and at notched specimens and in addition to explain the phenomenon of non-propagating cracks.

FATIGUE CRACK GROWTH IN SMOOTH SPECIMENS

Elastic Crack Propagation and the Threshold Stress Intensity

The following expression has been proposed [1] for the elastic stress intensity factor ΔK of a short crack having a length ℓ.

$$\Delta K = F \Delta S \sqrt{\pi(\ell + \ell_o)} \tag{1}$$

where ΔS is the applied nominal stress range, ℓ_o is a constant for a given material and material condition, and F is a geometry dependent constant which is of the order of unity for cracks of a size that is small compared to other geometric dimensions. The threshold stress at a very short crack length is shown to approach the fatigue limit of the material ($\Delta\sigma_e$) based on smooth specimen tests [1,13-15], and from Equation (1) the threshold stress intensity factor ΔK_{th} can be obtained as:

$$\Delta K_{th} = \Delta\sigma_e \sqrt{\pi(\ell_o)} \tag{2}$$

or

$$\ell_o = \left[\frac{\Delta K_{th}}{\Delta\sigma_e}\right]^2 \frac{1}{\pi} \tag{3}$$

where F is approximated as unity. At any crack length ℓ, the threshold stress is then obtained through Equation (1) as:

$$\Delta\sigma_{th} = \frac{\Delta K_{th}}{\sqrt{\pi(\ell+\ell_o)}} \qquad (4)$$

As the crack length decreases, the length ℓ_o constitutes an increasing fraction of the effective length, until at very short lengths it represents an effective crack length at which the fatigue limit stress can propagate a crack into the interior of the specimen. It is significant that, using long crack data as base line behavior, the stress intensity defined by Equation (1) will predict higher crack propagation rates for short cracks than would be the more usual representation of stress intensity which deletes the term ℓ_o.

A series of experiments on CSA G40.11 steel plates for fully reversed (R = -1) loading were performed to examine the accuracy of the term ℓ_o in correlating experimental threshold and short crack results. In these experiments short cracks were initiated at the ends of slits cut into the sides of a 2.5mm thick sheet specimens by applying a very small load. The slit was then machined away and the threshold stress levels for various crack lengths determined using a series of specimens. Considerable care was exercised to avoid any effect of previous loading on the behavior, with a stress relief treatment being applied in some cases. Figure 1 represents these experimental results together with the theoretical curve given by Equation (4). The predictions of the equation adequately describe the effect of crack size on threshold stress intensity. Figure 1 also indicates that, as the crack length decreases below 1mm, the term ℓ_o constitutes an increasingly important fraction of the effective length, resulting in an increasing deviation of the threshold stress from the more usual prediction of Equation (1) without ℓ_o. Figure 2, which reproduces the above experimental data on the basis of ΔK_{th} versis crack length, shows that while the threshold stress intensity factor ΔK_{th} calculated via Equation (1) is dependent on crack length if the term ℓ_o is deleted from the equation, ΔK_{th} remains constant if ℓ_o is included in the equation. This observation is also supported by the experimental results given in References [13-15]

To examine the ability of the term ℓ_o to predict the crack growth rates of short cracks, measured crack growth rates in CSA G40.11 steel plates as a function of stress intensity factor without including the term ℓ_o are given in Figure 3. Results for crack lengths less than 1mm have higher crack growth rates than predicted by the long crack trend as shown in the figure, but if Equation (1) is used to correlate short crack results the discrepancy between short and long crack results is eliminated, as shown in Figure 4. This observiation is also supported by the experimental results

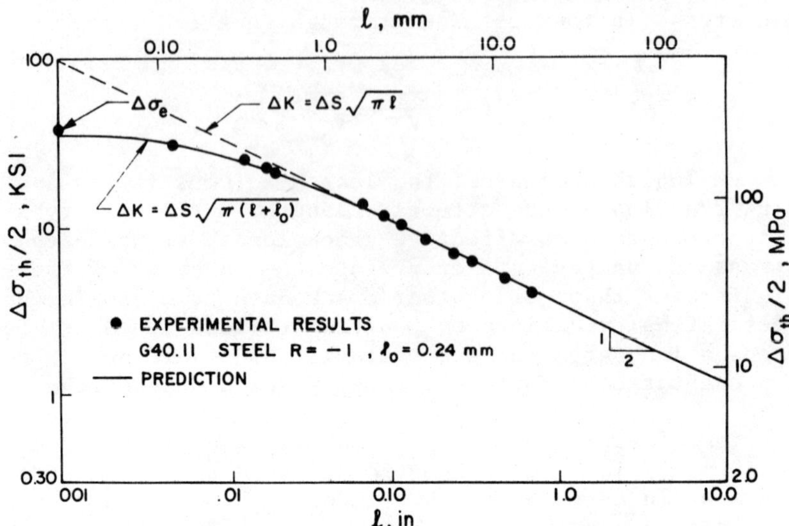

Figure 1. Effect of crack length on threshold stress amplitude for fatigue crack growth in G40.11 steel.

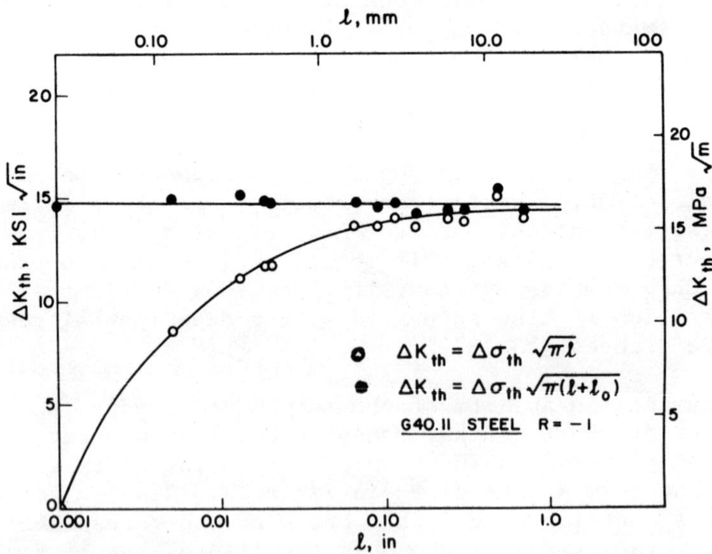

Figure 2. Estimation of threshold stress intensity factor.

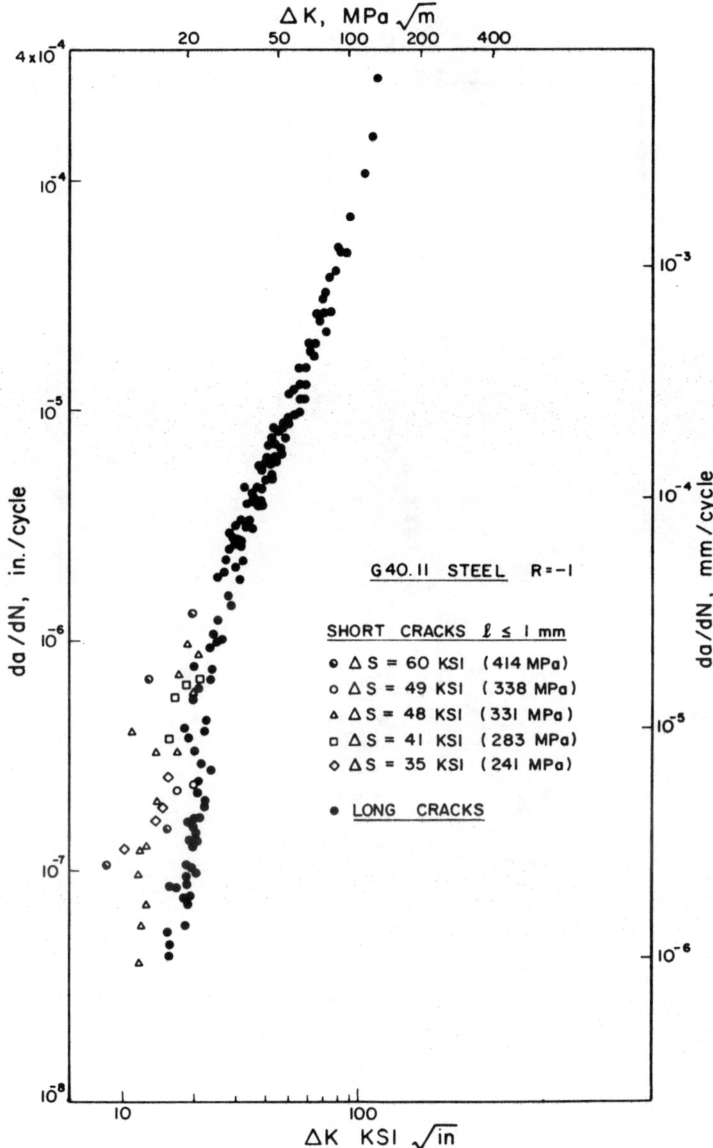

Figure 3. Short and long crack results for G40.11 steel without ℓ_o included in the ΔK solution.

The term ℓ_o which modifies the stress intensity appears to adequately account for two unique features of deformation near a free surface. The first of these is the effect of the boundary on the distribution of stresses around the crack tip. Talug and

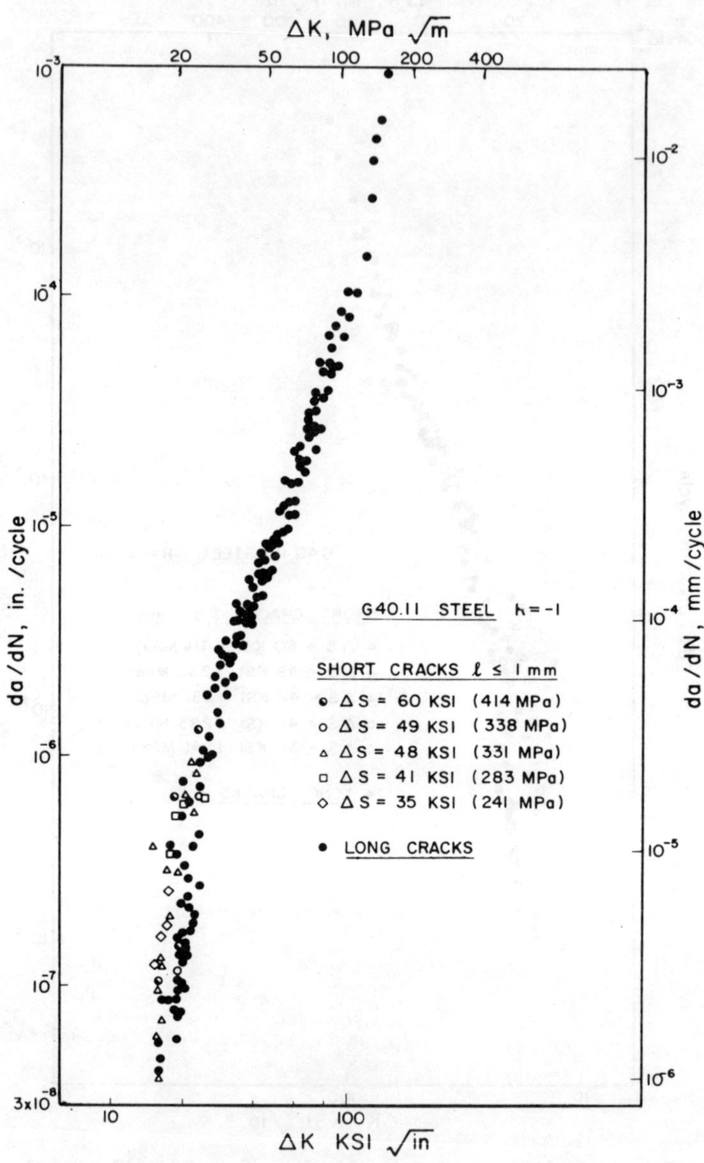

Figure 4. Fatigue crack propagation rate as a function of ΔK. Given by Equation (2).

Reifsnider [3] pointed out that when the distance from the crack tip to the surface becomes very small the conventional ΔK based only on the singular part of the stress field is inadequate to fully describe the near crack tip stress field. The second feature is the reduced flow resistance of the surface grains of a metal compared to interior grains due to their lack of constraint by surrounding grains [1]. As yet there is no analytical method for combining the effect of these features on crack tip plasticity but the empirical calculation of ℓ_o as the effective crack length necessary to start a crack by exceeding the threshold stress intensity when the applied stress is equal to the endurance limit stress provides accurate predictions of crack propagation behavior. The following paragraph examines its relationship to the grain size for low strength steels.

Recent observations [7-12] have shown that material strength and the scale of the microstructure can influence the threshold stress intensity and near threshold crack propagation rates. An increase in threshold values has been observed in lower-strength steels as the yield stress decreases [7-16]. Other investigators [7,17-19] have shown that the threshold for low carbon steels and titanium alloys increases with the grain size. This increase in threshold with grain size is the reverse of the decrease in yield stress and fatigue limit with grain size [7,18]. Since values for ℓ_o, given by Equation (3) are obtained based on values of ΔK_{th} and $\Delta\sigma_e$, a definite relationship between ℓ_o and grain size must exist. To establish such a relationship experimental data found in literature [7-10,12,20] are given in Figure 5. Values of ℓ_o are estimated using Equation (3) by substituting values of ΔK_{th} and the yield stress σ_y to approximate the fatigue limit (note that this approximation ignores cyclic hardening and softening and assumes that the fatigue limit is reasonable approximated by the cyclic yield stress). Figure 5 suggests that ℓ_o indeed varies linearly with grain size for the low strength steels given in the figure.

Plastic Crack Propagation

In recent years, several attempts have been made to extend linear elastic fracture mechanics into the elastic plastic regime by using the J integral method of analysis. As originated by Rice [22], the J-integral is analogous to the strain energy release rate, G, except that it is based on nonlinear, rather than linear, elasticity. Thus, for the special case of linear elasticity, J reduces the G, which is in turn directly and simply related to the stress intensity factor ΔK. For elastic-plastic materials, J loses its physical interpretation in terms of the potential energy available for crack extension, but retains physical significance as a measure of the intensity of the characteristic crack tip strain field, with the value of J being influenced by both load and plastic deformation. Dowling [5,22-23] attempted to analyze fatigue crack propagation rates in terms of the J integral for several kinds of specimens, specifically for centrally cracked specimens, compact specimens, and smooth axial specimens. He successfully correlated

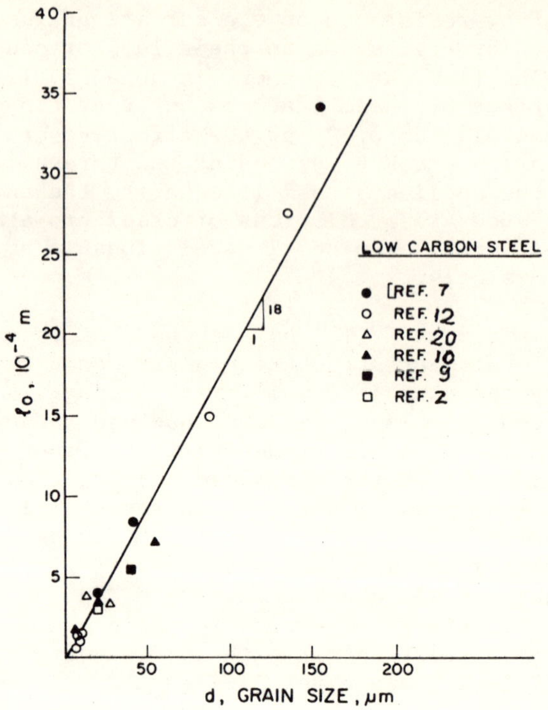

Figure 5. Relationship between ℓ_o and grain size for low strength steels.

elastic plastic data described using the J integral with elastic data based on linear elastic fracture mechanics. Results based on his solution [5] for the J integral for cracked smooth specimens are given in Figure 6, which shows that data for crack lengths less than 0.20mm indicate higher crack growth rates than those predicted by the long crack trend. To avoid this discrepancy between short and long crack results solutions for the J integral will be modified at short crack lengths.

In estimating a value of the J integral for a small crack under uniform strain in a smooth specimen, linear elastic and exponential hardening plastic cases may be considered separately and then combined to approximate elastic-plastic stress strain behavior [5,24]. The J integral was developed for nonlinear elastic materials, and for such materials can be expressed in terms of the potential energy available for crack extension. For the special case of a linear elastic material, J reduces to the strain energy release rate, G. Assuming plane stress, J, for the elastic case is simply related to

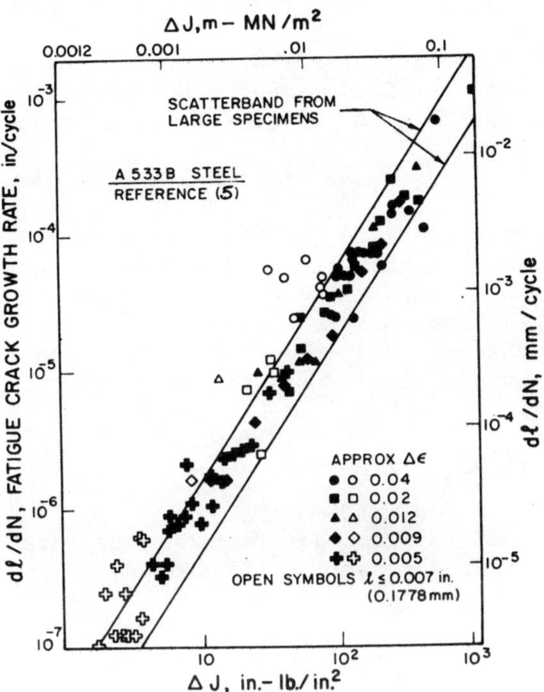

Figure 6. Short and long crack results for A533B steel without ℓ_o included in ΔJ solution.

K and E as follows:

$$\Delta J_e = G = \frac{\Delta K^2}{E} \tag{5}$$

For plastic deformation, J loses its potential energy interpretation, but retains physical significance as a measure of the crack tip strain field.

Consider the case of a smooth specimen containing a crack and having an applied stress range, ΔS; the linear elastic stress intensity is expressed by Equation (1). From Equations (1) and (5)

$$\Delta J_e = \frac{F^2 \, \Delta S^2 \, \pi(\ell+\ell_o)}{E} \tag{6}$$

or

$$\Delta J_e = 2\pi F^2 \, W_e(\ell+\ell_o) \tag{7}$$

where W_e is the elastic strain energy density, $(\Delta S^2/2E)$.

An approximate solution for J for the exponential hardening plastic case may be obtained based on an estimate that has been made by Shih and Hutchinson [24] for tension loaded cracked members:

$$\Delta J_p = 2\pi F^2 f(n) W_p (\ell + \ell_o) \tag{8}$$

where f(n) is a function [24] of the strain hardening exponent, n, which is defined by the stress-strain realtionship $\Delta S \alpha (\Delta e_p)^n$. The value of F is assumed to be the same as for the elastic case. The quantity W_p is the plastic strain energy density, which may be expressed [5] in terms of n, the stress (ΔS), and the plastic strain range (Δe_p), causing Equation (8) to yield:

$$\Delta J_p = 2\pi F^2 f(n) \frac{(\Delta S \Delta e_p)}{n+1} (\ell + \ell_o) \tag{9}$$

For combined elastic-plastic deformation, the total J may be approximated by adding Equations (7) and (9).

$$\Delta J = 2\pi F^2 (\ell + \ell_o) \{[\frac{\Delta S^2}{2E}] + [\frac{\Delta S \ \Delta e_p}{n+1}] f(n)\} \tag{10}$$

Noting that the plastic strain range, Δe_p, is equal to the total strain range, Δe minus the elastic strain range, $\Delta S/E$ and the last equation may be expressed as:

$$\Delta J = 2\pi F^2 (\ell + \ell_o) \{\frac{f(n)}{n+1} (\frac{\Delta S \Delta e E}{E}) - (\frac{2f(n)}{n+1} - 1) \frac{\Delta S^2}{2E}\} \tag{11}$$

Equation (11) gives the J integral estimate for a cracked smooth axial specimen. In this equation ℓ_o, n, and f(n) are known constants for a given material. The term ℓ_o which predicts the behavior of short cracks is assumed to be independent of the applied strain level. Note also that Equation (11) reduced to Equation (6) for elastic crack propagation and therefore for elastic material response Equations (1) and (11) will give the same result. To examine the accuracy of Equation (11), the data of Figure 6 are re-analyzed based on Equation (11) and plotted in Figure 7. For the half circular surface cracks found in these tests, a value of F = .71 is appropriate [5]. The firgue indicates that Equation (11) indeed accurately models the effects of plasticity and crack size on crack growth rate. Note that the effectiveness of the term ℓ_o in correctly predicting the higher crack growth rates for short cracks is independent of the applied strain level.

A strain based intensity factor using strain and length rather

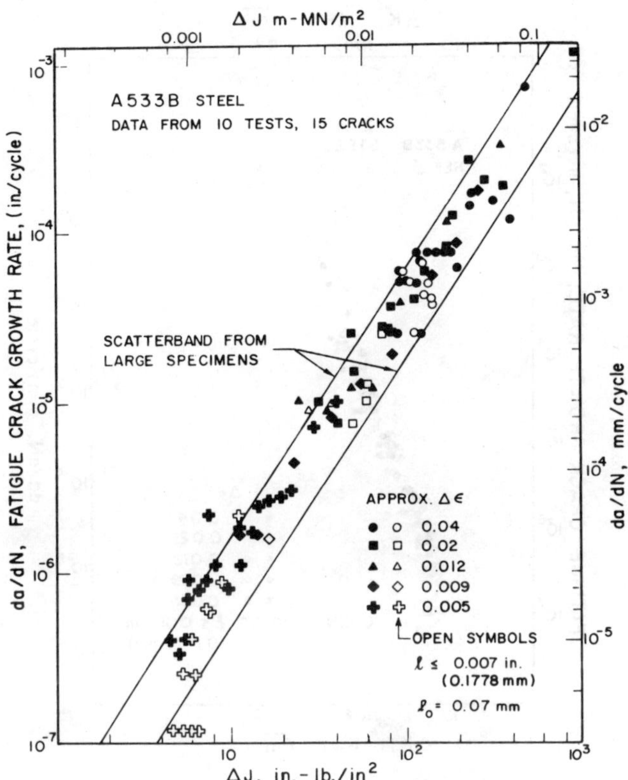

Figure 7. Fatigue crack growth rates as a function of ΔJ given by Equation (11).

than stress and length has been applied also to cracks in plastically strained smooth specimens [1-2,26-28]. Rewriting Equation (1) in terms of strains gives:

$$\Delta K = FE\ \Delta e\sqrt{\pi(\ell+\ell_o)} \qquad (12)$$

For the elastic case Equation (12) is identical to Equation (1) but as the strain increases to inelastic levels ΔK values given by Equation (12) exceed those of Equation (1). When plotted in terms of ΔK values given by Equation (12), short crack growth data for plastically strained smooth specimens correlated with elastic long crack data [1-2]. The data of Figures 6 and 7 are replotted in Figure 8 on coordinates of da/dN versus ΔK, where values of ΔK are based on Equation (12). Again, for the half circular surface cracks found in these tests a value of F = 0.71 is appropriate [5]. The figure indicates the accuracy of Equation (12) in modelling the

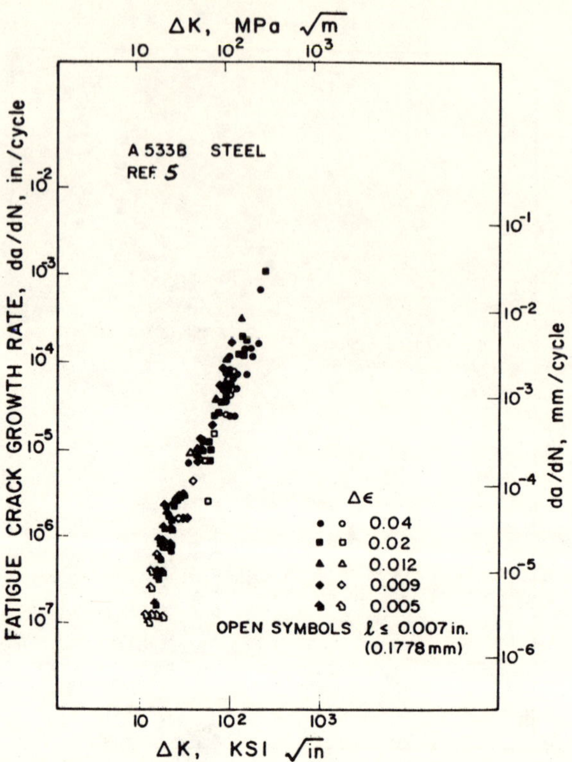

Figure 8. Fatigue crack growth rates as a function of ΔK given by Equation (12).

combined effects of plasticity and small crack size. Also, notice again that the effectiveness of the term ℓ_o in correctly predicting the higher crack growth rates for short cracks is independent of the applied strain level.

FATIGUE CRACK PROPAGATION IN NOTCHED SPECIMENS

Elastic Plastic Solutions

When applied stress levels are low enough that notch strains remain elastic, solutions for the elastic stress concentration factor k` for a crack is a notch may be used to determine the stress intensity factor as follows:

$$\Delta K = k` \; \Delta S\sqrt{\pi(\ell+\ell_o)} \tag{13}$$

where k' is a function of $\ell/(\ell+c)$ as given in Reference [29]. Here the crack length, ℓ_o, is measured from the notch root and c is the notch depth. The quantity ℓ_o is again the material constant defined by Equation (3). The stress concentration factor k' decreases from an initial maximum value equal to 1.12 k_t where k_t is the theoretical stress concentration factor, to a value of

$$\sqrt{\frac{\ell+\ell_o+c}{\ell+\ell_o}}$$

as the crack passes outside the field of influence of the notch. At stress levels causing local yielding Equation (13) is invalid [2,25]. However, Equations (11) and (12) remain valid and will now be used to estimate ΔJ and ΔK values for a short crack in the practical design situation in which yielding is restricted to the immediate vicinity of the notch root.

Replacing nominal stress and strain ranges ΔS and Δe with local stress and strain ranges in the vicinity of the crack tip $\Delta\sigma$ and $\Delta\varepsilon$ in Equations (11) and (12) gives:

$$\Delta J = 2\pi F^2 (\ell+\ell_o) \left\{\frac{f(n)}{n+1} \frac{(\Delta\sigma \cdot \Delta\varepsilon E)}{E} - \left(\frac{2f(n)}{n+1} - 1\right) \frac{\Delta\sigma^2}{2E}\right\} \quad (11a)$$

and

$$\Delta K = FE\Delta\varepsilon\sqrt{\pi(\ell+\ell_o)} \quad (12a)$$

Although neither exact nor finite element plastic solutions are generally available for cracks in notches, estimates of local strains may be derived from a relationship between stress and strain concentration factors proposed by Neuber [30]. For the practical case of elastic nominal stress levels ΔS Neuber's rule is obtained in the following form [2,25]

$$k'\Delta S = [\Delta\sigma\Delta\varepsilon E]^{1/2} \quad (14)$$

Values of the left hand side of Equation (14) for a given crack length and nominal stress may be computed using an elastic solutions for k' [29]. Since terms on the right hand side involve only material stress strain response, a base curve of $(\Delta\sigma\Delta\varepsilon E)^{1/2}$ versus $\Delta\varepsilon \cdot E$ and be constructed from stress strain data for a given material [2,25]. Values of $\Delta\varepsilon$ corresponding to a given value of $k'\Delta S$ obtained from this curve may be entered into a material stress-strain curve, to obtain values of $\Delta\sigma$. These $\Delta\sigma$ and $\Delta\varepsilon$ values can then be inserted into Equations (11a) and (12a) to determine values of ΔJ or ΔK corresponding to the particular crack length. If the crack front is straight a reasonable assumption for the thin plates considered in the next section, a value of unity is appropriate for F in Equation (11a) and (12a).

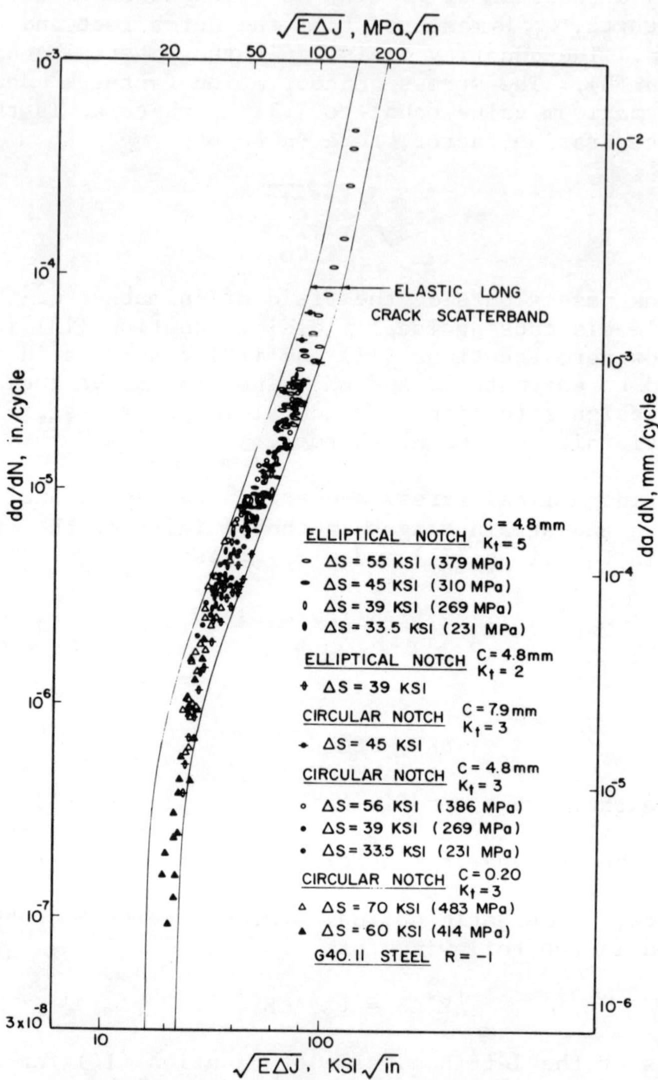

Figure 9. Fatigue crack growth rates as a function of ΔJ given by Equation (11a).

Fatigue Crack Growth Rates

Constant amplitude load controlled tests were performed at different stress levels in thin circular and elliptical notched plates fabricated from CSA G40.11 steel [2.25]. Crack propagation rates were determined from travelling microscope measurements of

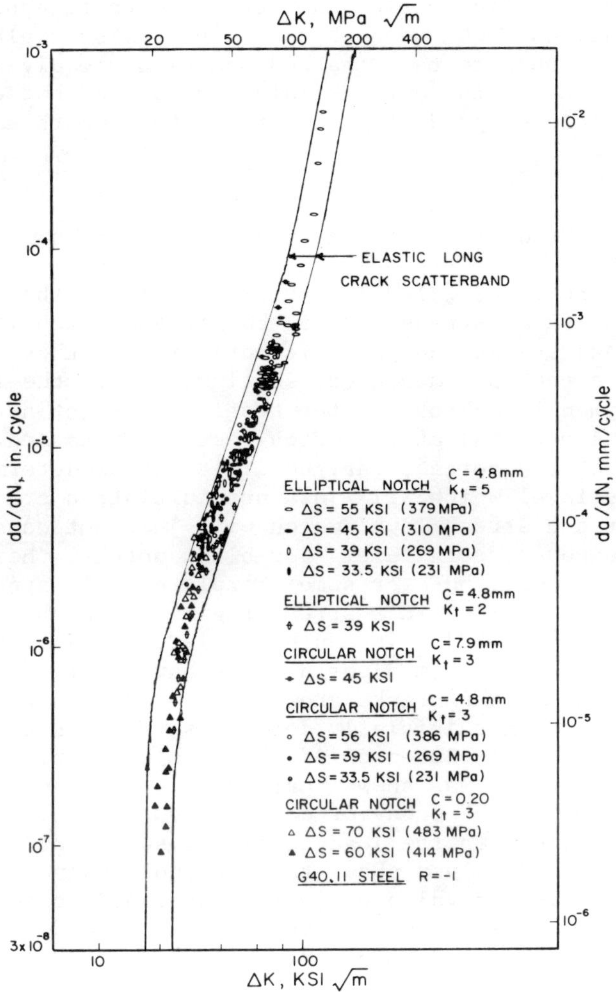

Figure 10. Fatigue crack growth rates as a function of ΔK given by Equation (12a).

crack length at various cycle numbers. Straightforward calculations of ΔJ and ΔK values using cyclic stress strain data to construct the curves suggested in the previous section and Equations (11a) and (12a) were performed and the results plotted versus da/dN in Figures 9 and 10. The data obtained for a variety of notches and load levels shows excellent agreement with the long crack data. The similar crack growth rate versus ΔJ and ΔK given in Figures 9 and 10 for elastic crack growth are explained by the simple elastic relationship between the variables $[(\Delta J = \Delta K^2)/E]$. Since Figures 9 and 10 also include inelastic crack growth data it appears that the same relationship between ΔK and ΔJ persists at plastic strain levels [1]. It appears then that ΔJ and ΔK are equivalent quantities for both elastic and inelastic strain levels and that either can be used to correlate or predict crack growth for smooth and notcted specimens [1-2,25].

Prediction of Non Propagating Cracks

Cracks were found [31-32] to form quickly at the roots of sharp notches, even at los stress levels, but if the notch is very sharp and the nominal stress range sufficiently small, they do not continue to grow across the specimen cross section. Thus, the fact that a notched specimen is unbroken after testing does not necessarily imply that the material at the notch root is uncracked. It follows, therefore, that the notched fatigue limit may be determined either by the stress level which will just not initiate a crack at the notch root or the stress level which will just not cause a crack to grow to cause complete failure. For blunt notches these two fatigue limits are identical, but for some sharply notched specimens, they are different, resulting in a stress regime in which non propagating cracks are found [31-32]. The authors [1-2,25] have shown that while cracks growning from blunt notches have continuously increasing crack growth rates, the cracks growing from sharp notches exhibit a crack growth rate that initially decreases, reaches a minimum and thereafter increases. Based on ΔK or ΔJ solution given in the previous sections, it was shown that [2,25] the change from "sharp notch" to "blunt notch" behavior depended both on notch geometry and notch size and when the load level for sharp notches results in an initial ΔK or ΔJ values above the threshold value for crack propagation but a minimum value below the threshold, crack growth started and then ceased when the applied K or J decreased to the threshold, resulting in dormant cracks [32].

Non propagating crack lengths and notch size effect can also be predicted in terms of the applied nominal stress ΔS_{th}. This stress may be obtained by substituting the value of ΔK_{th} into Equation (13). Rearranging Equation (13), we obtain:

$$\Delta S_{th} = \frac{\Delta K_{th}}{k' \sqrt{\pi(\ell+\ell_o)}} \tag{15}$$

Substituting for the value of ΔK_{th} using Equation (2), the last equation reduces to:

$$\Delta S_{th} = \frac{\Delta\sigma_e}{k'} \sqrt{\frac{\ell_o}{(\ell+\ell_o)}} \tag{16}$$

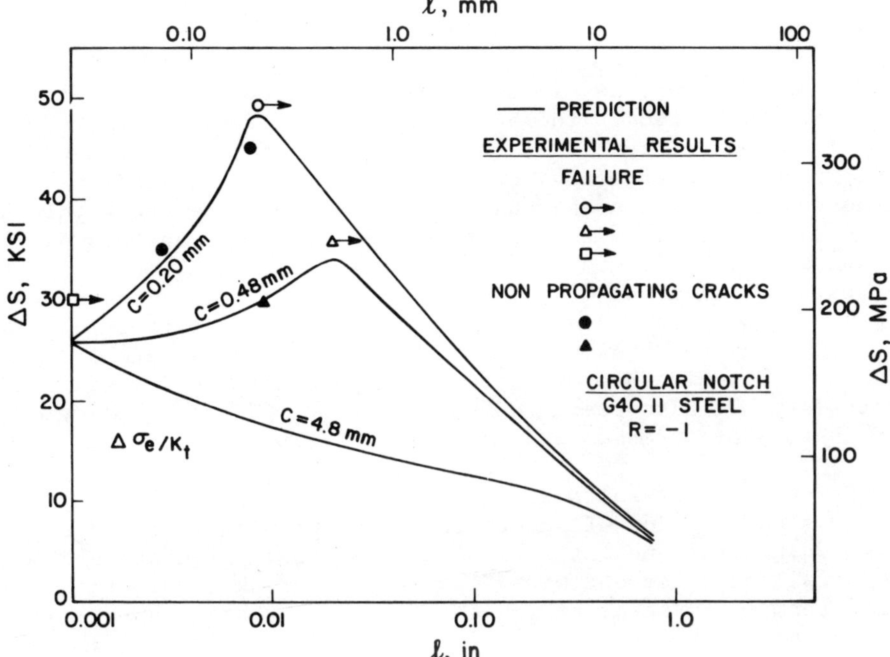

Figure 11. Prediction of non propagating cracks in circular notches.

Solutions for ΔS_{th} given by Equation (16) are shown in Figure 11 for cracks initiating from circular notches with different diameters. For the largest notch the threshold stress corresponding to failure is at initiation and once cracks initiate they will continue growing until final fracture. This initiation level is equal to $\Delta\sigma_e/k_t$. However, as the notch size decreases the peak value of the threshold stress versus length curve shifts to the right as shown in Figure 11. This peak value gives the minimum stress level required to fail the specimen. At stress levels below this value but above the value of $\Delta\sigma_e/k_t$ which corresponds to initiation, cracks will start but will not propagate. Therefore, for a given notch geometry, decreasing the size of the notch results in an increase in the stress level required for failure and a decrease in notch sensitivity. Experimental results for non propagating cracks and the stress levels corresponding to failure given in Figure 11 for a CSA G40.11 steel lie close to the curves predicted by Equation (15). Also, non propagating crack lengths and stress levels corresponding to failure were accurately predicted in notches of the same size but with different geometry [32].

CONCLUSIONS

Elastic as well as elastic plastic fracture mechanics solutions are modified to predict the behavior of short fatigue cracks. The lack of a physical continuum at small size scales is account for by introducing an effective crack length, ℓ_o, calculated in terms of the smooth specimen endurance limit and the long crack threshold stress intensity, into the solutions for intensity factors and the J integral. This length, ℓ_o, is added to the actual crack lengths to obtain correlations of crack growth rates for long and short cracks. It appears to vary linearly with grain size for low strength steels and may be considered at the surface as a measure of the reduced flow resistance of surface grains due to their lack of constraint. It also accounts for the effect of the boundary on the distribution of stresses around the crack tip.

Crack growth results for short cracks, in both elastic and plastic strain fields of smooth specimens when interpreted in terms of the J integral and the strain based intensity factor solutions, show excellent agreement with elastic long crack data. The accuracy of the term ℓ_o in predicting higher crack rates for short cracks is independent of the applied strain level. The J integral and the strain based intensity factor solutions successfully correlated data for the growth of short cracks in plastically strained notches with elastic long crack rsults. Conversely, short crack growth at inelastically strained notches was accurately predicted from long crack data. Also, non propagating crack lengths and the minimum stress levels required for failure are correctly predicted based on the fracture mechanics solutions.

REFERENCES

1. El Haddad, M.H., Smith, K.N. and Topper, T.H., 1978 ASME/CSME Joint Conference on Pressure Vessels and Piping, Nuclear Energy and Materials, Montreal, Juen 1978, Paper No. 78-Mat.-7.
2. El Haddad, M.H., Smith, K.N. and Topper, T.H., Eleventh National Symposium on Fracture Mechanics, ASTM, Blacksburgh, VA, June 1978.
3. Talug, A. and Reifsnider, K., ASTM STP 637, 1977, pp 81-96.
4. Pearson, S., Engineering Fracture Mechanics, Vol. 7, 1975, pp. 235-47.
5. Dowling, N.E., ASTM STP 637, 1977, pp 97-121.
6. Paris, P.C. and Erdogan, F., J. Basic Engineering, Series D of the Transactions of ASME, Vol. 85, 1963, pp. 528-34.
7. Masounave, J. and Bailon, J.P., Scr Metallurgica, Vol. 10, 1976, pp. 165-70.
8. Gerberich, W.W. and Moody, N.R., ASTM-NBS-NSF Symposium on Fatigue Mechanisms, Kansas City, MO, May 1978.
9. Kunio, T. and Yamada, K., ASTM-NBS-NSF Symposium on Fatigue Mechanisms, Kansas City, MO, May 1978.
10. Taira, S., Tanaka, K. and Hoshina, M., ASTM-NBS Symposium on Fatigue Mechanisms, Kansas City, MO, May 1978.
11. Clark, W.G., Metals Engineering Quarterly, August 1974, pp 16-22.
12. Masounava, J. and Bailon, J.P., Proc. of the 2nd Int. Conf. of the Mech. Behaviour of Materials, American Society for Metals, Cleveland, OH, 1976, p. 636.
13. Kitagawa, H. and Takahashi, S., 2nd Int. Conf. on Mechanical Behaviour of Materials, Boston, MA, August 1976, pp 627-30.
14. Frost, N.E., Proc. of the Institution of Mechanical Engineers, Vol. 173, No. 35, London, 1959, pp 811-35.
15. Frost, N.E., Marsh, K.J. and Pook, L.P., Metal Fatigue, Clarendon Press, Oxford, 1974.
16. Ritchie, R.O., Fatigue 1977 Conference, Univ. of Cambridge, March 1977.
17. Beevers, C.J., Fatigue 1977 Conference, Univ. of Cambridge, March 1977.
18. Robinson, J.L. and Beevers, C.J., Metal Sci. J., 7, 1973, p 153.
19. Irving, P.E. and Beevers, C.J., Material Sci., Eng. 14, 1974.

20. Cooke, P.J. and Beevers, C.J., Materials Science and Engineering, Vol. 13, 1974, p. 201.

21. Rice, J.R., ASME Journal of Applied Mechanics, June 1968, pp. 379-86.

22. Dowling, N.E. and Begley, J.A., ASTM STP 590, 1976, pp 82-103.

23. Dowling, N.E., ASTM STP 601, 1976, pp 19-32.

24. Shih, C.F. and Hutchinson, J.W., Division of Engineering and Applied Physics, Harvard Univ., Cambridge, MA 02138, Report No. DEAP. S. 14, July 1975.

25. El Haddad, M.H., Dowling, N.E., Topper, T.H. and Smith, K.N., J. Integral Appl. for Short Fatigue Cracks at Notches, submitted to Int. J. of Fracture, May 1978.

26. Boettner, R.C., Laird, C. and McEvily, A.J., Transactions of Metallurgical Society of AIME, Vol. 233, 1965, pp. 379-87.

27. Solomon, H.D., J. of Materials, Vol. 7, No. 3, Sept. 1972, pp 299-306.

28. Rau, C.A., Gemma, A.E. and Leverant, G.R., ASTM STP 520, 1972, pp 166-78.

29. Tada, H., Paris, P.C. and Irwin, G.R., The Stress Analysis of Cracks Handbook, Del Research Corp., Hellertown, PA, 1973.

30. Neuber, H., Journal of Applied Mechanics, ASME, 28, 1961, pp 544-50.

31. Frost, N.E., The Aeronautical Quarterly, Vol. 8, pp 1-20.

32. El Haddad, M.H., Smith, K.N. and Topper, T.H., "Prediction of Non Propagating Cracks", accepted for Publication - Eng. Fracture Mechanics, August 1978.

CHAPTER 8

REVIEW OF CONTEMPORARY APPROACHES TO FATIGUE DAMAGE ANALYSIS

D. F. Socie and JoDean Morrow

Department of Theoretical and Applied Mechanics

University of Illinois, Urbana, Illinois

PURPOSE AND SCOPE

A review of current methods for fatigue damage analysis employing smooth specimen materials data for predicting the service life of components and structures subjected to variable loading will be presented. It is written for the beginner in this area of fatigue damage analysis rather than for experienced practitioners. Special emphasis is placed on the detailed elements of the analysis, as well as the overall pattern for synthesizing these elements into a working computer-based program.

Most of the pertinent literature in this field is less than ten years old and often difficult for the novice to locate. For this reason, key papers are cited as a guide to publications and useful data sources. Our purpose is to provide the reader with a single document containing, as nearly as possible, all the information required to implement the current state of the art damage analyses.

After discussing the details of the analysis, application of the techniques will be illustrated by making life predictions for the SAE Cumulative Fatigue Damage Test Program.

Appendices are presented with flow charts and Fortran IV listings of computer programs of all the necessary elements in a complete fatigue damage analysis for crack initiation life.

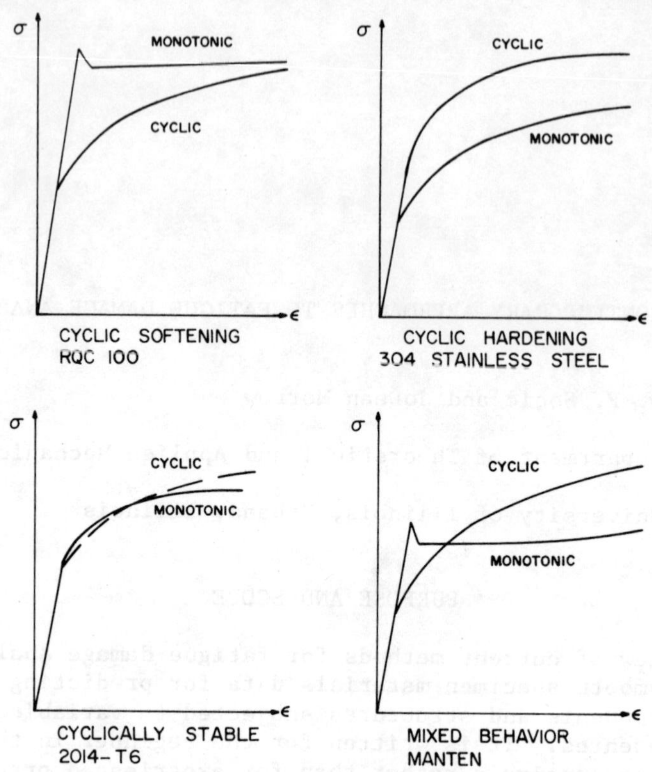

Figure 1. Cyclic and monotonic stress-strain behavior of metals.

FUNDAMENTAL OF FATIGUE DAMAGE ANALYSIS

Cyclic Stress-Strain Behavior of Metals

The monotonic stress-strain curve (often simply referred to as a tensile stress-strain curve) for metals is used to obtain design paraments for limiting stresses on structures and parts subjected to static loading. By analogy the cyclic stress-strain curve may be used by the designer to evaluate the durability of structures and parts subjected to cyclic loading. The stress-strain response of metals which are metastable under cyclic loading is usually altered due to repeated plastic strain. Depending on the initial condition of a metal (i.e., quenched and tempered, annealed, cold worked, etc.) and the test condition, it may cyclically harden, soften or remain relatively stable. The cyclic stress-strain curve can be compared with the monotonic stress-strain curve to quantitatively assess cyclically induced changes in material behavior, as shown in Figure 1. In the case of RQC-100, it is shown that cyclic yielding occurs well below the monotonic yield. This shows the falsity of designing to a monotonic yield strength rather than using the cyclic stress-strain curve for designing structures to resist fatigue loading.

Just as the monotonic stress-strain curve relates static applied stress and the resultant static strain, the cyclic stress-strain curve relates cyclic stress and strain. Of particular importance is the amount of cyclic plastic strain, since this quality is intimately related to fatigue damage, as will be discussed in the next section. An equation of the following form is generally used to express the cyclic stress-strain relationship:

$$\frac{\Delta\varepsilon}{2} = \frac{\Delta\varepsilon_e}{2} + \frac{\Delta\varepsilon_p}{2} = \frac{\Delta\sigma}{2E} + \left(\frac{\Delta\sigma}{2K'}\right)^{1/n'} \qquad (1)$$

The Δ, in the above equation, indicates completely reversed ranges of stresses and strains and subscripts e and p stand for elastic and plastic strain. The two material properties, n' and K', are the cyclic strain hardening exponent and cyclic strength coefficient, respectively. These are determined by subjecting smooth samples to completely reversed strain control, as illustrated in Figures 2(a) and 2(b). After the transient adjustment, essentially the same size and shape of hysteresis loop will be produced cycle after cycle. The stable, steady-state loop for each specimen tested is used to obtain one point on the cyclic stress-strain curve. Enough specimens are cycled at different strain ranges to provide the stable cyclic stress-strain behavior over a sufficient range to fit a function of the form of Equation (1) to the locus of tips of these stable loops, as shown in Figure 3.

A number of short-cut methods for approximating the cyclic stress-strain curve from only one specimen have been suggested. Perhaps the most popular of these is the incremental step strain test, illustrated in Figure 4. In this test one specimen is subjected to several large, completely reversed, strain cycles, and then the strain is gradually reduced in about forty steps to the point of zero stress and strain. The strain is then incrementally increased to the large level, then decreased, and so forth to failure. After the initial transient hardening or softening, the plot of stress versus strain furnishes an approximation to the locus of tips of stable loops. Once again an equation, of the form of Equation (1), can be fitted to this curve. However, the reader is cautioned that the values of K' and n' obtained in this way may be somewhat different from that obtained using the companion specimen method.

For the purpose of better understanding the meaning of the cyclic stress-strain curve, it should be noted that if a sample is pulled in tension from zero stress and strain after incremental step

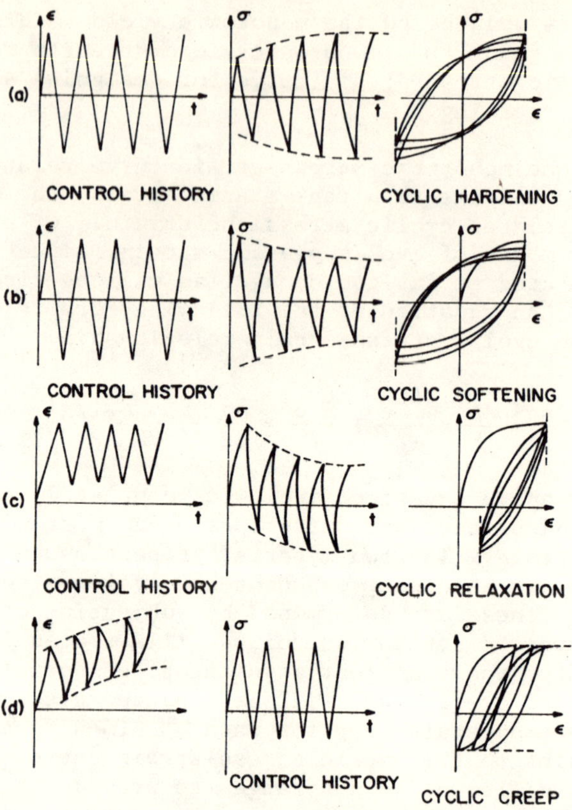

Figure 2. Schematic illustrations of cyclic transient phenomena.

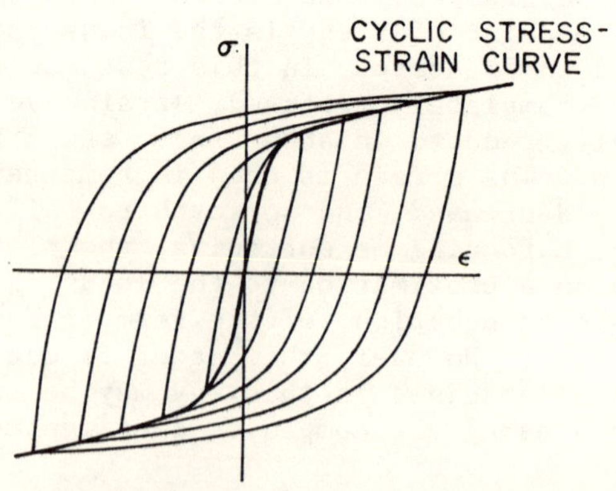

Figure 3. Cyclic stress-strain curve drawn through stable loop tips

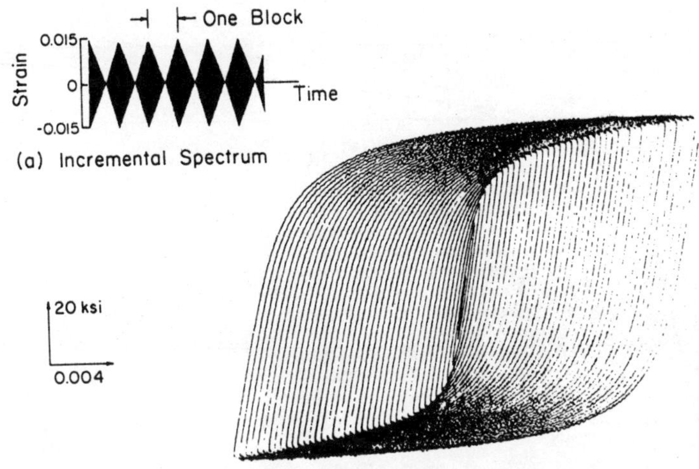

Figure 4. Incremental step straining of sheet steel.

straining, the stress-strain curve that is obtained will be very much like the stable cyclic stress-strain curve. This is illustrated in Figure 5. Once again the reader is cautioned that the values of n' and K' obtained from a tension test after incremental step straining may be somewhat different than those obtained from the companion specimens.

Fatigue Properties of Metals

Until the last decade the only fatigue property of metals determined by testing and reported in handbooks and literature was the so-called "endurance limit" or fatigue limit. This was the stress amplitude in a completely reversed test (usually rotating bending) which was considered to be safe - that is, below this stress amplitude the fatigue life was considered to be infinite. The pioneering work of Coffin and Manson in 1954, where the cyclic plastic strain and fatigue life were shown to be related by a simple power function, led to a more detailed formulation of fatigue properties with metals over the entire life range from only a half cycle or a tension test to millions of cycles. The form of the Coffin-Manson relation is as follows:

$$\frac{\Delta\varepsilon_p}{2} = {}'_f(2N_f)^c \qquad (2)$$

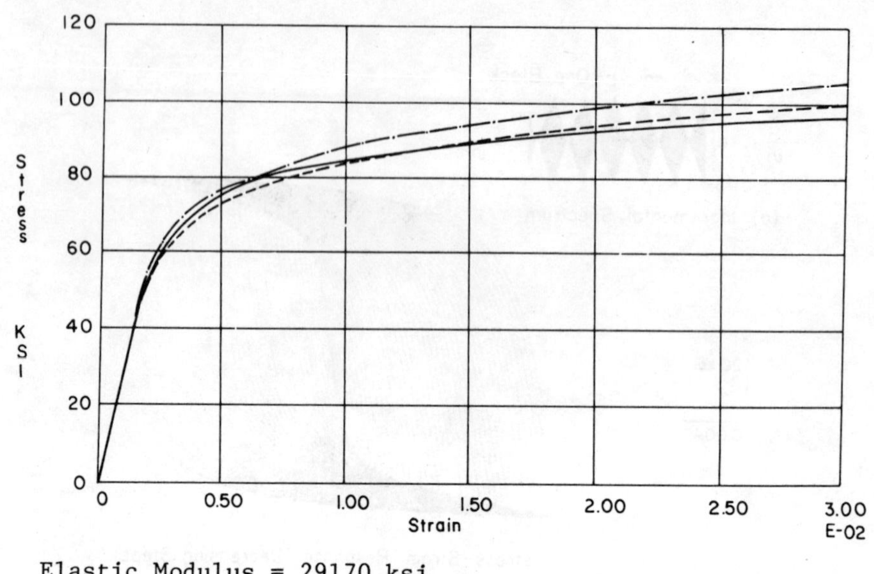

Elastic Modulus = 29170 ksi

n' = .092958	K' = 135.106	ksi	from LCF Tests
n' = .138	K' = 174.3	ksi	from Incremental Step
n' = .132	K' = 161.9	ksi	from Tension

——— Solid Line - Plot from LCF Test
—·—·— Marked Line - Plot from Incremental Step Test
— — — Marked Line - Plot from Tension After Cycling

Figure 5. Cyclic stress-strain curves for RQC-100 Steel.

The plastic strain amplitude, $\Delta\varepsilon_p/2$, is related to the life, $2N_f$, according to this equation in a log-log linear fashion, as shown in Figure 6. The symbol $2N_f$ is used to indicate reversals or half cycles, while N_f means number of cycles to failure (usually taken as the separation of the specimen into two pieces). For a constant amplitude test, the number of reversals is equal to twice the number of cycles. In a variable amplitude test it is not always a simple matter to identify a cycle. However, half cycles or reversals can always be evaluated as peak-valley or valley-peak excursions. The next section will be devoted to the matter of counting cycles during variable loading histories.

The representation of fatigue life in terms of plastic strain, while attractive from a fundamental viewpoint because of the intimate cause and effect relation between the two, is difficult to apply to practical engineering problems. This is because the

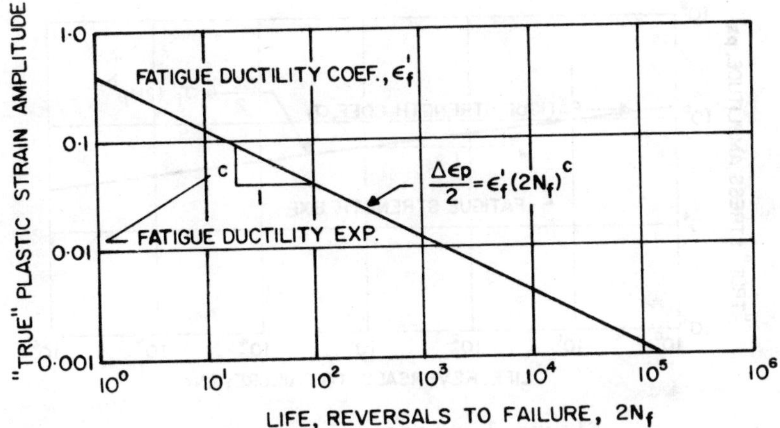

Figure 6. Plastic strain-life curve.

engineer and designer seldom know the amount of cyclic plastic strain present at the critical location in a member or part. Rather, they are more apt to be able to measure or estimate the total cyclic strain by using strain gages or methods of applied mechanics.

While it is common practice to change the strain to an equivalent elastic stress by using Hooke's law, it is well to remember that a protion of the total strain is not elastic whenever a fatigue problem exists.

The obsession that engineers have with the cyclic stress has been one of the major impediments to the acceptance and application of contemporary approaches to fatigue analysis. The endurance limit concept is ingrained in the thinking of many engineers, and the idea of the S-N curve to represent the finite life region is still used as a basis for cumulative damage analysis by many engineering groups. The S-N curve in the finite life region is often represented as a power function of stress in a form similar to Equation (2). This formulation was first suggest by H. O. Basquin in 1910 and is of the following form:

$$\frac{\Delta\sigma}{2} = \sigma_f'(2N_f)^b \tag{3}$$

Here the σ represents true stress. This equation implies a log-log linear relationship between stress and life, as shown in Figure 7. Note that the straight line continues to lower and lower stresses with longer and longer lives and that the "endurance limit" concept is not included in this equation.

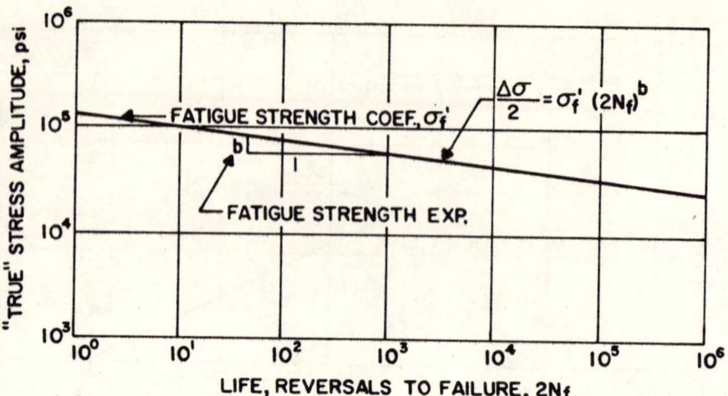

Figure 7. Stress-life curve.

This elimination of the "endurance limit" is appropriate for cumulative fatigue damage evaluation, since it is well known that large overstrains common in real life situations will cause the "knee" of the S-N curve to be eradicated. Cyclic stresses below the endurance limit will then produce finite fatigue lives.

In addition, many metals of engineering interest do not exhibit an "endurance limit" even in the absence of a periodic overstrain. In fact, principally ferrous-based metals and some other BCC metals are the only ones capable of developing an "endurance limit". The effect of overstrain in eliminating the endurance limit is illustrate for several steels in Figure 8.

Mean stress effects may be included in a variety of ways, as shown in Figure 9. For the purpose of this formulation of fatigue properties of metals, it is convenient to incorporate mean stress effects as an equivalent change in static strength. This consists of changing the coefficient in Equation (3) from σ_f' to $(\sigma_f' - \sigma_o)$. Equation (3) may then be modified as follows:

$$\frac{\Delta\sigma}{2} = (\sigma_f' - \sigma_o)\ (2N_f)^b \tag{4}$$

To obtain an expression relating total strain, mean stress and life, Equation (4) is divided by the Young's modulus, E, to obtain the elastic strain and added to Equation (2) for the plastic strain, as shown in Figure 10.

$$\frac{\Delta\varepsilon}{2} = \frac{\Delta\varepsilon_e}{2} + \frac{\Delta\varepsilon_p}{2} = \left(\frac{\sigma_f' - \sigma_o}{E}\right)\ (2N_f)^b + \varepsilon_f'(2N_f)^c \tag{5}$$

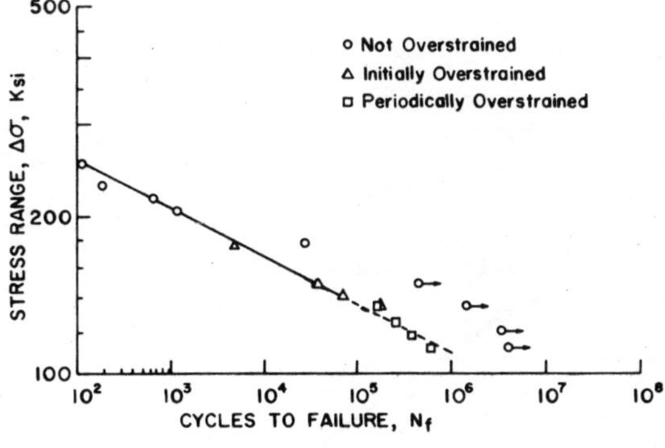

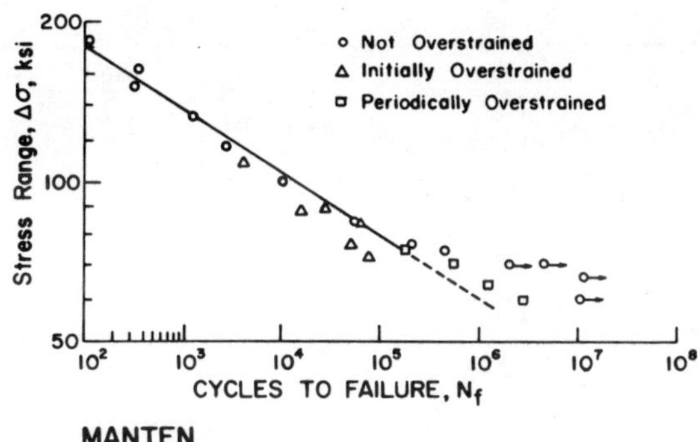

Figure 8. Effect of overstrain on the fatigue behavior of RQC-100 and Man-Ten.

The two exponents and the coefficients are regarded as fatigue properties of the metal, and they are designated as follows:

b = fatigue strength exponent
c = fatigue ductility exponent
σ_f' = fatigue ductility coefficient
σ_f' = fatigue strength coefficient

By manipulating Equation (2) and (3) to eliminate life and carrying the result to the form of Equation (1), it can be shown that the cyclic strain hardening exponent, n', is determined by the fatigue strength and ductility as follows:

$$n' = \frac{b}{c} \tag{6}$$

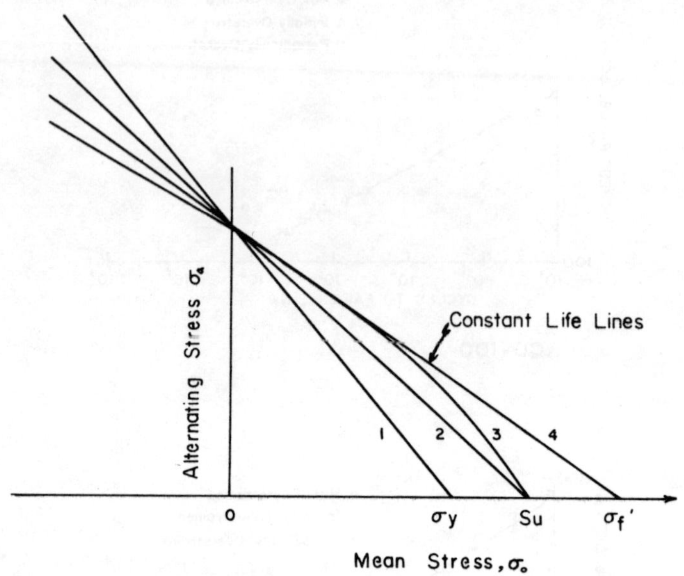

1 Soderberg

2 Goodman

3 Gerber

4 Morrow

Figure 9. Mean stress corrections.

Similarly, it can also be shown that the cyclic strength coefficient can be determined from the fatigue properties as follows:

$$K' = \frac{\sigma'_f}{(\;'_f)^{n'}} \tag{7}$$

When fatigue data are available so that the fatigue properties can be determined, it is advisable to use Equation (6) and (7) to determine n' and K' rather than using approximate values from the incremental step test. If appropriate fatigue data are not available the following approximate relationships are sometimes useful in estimating fatigue properties of ductile steels from tensile data.

$\sigma'_f \approx \sigma_f$ = true fracture strength

$\approx S_u$ + 50 ksi; S_u = ultimate tensile strength

$$b \approx -\frac{1}{6}\log\left(\frac{2\sigma_f}{S_u}\right)$$

$$\approx -0.1 \text{ for soft steels}$$

$$\approx -0.05 \text{ for hardest steels}$$

$$\varepsilon_f' \approx \varepsilon_f = \text{true fracture ductility}$$

$$\varepsilon_f = \ell n_{100} = \%RA \; ; \; \%RA = \text{percent reduction in area}$$

$$\approx -0.5 \text{ for soft steels to } -0.7 \text{ for hardened steels}$$

The above crude approximations of the fatigue properties of steels should be used only for preliminary design estimates when proper laboratory data are not yet available.

Cycle Counting

Some of the cycle counting methods in use today for fatigue analysis are peak, level crossing, range, range-mean, range-pair, and rainflow. Of these various methods, rainflow or its equivalent range-pair has been shown to yield superior fatigue life estimates [2].

Recall that fatigue properties are determined from a constant amplitude testing. An irregular load history must be reduced into

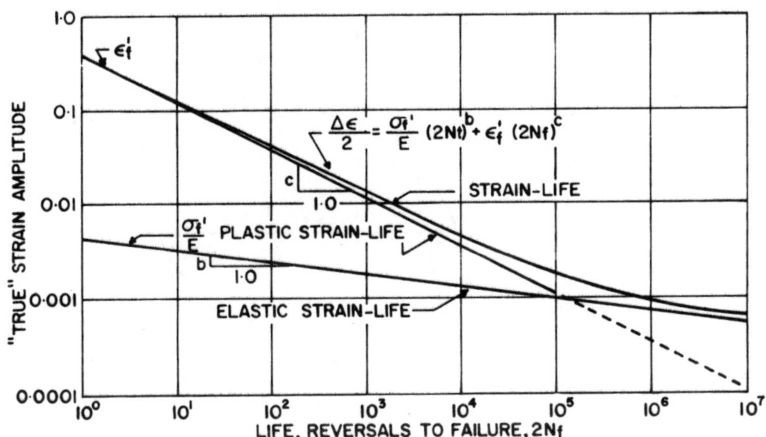

Figure 10. Total strain-life curve.

a series of constant amplitude events for comparison with the smooth specimen data. The concept of a cycle in an irregular history is difficult to define; however, a reversal can easily be defined as a change in sign in the loading. A constant amplitude sinusoidal cycle would contain two reversals. The apparent reason for the superiority of rainflow counting is that it combines load reversals in a manner that defines a cycle as a closed hysteresis loop. Each closed hysteresis loop has a strain range and mean stress associated with it that can be compared with the constant amplitude fatigue data in order to calculate fatigue damage.

In order to illustrate rainflow counting, consider the simple spectrum and the corresponding stress-strain response, shown in Figure 11. The four events in the stress-strain response that resemble constant amplitude cycles are easily identified as a-d-a, c-b-c, d-e-d, and f-g-f. Rainflow counting recognizes these events as closed hysteresis loops and counts them as cycles. In this

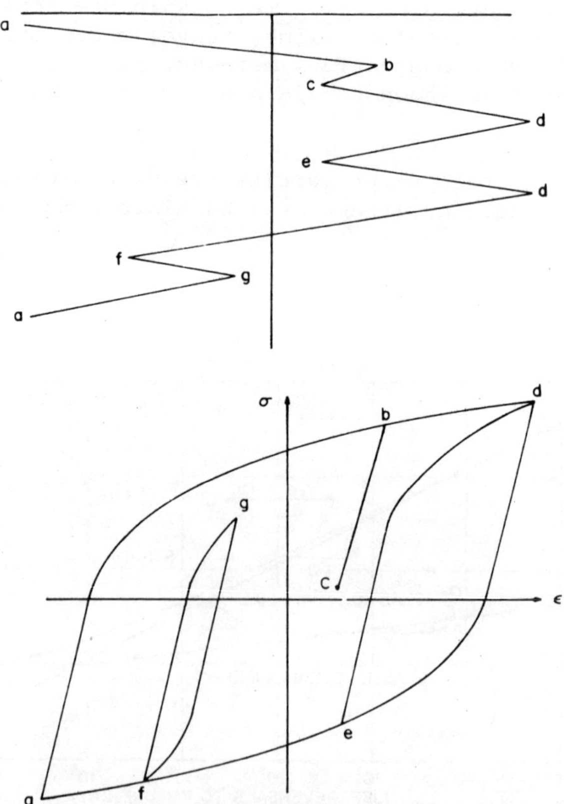

Figure 11. Stress-strain response during an irregular loading history.

procedure the small event b-c is treated as an interruption of the overall event a-d. Although the vents c-d and e-d appear identical in the strain time history, and would be counted as equally damaging by the range or range-mean method, the mean stresses and plastic strains are quite different even though the mean strain and strain ranges are equal. It should also be noted that mean stresses cannot be calculated directly from mean strains. The basic idea behind rainflow counting is to treat small events as interruptions of larger overall events and, in teh simplest terms, to match the highest peak and deepest valley, then the next largest and smallest together, etc., until peaks and valleys have been paired.

METHODS OF FATIGUE DAMAGE ANALYSIS

Simulation of Stress-Strain Response

The purpose of simulating the stress-strain response is to determine the parameters necessary for cumulative damage fatigue analysis. Information such as stress amplitude, mean stress, elastic and plastic strain can be determined for each reversal in the load history. The most important feature of the model is its ability to correctly describe the history dependence of cycle deformation. This so-called memory effect can best be illustrated in the stress-strain response, shown in Figure 11. As the material deforms from a to b, it follows a path described by the cyclic stress-strain curve [Equation (1)] magnified by a factor of two, because it is a hysteresis loop rather than a monotonic loading. At point b the load is reversed and the material elastically unloads to point c. When the load is reapplied from c to d, the material elastically deforms to point b where the material remembers its prior history (i.e., from a to b) and deformation continues along path a to d as if event b-c had never occurred.

To be exact, the model should also have the capability of including the effects of cyclic hardening or softening and of cycle-dependent relaxation of mean stress, as shown in Figure 12. In analyzing real structures, transient behavior is usually ignored and stable cyclic materials properties are used, thus eliminating the need to model cyclic hardening or softening. In a stationary random load history, cycle-dependent mean stress relaxation can usually be ignored, because it contributes little to the total fatigue damage. In certain types of block loading, particularly those with occasional large shifts in the mean, the presence of mean stresses and their relaxation may have a significant influence and should be included.

Martin [3] developed a computer model for cyclic deformation, based on a series of springs and frictional sliders, which is used to describe the hysteresis and cyclic memory effects exhibited by

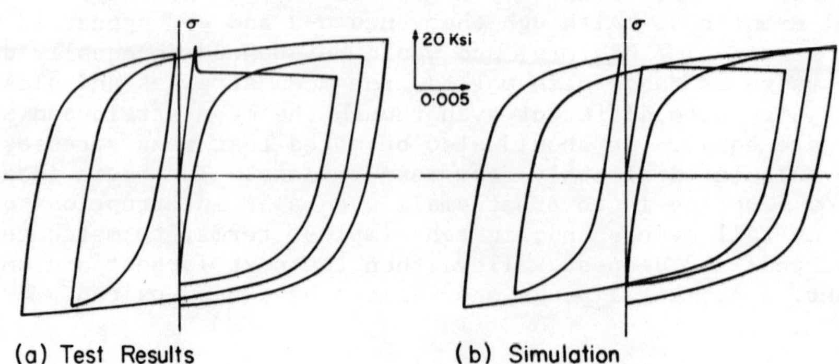

Figure 12. Actual and simulated stress-strain response of A36 steel [3].

metals. This type of model requires a great deal of computation because of the number of equations involved. Wetzel [4] developed a similar model, based on an "availability concept", which reduces the computational aspects of the analysis and, as such, is much easier to program for a digital computer.

The cyclic stress-strain curve [Equation (1)] is approximated by a series of straight line segments or elements. The number and size of the elements is arbitrary, depending on the manner in which elements are used. A large number of elements (50-100) can be used where each element is used to its fullest extent. When a smaller number of elements is used, they are usually interpolated to obtain midrange values. The following rules govern the manner in which elements are used: (1) start with the first element, (2) use them in order, (3) skip those elements unavailable for deformation, and (4) continue until the control condition is reached. For the purpose of illustration, a model consisting of ten elements will be used to simulate the stress-strain response of the spectrum shown in Figure 11.

First, the largest strain (manimum or minimum) in the history must be determined. This point will lie on the cyclic stress-strain curve and serve as a reference for determining the stress associated with each reversal. The spectrum is arranged in such a manner that the largest strain is the first and last value in the history.

The stress-strain elements are then formed using Equation (1), as illustrated in Figure 13. Starting at point a, corresponding to a strain of -0.015 and a stress of -70 ksi, as determined from the cyclic stress-strain curve, the stress and strain elements are doubled, so that the cyclic loading curve is similar to the initial

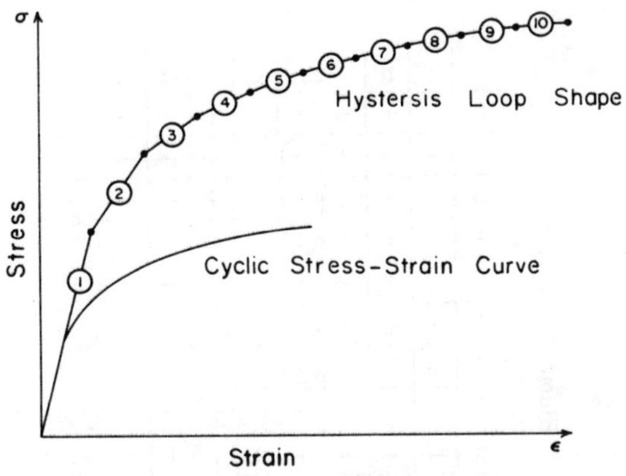

Element	Stress, $\Delta\sigma$	Strain, $\Delta\epsilon$
1	30	0.0015
2	12	0.0015
3	9	0.0015
4	7	0.0015
5	5	0.0015
6	3	0.0015
7	2	0.0015
8	2	0.0015
9	1	0.0015
10	1	0.0015

Figure 13. Elemental cyclic stress-strain curve.

loading curve but magnified by a factor of two. The availability coefficient of all the elements is set to -1, because the maximum strain was compressive. An availability of -1 means that all elements with an availability of -1 are available for tensile deformation but not compressive. Similarly, all elements with an availability of +1 are available for compressive deformation but not tensile. Once an element has been used for deformation, its availability is changed to the opposite sign. Simply stated this means that after an element is used in tension it must be used in compression before it can be used again in tension.

In traversing from a to b, six and one-half elements would be required but, since elements can only be used to the fullest extent, seven full elements are used. The stress and strain are incremented by seven elements with the resulting stress of 62 ksi and strain of 0.006 at b. The error between the calculated and observed strain is, of course, reduced when a greater number of line segments are used to describe the cyclic stress-strain curves. Since the first seven

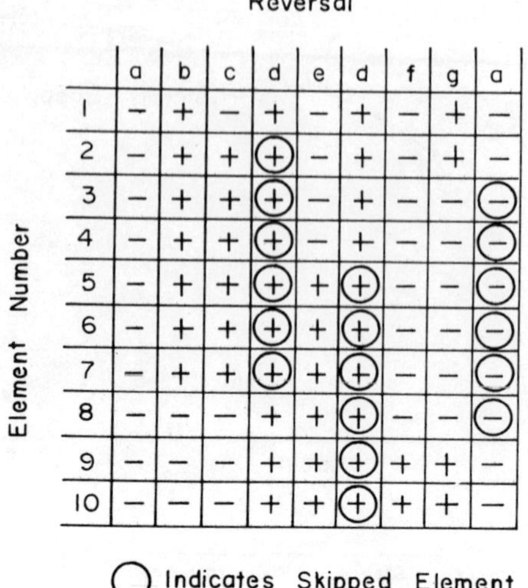

Reversal

Element Number	a	b	c	d	e	d	f	g	a
1	−	+	−	+	−	+	−	+	−
2	−	+	+	(+)	−	+	−	+	−
3	−	+	+	(+)	−	+	−	−	(−)
4	−	+	+	(+)	−	+	−	−	(−)
5	−	+	+	(+)	+	(+)	−	−	(−)
6	−	+	+	(+)	+	(+)	−	−	(−)
7	−	+	+	(+)	+	(+)	−	−	(−)
8	−	−	−	+	+	(+)	−	−	(−)
9	−	−	−	+	+	(+)	+	+	−
10	−	−	−	+	+	(+)	+	+	−

○ Indicates Skipped Element

Figure 14. Availability matrix used for material response model.

elements were used in tension, their availability sign is changed to +1, as shown in Figure 14. One element is required to go from b to c. The availability is changed to -1, because the elements were used in compression. The stress and strain are also incremented by one element with the result of 2 ksi and 0.003, respectively.

Four elements are required to go from c to d. The availability matrix shows that elements 2 through 7 are unavailable for tensile deformation, therefore, elements 1, 8, 9 and 10 must be used to makeup the four elements required to reach a strain of 0.015. The strain is incremented by 0.012, the stress by 68 ksi, the sum or elements 1, 8, 9 and 10. At this point all elements are available for compressive deformation. Reversal d-e requires four elements in compression with the resulting strain and stress of 0.003 and -46 ksi. The availability sign on the first four elements is changed to -1, indicating that they are available for tensile deformation for reversal e-d, which uses four elements in tension. All of the elements are once again available for compressive deformation. The last three reversals are treated in the same manner as the first three except that the loading direction is changed, i.e., a-b is tensile, while d-f is compressive. The completed element matrix, Figure 14, shows all of the elements that were used for each reversal. Figure 15 shows the simulated stress-strain response for the spectrum, which is in excellent agreement with the measured response of Figure 11.

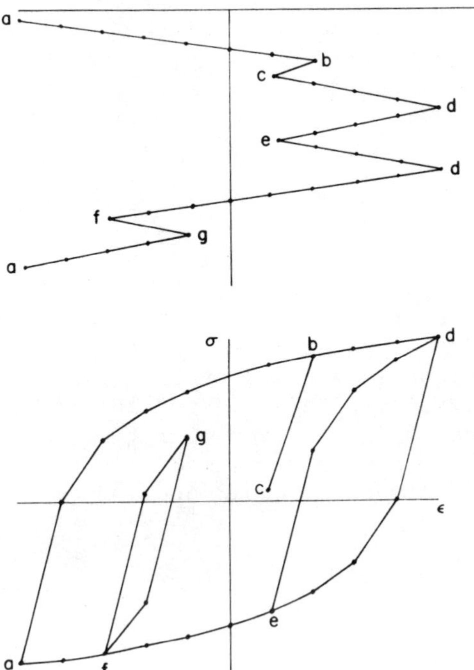

Figure 15. Simulation of stress-strain response for an irregular load history.

Besides simulating the stress-strain response of the material, the model provides a convenient and efficient method of cycle counting. There are four closed hysteresis loops in this spectrum: a-d-a, b-c-b, d-e-d and f-g-f. A close examination of the element matrix will show that when a hysteresis loop is closed, one or more elements has been skipped.

For example, in reversal c-d, elements 3 through 8 were skipped indicating that a closed hysteresis loop consisting of elements 1 and 2 was formed. Similarly, in reversal e-d, elements 5 through 10 would be skipped if they were required, indicating a closed hysteresis loop of elements 1 through 4 was formed. The key to cycle counting is that a closed hysteresis loop is formed whenever the next element in the availability matrix is unavailable for deformation. Also, there are no skipped elements in a hysteresis loop, which means that every time element 5 was skipped a loop consisting of elements 1 through 4 was formed.

Once the cycles have been defined and the stress-strain

response determined, the appropriate fatigue parameters can be determined, so that a damage analysis can be performed. Also, it should be noted that a cycle counting algorithm, completely independent of the material response model, can be developed along the lines of an availability concept as suggested by Richards et al. [5].

Component Calibration Techniques

In many practical problems engineers and designers are required to evaluate the fatigue resistance of new components while they are at the drawing board or prototype state of development. One method for performing this type of analysis is the so-called "component calibration" technique, described by Landgraf [6], which requires a relationship between applied load and local strains, such as the one shown in Figure 16. This type of information can be obtained analytically with a finite element model or experimentally by testing the component. The component would normally be tested by attaching

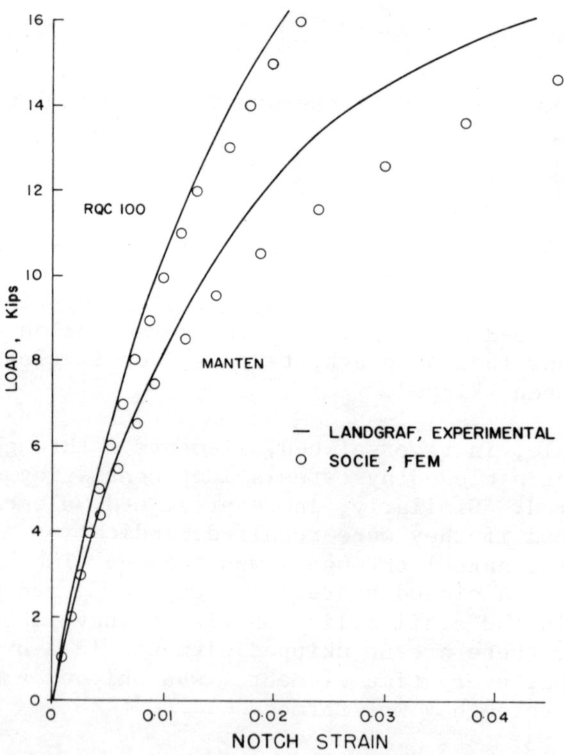

Figure 16. Load-strain curve from finite element analysis.

strain gages to the critical locations and applying one load and unload cycle while measuring the load-strain response. However, this type of test will produce inaccurate data because of the cyclic hardening or softening characteristics of the material. For this reason, an incremental step type of test should be used in obtaining load-strain curves from a single component. Similarly, cyclically stable material properties should be used in any analytical calculations.

The conversion of applied load into strain is accomplished in exactly the same manner as strain was converted into stress. The load-strain response has all of the features normally associated with stress-strain response, i.e., hysteresis effects, memory, and cyclic hardening and softening. The transient response is normally neglected, so that the load-strain response model only accounts for hysteresis and memory effects. From a computational viewpoint, this technique is exactly the same as those described in the last section for stress-strain response. In fact, the load-strain and stress-strain response models can be combined, so that the applied load can be converted to both the local stress and strain with one simple computer algorithm.

Notch Analysis

Neuber's Rule. In dealing with real components, it is often necessary to relate the nominal loads or strains to the maximum stresses and strains at the critical location. Neuber derived a rule which applies when the material at the notch root deforms non-linearly. The theoretical stress concentration, K_t, is equal to the geometric mean of the actual stress and strain concentration factors, K_ε and K_σ.

$$K_t = (K_\sigma K_\varepsilon)^{1/2} \tag{8}$$

Topper, et al. [7] modified Neuber's rule for use in cyclic loading applications by substituting the fatigue notch factor, K_f, for the stress concentration factor and rewriting Equation (8) in the following form:

$$K_\sigma = \frac{\Delta\sigma}{\Delta S}$$

$$K_\varepsilon = \frac{\Delta\varepsilon}{\Delta e} \tag{9}$$

$$K_f = \left(\frac{\Delta\sigma}{\Delta S}\frac{\Delta\varepsilon}{\Delta e}\right)^{1/2}$$

where $\Delta\sigma$ = stress range at notch root
ΔS = nominal stress range

$\Delta\varepsilon$ = strain range at notch root
Δe = nominal strain range

This relationship is conveniently used in the following form:

$$K_f^2 \Delta S \Delta e = \Delta\sigma \Delta\varepsilon \tag{10}$$

All terms on the left side are determinable for each reversal from the load history and cyclic stress-strain response of the material, and those terms on the right side represent the local stress-strain behavior of the material at the notch root. The terms on the left side are a determinable constant for each reversal and the result is and equation of the form xy = c, which is a rectangular hyperbola. When the nominal strains are elastic, Equation (10) may be used in the following form:

$$\Delta\sigma \Delta\varepsilon = \frac{(K_f \Delta S)^2}{E} \tag{11}$$

Combining this form with Equation (1), an expression for the notch stress becomes

$$\frac{\Delta\sigma^2}{E} + \Delta\sigma\left(\frac{\Delta\sigma}{K'}\right)^{1/n'} = \frac{(K_f \Delta S)^2}{E} \tag{12}$$

This equation is easily solved using the Newton-Raphson iteration technique. Once the notch stress is obtained, it can be used in Equation (1) to solve for the elastic and plastic strains at the notch root. After each reversal, a new axis is defined and the right hand side of Equation (12) recalculated in order to solve for the new stress and strain range. This process is illustrated in Figure 17.

One of the difficulties in applying this type of analysis is the constant accrual of error in the stress and strain, because each point is referenced to the end of the previous reversal rather than the original starting stress and strain. One method employed to overcome numerical difficulties is to use a straight line approximation developed by Stadnick [8], as shown in Figure 18. The slope, m, of the line is given by

$$m = \frac{K_f S_{max} - \sigma_{max}}{K_f \frac{S_{max}}{E} - \varepsilon_{max}} \tag{13}$$

and the stresses calculated from

$$S = \frac{\sigma - m\varepsilon}{K_f(1 - \frac{m}{E})} \tag{14}$$

While this approach is not exact, it has the advantage of referencing all deformations to the original stresses and strains. This results in less total error than a Neuber analysis that defines a new origin at each reversal.

An alternate approach is to directly incorporate Neuber's rule into the materials response model. There is a one-to-one correspondence between nominal and notched closed hysteresis loops, i.e., for each closed hysteresis loop of amplitude, ΔS and Δe, there is a closed hysteresis loop of amplitude, $\Delta\sigma$ and $\Delta\varepsilon$, at the notched root. The elements, ΔS and Δe, have already been determined for the stress-strain response model. Notch stress-strain elements, $\Delta\sigma$ and $\Delta\varepsilon$, are then determined for each nominal stress-strain element using a combination of Equations (1) and (10). The new notch elements are then used in exactly the same manner as the original stress-strain elements described in a preceding section on Simulation of Stress-Strain Response. Whenever an element is used in the nominal stress-

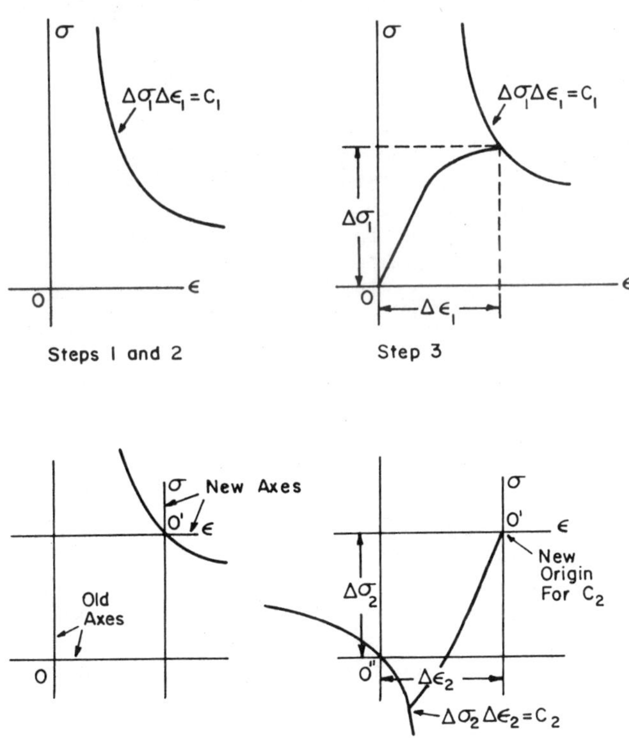

Figure 17. Steps of Neuber control.

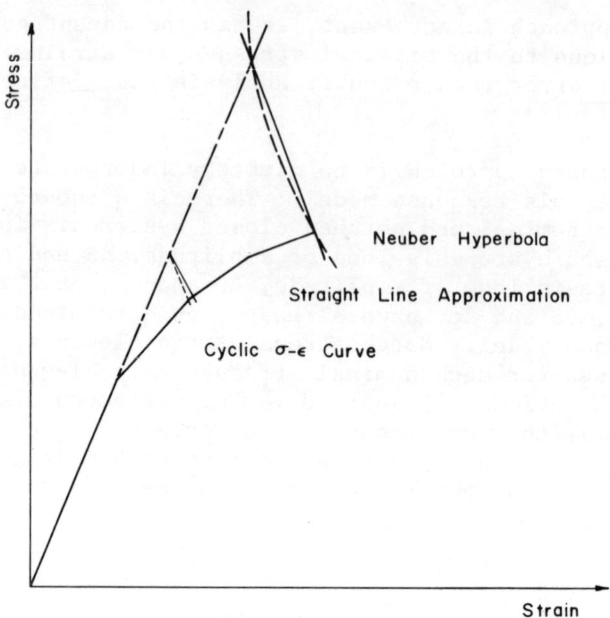

Figure 18. Construction for determining notch root stress and strain.

strain behavior, it is also used in the notch stress and strain response; conversely, whenever an element is skipped in the nominal behavior, it is also skipped in the notch response. The significance of this is that, once the nominal stress-strain response is obtained, the notch response is automatically determined. For example, in Figure 15 reversal c-d uses elements 1, 8, 9 and 10 to obtain the nominal stress-strain response. Notch elements 1, 8, 9 and 10 would be used to obtain the notch response. Material response, cycle counting, component calibration and notch response can be combined into a single model for efficient computation of fatigue damage similar to that of Socie [9].

<u>Fatigue Notch Factor</u>. The fatigue notch factor, K_f, is always less than or equal to the theoretical stress concentration factor, K_t, and its determination is an important part of the notch analysis. The appropriate value of K_f depends not only on the geometry but also on the material, thickness, surface finish, and stress gradient. For notched parts, the appropriate value of K_f can be estimated from Peterson's equation

$$K_f = 1 + \frac{K_t - 1}{1 + \frac{a}{r}} \tag{15}$$

where r is notch root radius and a is material constant, depending on strength and ductility.

For heat treated steels the following equation may be used to estimate a

$$a(\text{in}) = 10^{-3} \left(\frac{300}{S_u}\right)^{1.8} \quad (16)$$

where S_u is ultimate tensile strength, ksi.

Experimentally, K_f can be determined for any component by performing fatigue tests and plotting for the product, $\Delta\sigma\ \Delta\varepsilon$, for the smooth specimen and, $\Delta S\ \Delta e$, for the component, as will be discussed lated in connection with Figure 31. The fatigue notch factor is the difference between the two curves shown in the figure.

Cumulative Damage Analysis

Cumulative damage fatigue analysis is usually based on the Palmgren-Miner linear damage rule. Fatigue damage is computed by linearly summing cycle ratios for the applied loading history, as indicated in the following equation

$$\text{Damage} = \Sigma \frac{n_i}{N_{fi}} \quad (17)$$

where n_i is the observed cycles at amplitude i, and N_{fi} is fatigue life at constant amplitude i.

After the fatigue damage for a representative segment or block of load history has been determined, the fatigue life in blocks is calculated by taking the reciprocal. Other, more complex, procedures for summing damage such as the double linear damage concept and Corten-Dolan approach are not necessary for accurate life predictions for irregular loading provided the local σ-ε at the point of crack initiation is known. Special applications such as high temperature fatigue, where there is an interaction between creep and fatigue may require additional complications in the damage summation portion of the analysis.

The fatigue life for any cycle or reversal can be determined from Equation (5).

$$\frac{\Delta\varepsilon}{2} = \left(\frac{\sigma'_f - \sigma_o}{E}\right) (2N_f)^b + \varepsilon'_f (2N_f)^c \quad (5)$$

The mean stress, σ_o, and cyclic strain range, $\Delta\varepsilon$, has been determined from the material response model, and the four fatigue properties have been determined from constant amplitude life tests. Equation (5) cannot be explicitly solved for life because of the

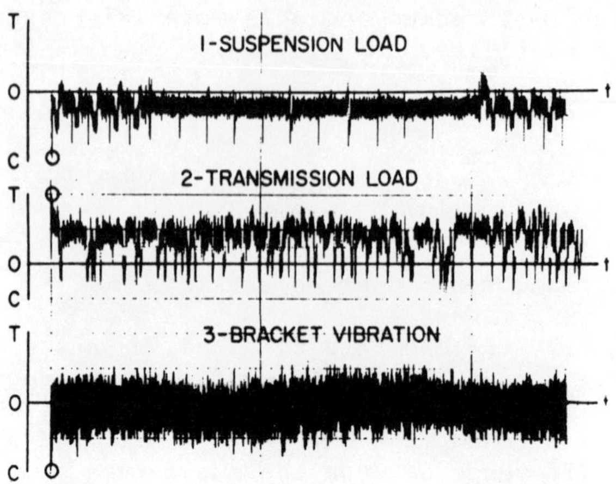

Figure 19. SAE load histories.

negative fractional exponents involved, but can easily be solved using Newton-Raphson or interval halving iteration techniques. The main advantage of this formulation of fatigue damage is that it eliminates the need to determine elastic and plastic components of the total strain. In other formulations of fatigue damage, the relative amounts of elastic and plastic strain are dependent on the choice of the cyclic stress-strain properties, where a small difference can produce a large difference in life estimates, especially at low strain amplitudes and long lives. Also, mean stress has a greater effect at longer lives because the elastic term dominates, which is in agreement with observed behavior.

APPLICATION OF A STRAIN BASED ANALYSIS TO THE SAE CUMULATIVE FATIGUE DAMAGE TEST PROGRAM

Review of Test Program

The test program [10] will be briefly reviewed for those readers unfamiliar with it. Three load histories, Figure 19, were selected from magnetic tape recordings of nominal strain from actual components under operating conditions. These load histories were designated (a) suspension with a dominant compressive mean, (b) transmission which has a tensile mean, and (c) bracket which is a random narrow band vibration. The analog signals were digitalized and filtered to provide a sequence of peaks and valleys that were normalized, so that the largest value in the spectrum was set equal to ± 999. No attempt was made to preserve the frequency content of these data.

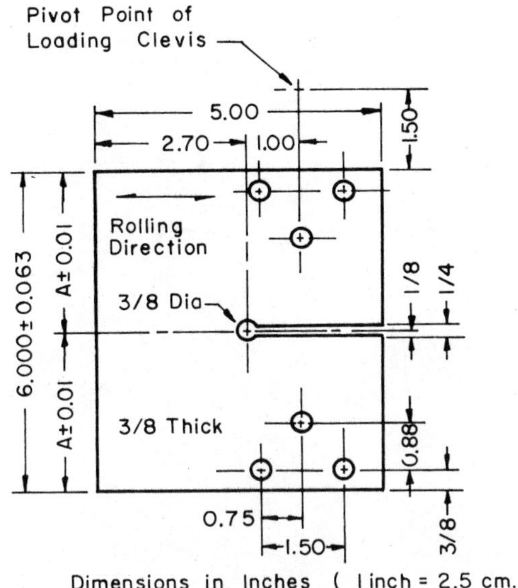

Figure 20. Sketch of SAE keyhole specimen.

The specimen, Figure 20, provided both axial and bending stresses and strain at the notch root, which is typical of many production components. Loads were applied to the specimen (comonent) through a monoball fixture, that allowed both tensile and compressive loads to be applied. Variable amplitude tests consisting of 57 specimens were performed, using the three load histories scaled to different maximum loads to produce fatigue lives ranging from 10^4 to 10^8 reversals, as shown in Table 1. Also, constant amplitude fully reversed data were obtained for this specimen.

Two commonly used structural steels, U. S. Steel's Man-Ten and Bethlehem's RQC-100, were used in this program. Mechanical properties of these steels are shown in Table 2. Both the monotonic and cyclic stress-strain curves for these materials are shown in Figures 21 and 22. The strain-life curves are shown in Figures 23 and 24. It should be noted that there is plastic deformation in Man-Ten at 10^6 reversals, which is normally thought to be in the elastic range.

Nominal Strain-Life Analysis

A nominal strain-life analysis will be used to predict the crack initiation life of the specimen, shown in Figure 20. The

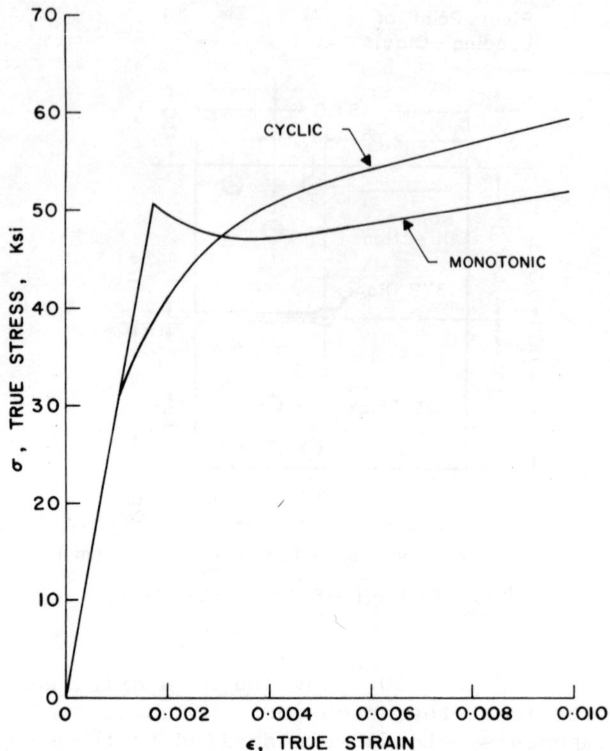

Figure 21. Stress-strain curves for Man-Ten.

following steps are required for performing the analysis using smooth specimen material properties.

Material properties. Material properties required for this analysis have been determined from a series of fatigue tests and are listed in Table 2. Specifically, the required fatigue properties are: fatigue strength coefficient and exponent, fatigue ductility coefficient and exponent, and the elastic modulus. If the material properties were unknown, they would have to be determined by a series of smooth specimen fatigue tests or estimated from tensile data. A discussion of this procedure can be found in the preceding section, Fundamentals of Fatigue Damage Analysis, Fatigue Properties of Metals.

Theoretical stress/strain concentration factor. The elastic stress concentration factor for the circular notch was estimated to be 3.0. Actual strain gage measurements at the notch root show the strain concentration factor to be approximately 2.93 when the

Table 1
Mechanical Properties of Man-Ten and RQC-100

Monotonic Properties		Man-Ten	RQC-100
Elastic Modulus,	E	30×10^3 ksi	30×10^3 ksi
Yield Strength,	S_y	47 ksi	120 ksi
Tensile Strength,	UTS	82 ksi	125 ksi
Reduction in Area,	%RA	65%	55%
True Fracture Strength,	σ_f	145 ksi	173 ksi
True Fracture Ductility,	ε_f	1.19	0.78
Strength Coefficient,	K	140 ksi	174 ksi
Strain Hardening Exponent,	n	0.21	0.08
Cyclic Properties			
Fatigue Ductility Coefficient,	ε_f'	0.26	1.06
Fatigue Ductility Exponent,	c	-0.47	-0.75
Fatigue Strength Coefficient,	σ_f'	133 ksi	168 ksi
Fatigue Strength Exponent,	b	-0.095	-0.075
Cyclic Strength Coefficient,	K'	174 ksi	167 ksi
Cyclic Strain Hardening Exponent,	n'	0.20	0.10
Cyclic Yield Strength,	S_y'	48 ksi	85 ksi

Table 2
Summary of SAE Cumulative Fatigue Damage Test Program

Man-Ten			RQC-100		
Test	Maximum Load	Blocks to Initiation	Test	Maximum Load	Blocks to Initiation
Suspension History					
SM 1	-16,000	7.7-28	SR 1	-16,000	19.9-64
SM 2	- 9,000	162-430	SR 2	- 9,000	1,710
SM 3	- 6,000	1,410-2,240	SR 3	- 7,000	11,200
SM 4	- 4,500	4,700	SR 4	- 6,000	48,000
SM 5	- 3,000	>85,370			
Bracket History					
BM 1	-16,000	1.5-2	BR 1	-16,000	3.3-5.1
BM 2	- 8,000	11.5-23	BR 2	- 8,000	47-113
BM 3	- 3,500	270-1,588	BR 3	- 3,500	2,673-5,020
BM 4	- 3,000	2,666			
BM 5	- 2,000	>20,630			
Transmission History					
TM 1	+16,000	8.4-12.8	TR 1	+16,000	22.2-29.9
TM 2	+ 8,000	74-420	TR 2	+ 8,000	269-460
TM 3	+ 3,500	3,755-5,800	TR 3	+ 3,500	>88,020

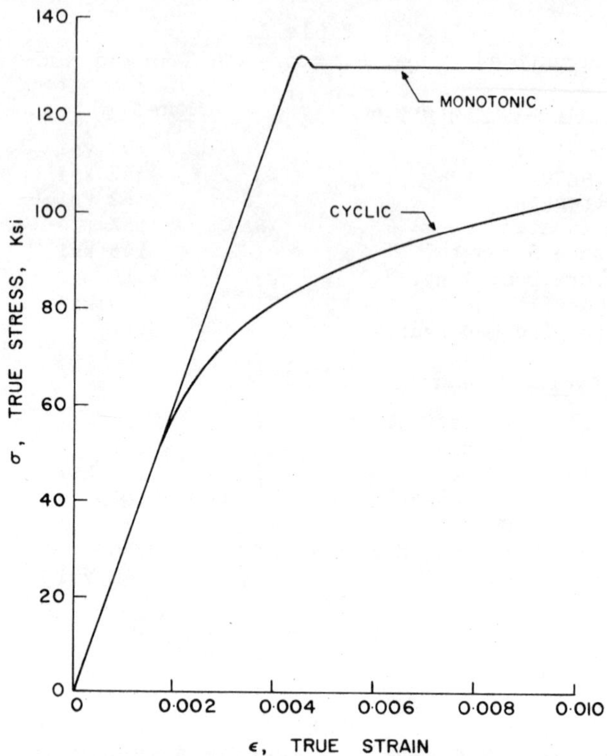

Figure 22. Stress strain curves for RQC-100.

reduced section of the specimen is treated as a simple beam. In actual practice the stress concentration factor can be determined from stress concentration table [11], finite element analysis, or from experimental stress analysis methods applied to actual components.

Nominal strain history. A difficult and expensive part of the analysis is to obtain the field loading for a component. The loads are usually recorded on magnetic tape, and then must be filtered and digitalized to obtain a sequence of peaks and valleys. A digitalized sequence of load peaks and valleys was obtained from the SAE Fatigue Design and Evaluation Committee and are listed in Reference [10]. Loads were converted to nominal strains by assuming that the reduced section of the specimen acts as a simple beam subjected to axial and bending stresses.

The following relationship from elementary beam theory was used:

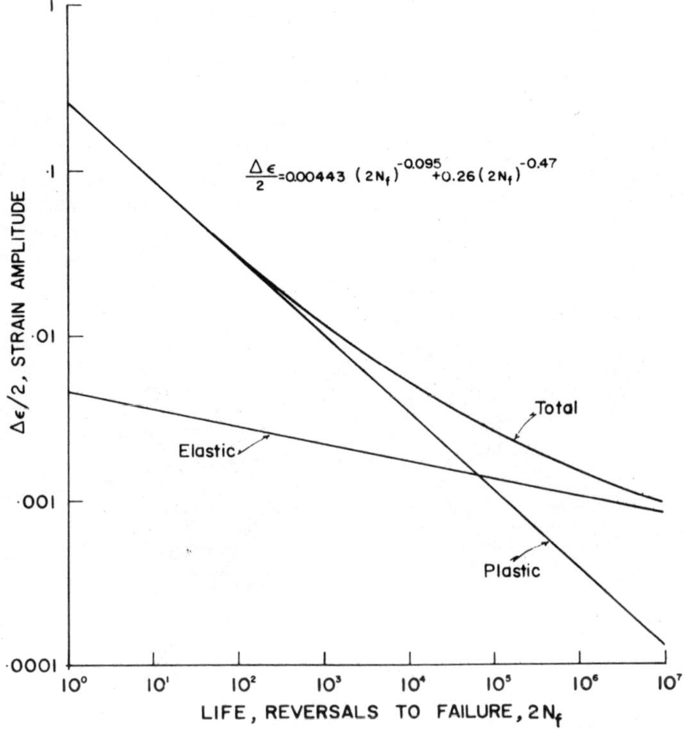

Figure 23. Strain-life curve for Man-Ten.

$$e = \frac{Mc}{IE} + \frac{P}{AE} \tag{18}$$

for the keyhole specimen:

$$e = 7.26\ \frac{P}{E}$$

where e equals nominal strain, P is applied load, kips, and E is elastic modulus, ksi.

Of course, if nominal strains were available, this conversion would be unnecessary.

Convert nominal strains to notch strains. The nominal strains through the simple relationship:

$$\varepsilon = K_t\ e \tag{19}$$

for the keyhole specimen:

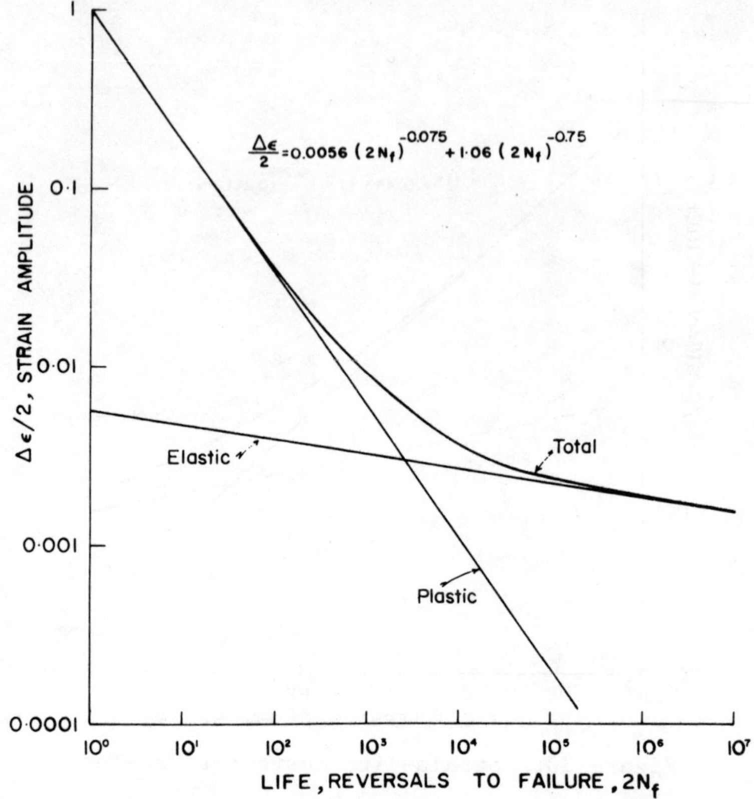

Figure 24. Strain-life curve for RQC-100.

$$\varepsilon = 3.0\ e$$

$$\varepsilon = \frac{21.8\ P}{E}$$

This type of conversion, as well as the load-strain conversion, assumes elastic behavior both nominally and at the notch root.

Rainflow count notch strains. Computational aspects of rainflow counting are described in Appendix A and will not be elaborated upon here.

Calculate and sum fatigue damage. Fatigue damage can be calculated during the rainflow counting procedure on a cycle-by-cycle basis or from a histogram of rainflow counted ranges. The fatigue damage for each cycle is calculated from Equation (17). Equation (5), with the mean stress set equal to zero, is used to calculate the fatigue life of each range.

The results of the nominal strain-life analysis are shown both in Table 3 and Figure 25. The computer algorithm used for these predictions is listed in Appendix A. In this approach the effect of mean stress is not included. By assuming elastic behavior, the mean stress could be calculated from mean strain; however, this could lead to considerable errors, as was shown in Figure 11. Sequence effects are listed only in the rainflow counting procedure.

An alternate approach to the above analysis would be to determine the nominal strain-life curve for the individual component through a series of constant amplitude fatigue tests. Then this strain-life curve would be used in place of the actual material properties, and the notch strains need not be determined. This type of approach is particularly useful in evaluating welded structures, where the stress concentration factors and mean stresses are difficult to obtain.

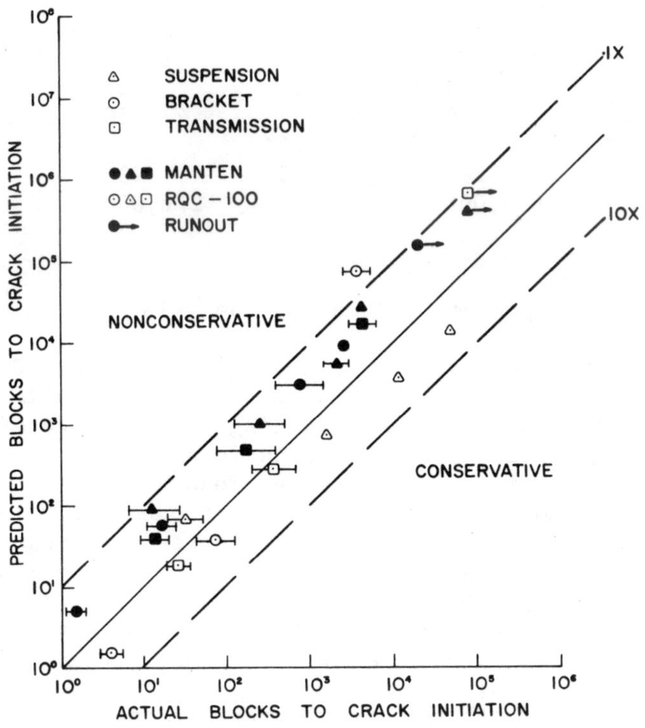

Figure 25. Nominal strain-life analysis.

Table 3

Strain Life Analysis SAE Keyhole Specimen

Test	Actual	Strain Life	Load-Strain Analysis	Load-Stress Strain Analysis	Neuber Notch Analysis $K_f = 2.6$
SM 1	7.7-28	93	33	34	27
SM 2	162-430	940	480	506	455
SM 3	1,410-2,240	5,890	3,642	3,988	4,590
SM 4	4,700	25,940	18,010	20,750	32,100
SM 5	>85,370	335,000	270,000	347,000	915,000
BM 1	1.5-2	5	1.3	1.2	1.0
BM 2	11.5-23	54	27	25	22
BM 3	270-1,588	3,053	2,225	2,000	3,335
BM 4	2,666	8,110	6,230	5,524	10,790
BM 5	>20,630	166,000	146,000	123,000	356,000
TM 1	8.4-12.8	42	10	10	8
TM 2	74-420	468	218	200	168
TM 3	3,755-5,800	21,470	15,422	13,410	21,440
SR 1	19.9-64	56	42	42	38
SR 2	1,710	687	599	689	1,210
SR 3	11,200	3,590	3,366	4,564	16,920
SR 4	48,000	14,310	14,283	22,920	123,000
BR 1	3.3-5.1	2	1.4	1.3	1.1
BR 2	47-113	39	34	30	49
BR 3	2,673-5,020	73,000	82,547	47,750	233,000
TR 1	22.2-29.9	19	13	12	10
TR 2	269-460	261	225	200	301
TR 3	88,020	650,000	753,000	276,000	1.4E7

Component Calibration Analysis

Experimental determination of load-strain curve. Component calibration is the name given to the method where applied loads are converted to notch root strains retaining all sequence effects. This method is applicable for both elastic and plastic strains and requires a relationship between applied load and notch root strain. A strain gage was mounted in the notch root, and the relationship between applied load and notch root strain was obtained from a monotronic tension test of specimens/components made from Man-Ten and RQC-100. Then the specimen was subjected to an incremental step type of test in order to cyclically stabilize the material. Monotonic and cyclic load-strain curves are shown in Figures 26 and 27 for Man-Ten and RQC-100, respectively. The same type of cyclic behavior is noted in both the smooth specimen and component, i.e., cyclic hardening for Man-Ten and cyclic softening for RQC-100. Cyclic plasticity and, hence, fatigue damage in the cyclically

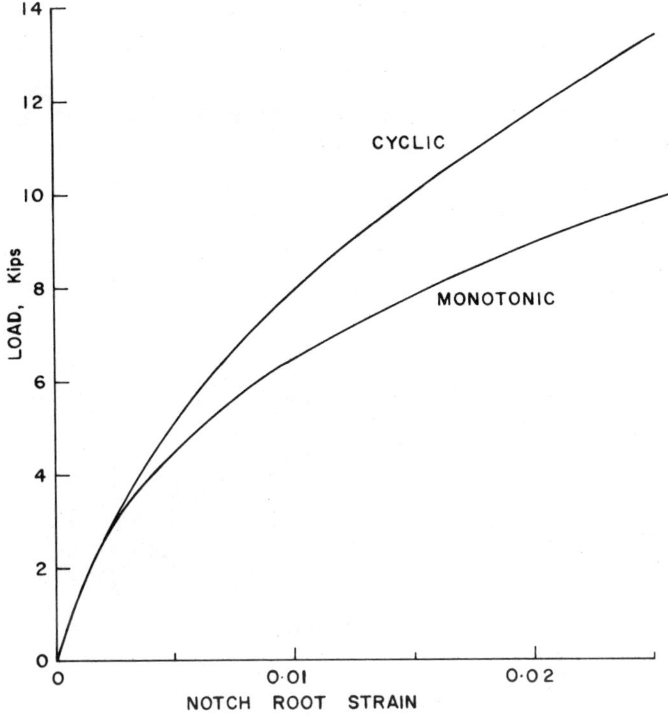

Figure 26. Load-notch root strain for Man-Ten.

stable component is considerably different from that calculated using the monotonic load-strain response. For this reason, only cyclically stable load-strain curves should be used in the analysis.

Finite element determination of load-strain curve. The specimen was a finite element modeled with the grid pattern shown in Figure 28. A program developed at the University of Illinois for the cyclic plastic analysis of notched plates was used to generate the load-strain curves shown in Figure 16. This program uses constant strain triangular elements with kinematic hardening. A piecewise linear solution is obtained for each load increment. Cyclically stable properties for the cyclic strength coefficient and strain hardening exponent were used in the analysis. The correlation between the experimental finite element analysis is quite good.

Load-strain conversion and fatigue analysis. An advantage of the component calibration techniques is that it accounts for notch root plasticity, and it does not require the determination of the stress concentration factor. The method, outlined below, does not include mean stress. The effect of mean stress is discussed in the next section.

(a) Material properties. Same as those used in nominal strain-life analysis.

(b) Load-strain curve. An equation of the form of Equation (1) was fit to the load-strain curves in Figure 26 and 27.

$$e = \frac{P}{C_1} + \left(\frac{P}{C_2}\right)^{1/D} \tag{20}$$

with the following results [6]:

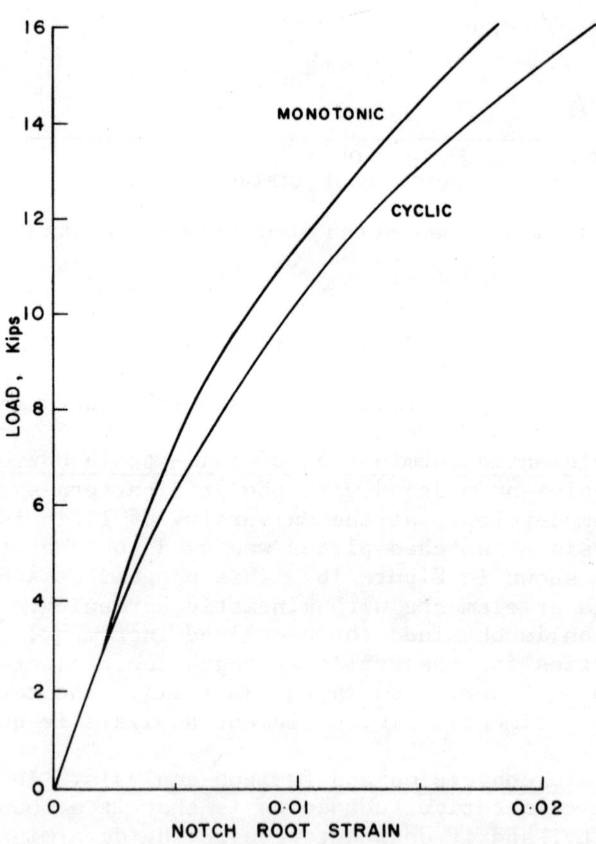

Figure 27. Load-notch root strain for RQC-100.

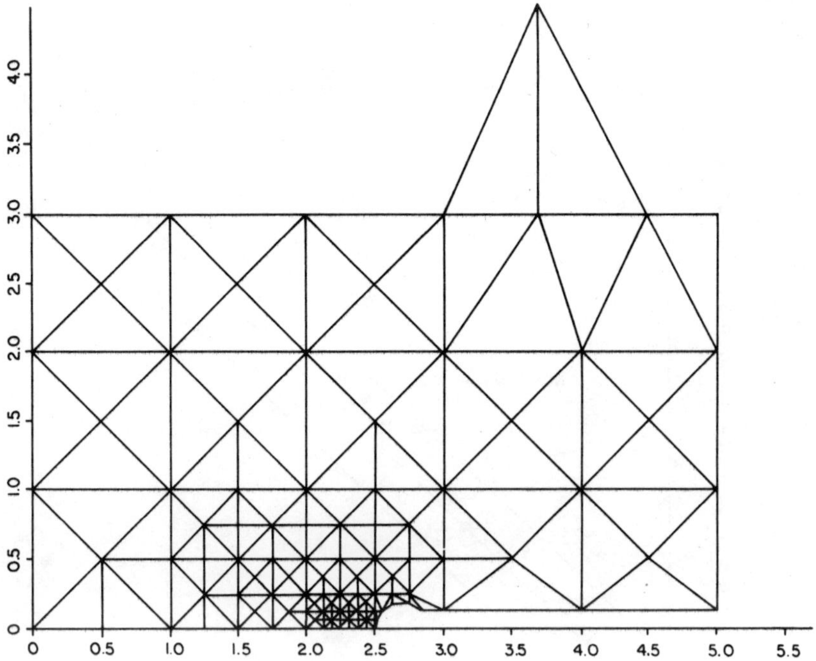

Figure 28. Element mesh for a keyhole notched member (SAE specimen).

	Man-Ten	RQC-100
C_1	1.41×10^6	1.41×10^6
C_2	6.67×10^4	6.55×10^4
D	0.39	0.31

(c) Load history. Same as nominal strain analysis.

(d) Rainflow count load history. Rainflow counting is used to obtain the load ranges for conversion to notch root strains.

(e) Convert load ranges to strain amplitude. Divide each load range by two to obtain load amplitude. For each load amplitude, identified by rainflow counting, enter the load-strain curve to obtain notch root strain amplitude. This strain amplitude is then used to calculate fatigue damage.

(f) Calculate and sum fatigue damage. Same as nominal strain-life analysis.

Results of this analysis are shown in Table 3 and Figure 29. Correlation between predicted lives and test data is somewhat better than the nominal strain-life analysis primarily because notch root

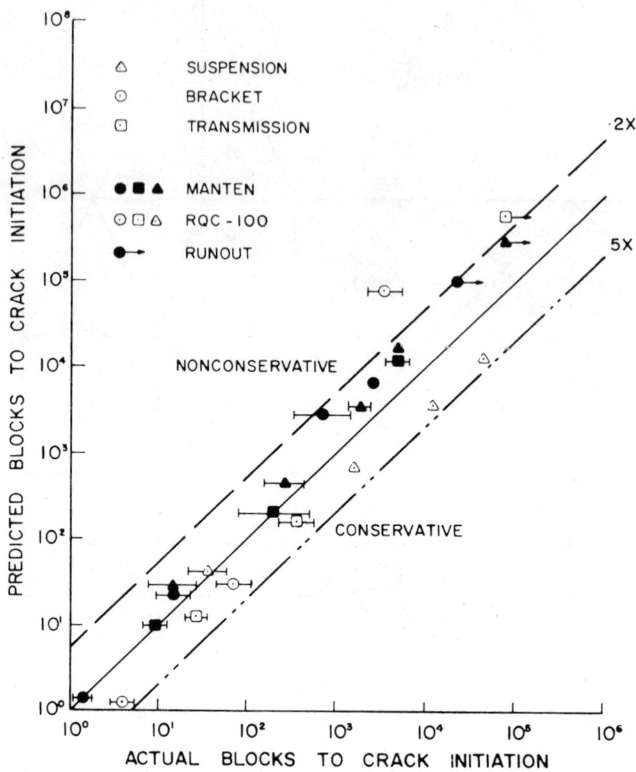

Figure 29. Load-strain analysis.

plasticity is accounted for. The only difference between the two analyses is the manner in which notch root strains are obtained from applied loads. The computer algorithm used for this analysis is found in Appendix B.

Load-stress-strain conversion and fatigue analysis. The analysis, outlined below, is a refinement of the previous one by accounting for the mean stress of each load reversal. The stress-strain response of each reversal much be determined on a reversal-by-reversal basis retaining all sequence effects. A detailed description of this procedure may be found in Reference [9].

(a) Material properties. In addition to the fatigue properties used previously, the cyclic stress-strain response of the material must be known. The cyclic strength coefficient and strain hardening exponent may be determined from the fatigue properties using Equations (6) and (7).

(b) Load-strain curve. Determined by methods used in the

first two sections, Purpose and Scope and Fundamentals of Fatigue Damage Analysis.

(c) Load history. Same as nominal strain analysis.

(d) Material response model. A model for following the load-strain and stress-strain response must be programmed for a digital computer using the concepts discussed in previous sections.

(d) Follow the load-stress-strain response of the specimen on a reversal-by-reversal basis. The strain range and mean stress of each reversal in the load history is obtained from the material stress-strain response model. The mechanics of the computer algorithm used for these calculations are discussed in Appendix C.

(f) Rainflow count applied loads or notch root strains. The simplest method of rainflow counting is to use the material response model, as illustrated in Appendix C.

(g) Calculate and sum fatigue damage. Fatigue damage is calculated for each closed hysteresis loop identified by rainflow counting. Equation (5) is solved for fatigue life and damage calculated from Equation (17).

The results of this analysis are shown in Figure 30 and Table 3. The only difference between this analysis and the previous one is the inclusion of mean stress effects. Computational aspects of the analysis are increased because of the material response model calculations. However, once the procedure is programmed for a digital computer, the analysis can easily be performed. Fairly accurate life predictions can be obtained when notch root plasticity and mean stress effects are properly accounted for.

Notch Analysis

Experimental determination of the fatigue notch factor. The fatigue notch factor was experimentally determined from a series of constant amplitude fully reversed fatigue tests of the specimen. A Neuber curve for the smooth specimen properties of Man-Ten and RQC-100 was constructed by plotting the product of stress and strain range versus fatigue life, as shown in Figures 31 and 32. These curves are obtained from the cyclic stress-strain and strain-life curves for each material.

Nominal stresses were obtained from the applied loads using the equation developed in Nominal Strain-Life Analysis of this section.

$$e = 7.26 \frac{P}{E} \tag{18}$$

$$S = 7.26 P$$

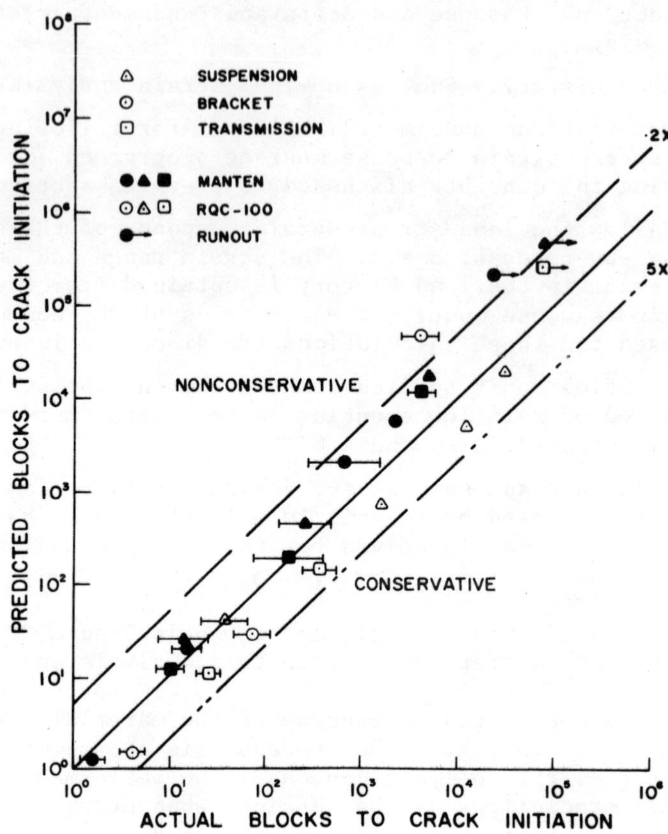

Figure 30. Load-stress-strain analysis.

Since the nominal stresses are assumed to be elastic, the Neuber parameter reduced to

$$\frac{\Delta S^2}{E}$$

and is plotted on the same curve as the smooth specimen data. The fatigue notch factor is then the square root of the difference between the two curves and is equal to 2.6 for this specimen configuration. This method requires several component fatigue tests and, in general, would not be applicable to prototype work.

Analytical determination of the fatigue notch factor. The theoretical stress concentration factor was estimated to be that of a circular hole in a semi-infinite plate for which $K_t = 3.0$. Finite element analysis used to develop the load-strain curve shows that K_t should be 3.08 and, experimentally, strain gage measurements show it to be 2.93. The fatigue notch factor is then calculated

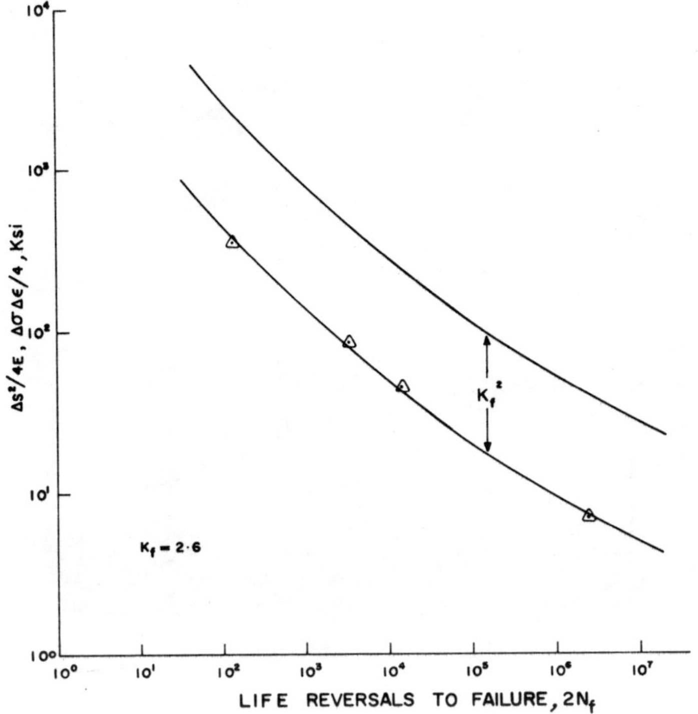

Figure 31. Neuber plot for Man-Ten specimen.

from Peterson's equation

$$K_f = 1 + \frac{K_t - 1}{1 + \frac{a}{r}} \tag{15}$$

for the keyhole specimen

$$a = \begin{cases} 0.01 & \text{for Man-Ten} \\ 0.004 & \text{for RQC-100} \end{cases}$$

$$r = 0.1875 \text{ in.}$$

$$K_f = \begin{cases} 2.90 & \text{for Man-Ten} \\ 2.96 & \text{for RQC-100} \end{cases}$$

Neuber notch analysis. Neuber notch analysis is most often used when the service history is in the form of nominal strains. The analysis for the keyhole specimen proceeds as follows:

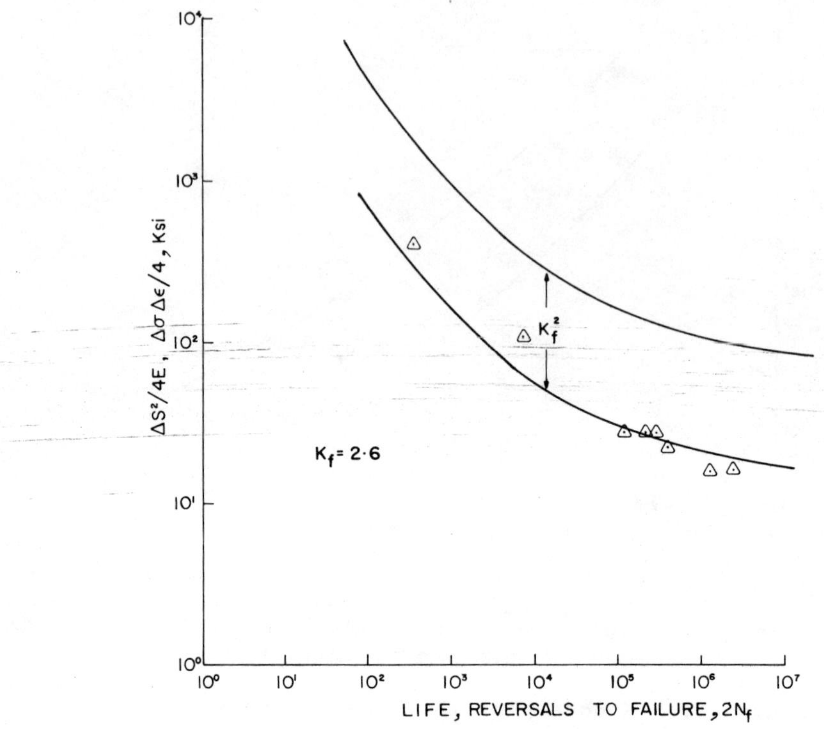

Figure 32. Neuber plot for RQC-100 specimen.

(a) Material properties. Same as load-stress-strain analysis.

(b) Load history. Applied loads were converted to nominal stresses from Equation (19).

(c) Fatigue notch factor. Determined by methods of Purpose and Scope and Fundamentals of Fatigue Damage Analysis.

(d) Material response model. A model for following the stress-strain response of the notch root and Neuber's rule must be programmed for a digital computer.

(e) Rainflow count stress. The stress history is rainflow counted to obtain nominal stress hysteresis loops (stress ranges). Mean stress at the notch root is obtained from the material response model.

(f) Determine notch root stresses and strains on a reversal-by-reversal basis. Equation (12) is solved for each hysteresis loop to obtain notch root stress amplitude. This stress is then used to obtain strain amplitude for damage calculations.

(g) Calculate and sum fatigue damage. Same as load-stress-strain analysis.

The results of this analysis for a fatigue notch factor equal to 2.6 are shown in Figure 33 and Table 3. The computational aspects are discussed in detail in Appendix D, which also contains a computer listing.

DISCUSSION OF LIMITATIONS OF APPROACH

The reader is reminded that the analysis described above is limited to predicting the fatigue crack initiation life of members subjected to variable loading histories. Analogous techniques for estimating the rate at which fatigue cracks will grow in service are not yet well developed. Fatigue crack propagation under variable loading history is the subject of current research. In this connection it is important to utilize non-arbitrary definition for

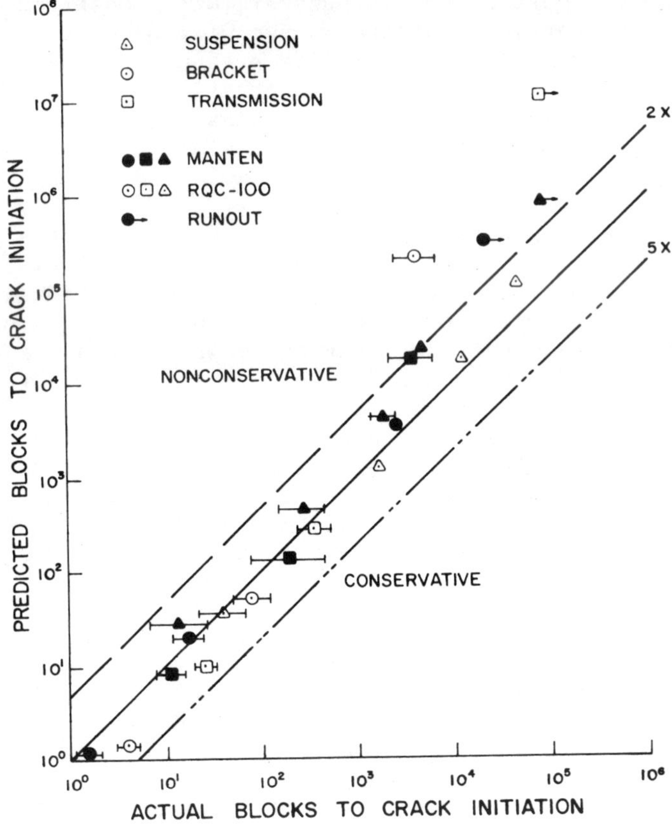

Figure 33. Neuber notch analysis.

fatigue crack initiation. In the work described in this chapter, it has been assumed that a crack of engineering significance will form at the most highly strained region at the same time as a smooth specimen subjected to the same stress-strain history would fracture into two pieces. While this is satisfactory for engineering estimates, no quantitative definition of the size of an initiated crack has been given although current research is being done in this area. Both crack initiation life and crack propagation life must be estimated in order to optimize the material and geometry for the application at hand. This is important to the efficient design for routine applications and essential for the design of advanced machines.

Also, note that a uniaxial state of stress has been assumed to exist in the most highly strained region, so that life can be estimated on the basis of smooth specimen data obtained by uniaxial testing. This assumption is reasonable valid for a number of engineering structural elements, but is not generally true. For example, if thick sections are involved, biaxial stresses will be present at notch roots which, in general, will constrain plastic deformation, so that less fatigue damage will occur than predicted using uniaxial assumptions. Thus life preductions, using the methods described herein, will tend to be conservative. A first-order correction for biaxial stresses and strains might be to use an equivalent strain calculated in a manner similar to that used for yielding.

The methods suggested here work best for reasonably ductile, homogeneous, isotropic materials. Applications of these methods in the case of metals with controlling internal discontinuities, such as present in cast metals or high hardness steels with non-metallic inclusions, requires that one treat the case material as micro-notched metal, as described by Mitchell [12]. A similar approach can be employed on weldments as suggested by Lawrence and his co-workers [13]. The weldment problem is further complicated by the presence of three distinct zones of metal, i.e., base metal, weld metal, and the heat affected zone. In this case it may be necessary to characterize the mechanical resistance of the separate zones. A similar difficulty arises in applying these techniques to case hardened parts.

Other factors that have been ignored in the analysis include the influence of the environment, temperature, frequency, hold times, wave-shape effects, dynamic strain aging, etc. These factors are particularly crucial when the member is subjected to elevated temperature during operation and when the design life is long. For long lives, where the designed stresses are kept low enough to avoid significant cyclic plasticity even at the notch root, factors such as the initial residual stresses and assembly stresses are of prime importance in determining fatigue performance. Effects of design and fabrication details are much more important in determining the fatigue performance than are the bulk properties of the metal. Thus, the predictive techniques described here are more appropriate for applications to low cycle fatigue problems (i.e., < 10^6 cycles) where significant cyclic plasticity occurs, at least occasionally, in the most highly strained region.

In a more optimistic vein, the approach suggested here can be a powerful tool for reaching engineering decisions concerning the low cycle fatigue crack initiation life of members subjected to virtually any history of loading. Of prime importance is the utilization of a correct cycle counting procedure that identifies close hysteresis loops, so that the dame done can be related to constant amplitude baseline data. Even the crude approaches, wherein the S-N curve of a component (e.g., weld joint) is sued in conjuntion with linear damage summation methods, are reasonably accurate provided the correct cycle counting method is used.

A knowledge of proper methods for cycle counting also permits the test engineer to edit service load spectra, so that laboratory test time can be greatly reduced without significantly altering the basic damage content of the spectrum. The spectrum editing approach is superior to the technique of increasing maximum operating load levels in order to shorten the test time, because such large loads may eliminate or induce residual stresses in the highly strained region that can have a large influence on the fatigue life of the member.

In this connection the approach is also useful for performing "sensitivity studies" as illustrated by Socie [9]. One can quickly evaluate the relative merits of one material versus another for a given part geometry subjected to the same load history. Likewise, the influence of minor modification in the part geometry on its fatigue performance may be quickly evaluated. Finally, the relative fatigue lives of two parts made of a particular metal subjected to different load histories can be evaluated.

APPENDIX A

Computer Program for Nominal Strain-Life Analysis and Rainflow Counting

The following computer program was used to make the fatigue life predictions based on a nominal strain-life approach.

Input Variables

EF	Fatigue Ductility Coefficient
C	Fatigue Ductility Exponent

SF	Fatigue Strength Coefficient
B	Fatigue Strength Exponent
E	Elastic Modulus
SCALE	Scale Factor
AKT	Fatigue Notch Factor
NPEAKS	Number of Data Points
DATA	Sequence of Peaks and Valleys

Major Calculation Variables

AV	Availability Matrix
RANGE	Maximum Range of Data
SEG	Segment or Element Size for Rainflow Counting
RUN	Running Total of Input Data
SGN	+ 1 if Tensile Load; -1 if Compressive Load

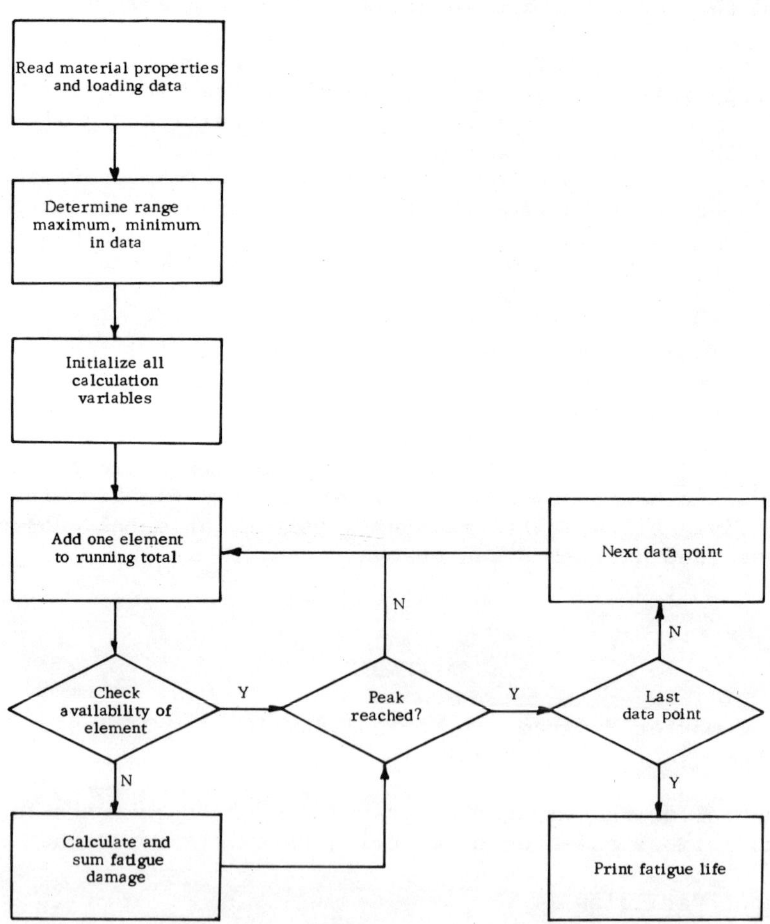

Figure A1. Overall flow diagram.

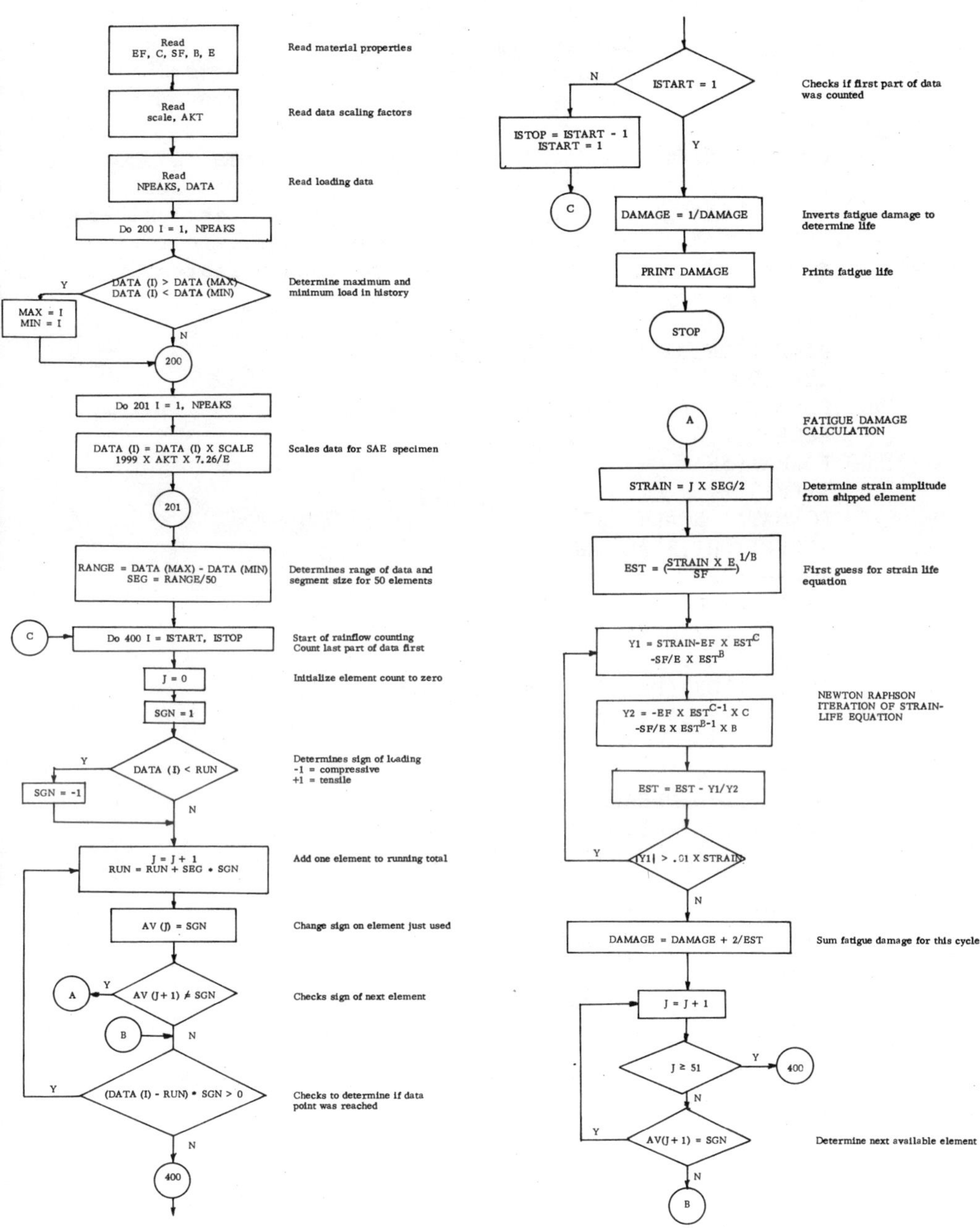

Figure A2. Detailed flow diagram.

Major Calculation Variables (Cont'd.)

DAMAGE	Fatigue Damage
STRAIN	Strain Amplitude
EST	Cycles to Failure for Strain Range

The overall flow diagram for all of the programs is shown in Fig. A1 with a detailed flow diagram for this analysis shown in Fig. A2. If a histogram of rainflow counted ranges is desired, the following statement is added to the fatigue damage calculation: HIST (J) = HIST (J) + 1. Of course, the dimension statement must be changed and the array initialized.

```
      DIMENSION DATA(6000),AV(51)
      TYPE 90
90    FORMAT(' EF,C,SF,E,E')
      ACCEPT 100,EF,C,SF,B,E
100   FORMAT(5E)
      TYPE 91
91    FORMAT(' SCALE, KT')
      ACCEPT 101,SCALE,AKT
101   FORMAT(2F)
      CALL IFILE(1,'BRACK')
      READ(1,102)NPEAKS
102   FORMAT(6X,I5)
      READ(1,103)(DATA(I),I=1,NPEAKS)
103   FORMAT(6X,14F5.0)
      MAX=1
      MIN=1
      DO 200 I=1,NPEAKS
      IF(DATA (I).GT.DATA(MAX)) MAX=I
      IF(DATA (I).LT.DATA(MIN)) MIN=I
200   CONTINUE
      DO 201 I=1,NPEAKS
      DATA(I)=DATA(I)*SCALE/999.*AKT*7.26/E
201   CONTINUE
              RANGE=DATA(MAX)-DATA(MIN)
      SEG=RANGE/50.
      ISTART=MAX
      IF(ABS(DATA(MIN)).GT.ABS(DATA(MAX))) ISTART=MIN
      RUN=DATA(ISTART)
      SGN=RUN/ABS(RUN)
      DO 202 I=1,51
      AV(I)=SGN
202   CONTINUE
      ISTART=ISTART+1
      IF(ISTART.GT.NPEAKS) ISTART=1
      ISTOP=NPEAKS
      DAMAGE=0.
```

```
300 DO 400 I=ISTART,ISTOP
    J=0
    SGN=1
    IF(DATA(I).LT.RUN) SGN=-1.
301 J=J+1
    RUN=RUN+SEG*SGN
    AV(J)=SGN
    IF(AV(J+1).NE.SGN) GO TO 350
    STRAIN=FLOAT(J)*SEG/2.
    EST=(STRAIN*E/SF)**(1./B)
302 Y1=STRAIN-EF*EST**C-SF/E*EST**B
    Y2=-C*EF*EST**(C-1.)-B*SF/E*EST**(B-1.)
    EST=EST-Y1/Y2
    IF(ABS(Y1).GT.0.01*STRAIN) GO TO 302
    DAMAGE=DAMAGE+2./EST
303 J=J+1
    IF(J.GE.51) GO TO 400
    IF(AV(J+1).EQ.SGN) GO TO 303
350 IF((DATA(I)-RUN)*SGN+0.01*SEG.GE.0.) GO TO 301
400 CONTINUE
    IF(ISTART.EQ.1) GO TO 401
    ISTOP=ISTART-1
    ISTART=1
    GO TO 300
401 DAMAGE=1./DAMAGE
    TYPE 104,DAMAGE
104 FORMAT(' BLOCKS TO FAILURE ',E)
    STOP
    END
```

APPENDIX B

Computer Program for Load-Notch Strain Analysis

Only three lines of the preceding program are changed to use a load-strain curve in the analysis.

First, the constants describing the load-strain curve must be read into the program.

```
ACCEPT 100, SCALE, C1, C2, D
```

Second, the load is calculated from the segment number and the strain is then calculated from the load.

```
ALOAD = J* SEG/2
STRAIN = ALOAD/C1 + (ALOAD/C2)** 1/D
```

```
      DIMENSION DATA(6000).AV(51)
      TYPE 90
90    FORMAT(' EF,C,SF,B,E')
      ACCEPT 100,EF,C,SF,B,E
100   FORMAT(5E)
      TYPE 91
      ACCEPT 100,SCALE,C1,C2,D
91    FORMAT(' SCALE,C1,C2,D')
      CALL IFILE(1,'SUSP')
      READ(1,102)NPEAKS
102   FORMAT(6X,I5)
      READ(1,103)(DATA(I),I=1,NPEAKS)
103   FORMAT(6X,14F5.0)
      MAX=1
      MIN=1
      DO 200 I=1,NPEAKS
      IF(DATA(I).GT.DATA(MAX)) MAX=I
      IF(DATA(I).LT.DATA(MIN)) MIN=I
200   CONTINUE
      DO 201 I=1,NPEAKS
      DATA(I)=DATA(I)*SCALE/999.
201   CONTINUE
      RANGE=DATA(MAX)-DATA(MIN)
      SEG=RANGE/50.
      ISTART=MAX
      IF(ABS(DATA(MIN)).GT.ABS(DATA(MAX))) ISTART=MIN
      RUN=DATA(ISTART)
      SGN=RUN/ABS(RUN)
      DO 202 I=1,51
      AV(I)=SGN
202   CONTINUE
      ISTART=ISTART+1
      IF(ISTART.GT.NPEAKS) ISTART=1
      ISTOP=NPEAKS
      DAMAGE=0.
300   DO 400 I=ISTART,ISTOP
      J=0
      SGN=1
      IF(DATA(I).LT.RUN) SGN=-1.
301   J=J+1
      RUN=RUN+SEG*SGN
      AV(J)=SGN
      IF(AV(J+1).NE.SGN) GO TO 350
      ALOAD=FLOAT(J)*SEG/2.
      STRAIN=ALOAD/C1+(ALOAD/C2)**(1./D)
      EST=(STRAIN*E/SF)**(1./B)
302   Y1=STRAIN-EF*EST**C-SF/E*EST**B
      Y2=-C*EF*EST**(C-1.)-B*SF/E*EST**(B-1.)
      EST=EST-Y1/Y2
      IF(ABS(Y1).GT.0.01*STRAIN) GO TO 302
      DAMAGE=DAMAGE+2./EST
```

```
303 J=J+1
    IF(J.GE.51) GO TO 400
    IF(AV(J+1).EQ.SGN) GO TO 303
350 IF((DATA(I)-RUN)*SGN+0.01*SEG.GE.0.) GO TO 301
400 CONTINUE
    IF(ISTART.EQ.1) GO TO 401
    ISTOP=ISTART-1
    ISTART=1
    GO TO 300
401 DAMAGE=1./DAMAGE
    TYPE 104,DAMAGE
104 FORMAT(' BLOCKS TO FAILURE ',E)
    STOP
    END
```

APPENDIX C

Computer Program for Load-Stress-Strain Analysis

The preceding program is modified to include mean stress effects by adding a section to calculate the stress that corresponds to each strain element. Three new variables are introduced;

STRESS	An Array Containing the Stress for Each Element
SM	Mean Stress
STRS	Mean Stress Reference

```
    DIMENSION DATA(6000),AV(51),STRESS(52),STRAIN(51)
    TYPE 90
90  FORMAT(' EF,C,SF,B,E')
    ACCEPT 100,EF,C,SF,B,E
100 FORMAT(5E)
    SHE=B/C
    CSC=SF/EF**SHE
    TYPE 91
    ACCEPT 100,SCALE,C1,C2,D
91  FORMAT(' SCALE,C1,C2,D')
    CALL IFILE(1,'BRACK')
    READ(1,102)NPEAKS
102 FORMAT(6X,I5)
    READ(1,103)(DATA(I),I=1,NPEAKS)
103 FORMAT(6X,14F5.0)
    MAX=1
    MIN=1
    DO 200 I=1,NPEAKS
    IF(DATA(I).GT.DATA(MAX)) MAX=I
    IF(DATA(I).LT.DATA(MIN)) MIN=I
```

```
 200 CONTINUE
     DO 201 I=1,NPEAKS
     DATA(I)=DATA(I)*SCALE/999.
 201 CONTINUE
     RANGE=DATA(MAX)-DATA(MIN)
     SEG=RANGE/50.
     ISTART=MAX
     IF(ABS(DATA(MIN)).GT.ABS(DATA(MAX))) ISTART=MIN
     RUN=DATA(ISTART)
     SGN=RUN/ABS(RUN)
     DO 202 I=1,51
     AV(I)=SGN
 202 CONTINUE
     DO 500 I=1,50
     ALOAD=FLOAT(I)*SEG/2.
     STRAIN(I)=ALOAD/C1+(ALOAD/C2)**(1./D)
 500 CONTINUE
     STRAIN(51)=ABS(RUN)/C1+(ABS(RUN)/C2)**(1./D)
     DO 510 I=1,51
     SIG=STRAIN(I)*E
 511 Y1=STRAIN(I)-SIG/E-(SIG/CSC)**(1./SHE)
     Y2=-1./E-(1./SHE)*(SIG/CSC)**(1./SHE-1.)/CSC
     SIG=SIG-Y1/Y2
     IF(ABS(Y1).GT.0.C1*STRAIN(I)) GO TO 511
     STRESS(I+1)=SIG
 510 CONTINUE
     STRESS(1)=0.
     STRS=STRESS(52)*SGN
     ISTART=ISTART+1
     IF(ISTART.GT.NPEAKS) ISTART=1
     ISTOP=NPEAKS
     DAMAGE=0.
 300 DO 400 I=ISTART,ISTOP
     J=0
     SGN=1
     IF(DATA(I).LT.RUN) SGN=-1.
 301 J=J+1
     RUN=RUN+SEG*SGN
     STRS=STRS+(STRESS(J+1)-STRESS(J))*SGN*2.
     AV(J)=SGN
     IF(AV(J+1).NE.SGN) GO TO 350
     SM=STES-STRESS(J+1)*SGN
     EST=(STRAIN(J)*E/(SF-SM))**(1./B)
 302 Y1=STRAIN(J)-EF*EST**C-(SF-SM)/E*EST**B
     Y2=-C*EF*EST**(C-1.)-B*(SF-SM)/E*EST**(B-1.)
     EST=EST-Y1/Y2
     IF(ABS(Y1).GT.0.01*STRAIN(J)) GO TO 302
     DAMAGE=DAMAGE+2./EST
 303 J=J+1
     IF(J.GE.51) GO TO 400
     IF(AV(J+1).EQ.SGN) GO TO 303
```

```
350 IF((DATA(I)-RUN)*SGN+C.01*SEG.GE.0.) GO TO 301
400 CONTINUE
    IF(ISTART.EQ.1) GO TO 401
    ISTOP=ISTART-1
    ISTART=1
    GO TO 300
401 DAMAGE=1./DAMAGE
    TYPE 104,DAMAGE
104 FORMAT(' BLOCKS TO FAILURE ',E)
    STOP
    END
```

APPENDIX D

Computer Program for Neuber Notch Analysis

The changes made to the nominal strain analysis program are the same as those made in the preceding section except the stresses are calculated using Neuber's rule.

```
    DIMENSION DATA(6000),AV(51),STRESS(52),STRAIN(51)
    TYPE 90
90  FORMAT(' EF,C,SF,B,E')
    ACCEPT 100,EF,C,SF,B,E
100 FORMAT(5E)
    SHE=B/C
    CSC=SF/EF**SHE
    TYPE 91
    ACCEPT 100,SCALE,AKF
91  FORMAT(' SCALE,KF')
    CALL IFILE(1,'BRACK')
    READ (1,102)NPEAKS
102 FORMAT(6X,I5)
    READ(1,103)(DATA(I),I=1,NPEAKS)
103 FORMAT(6X,14F5.0)
    MAX=1
    MIN=1
    DO 200 I=1,NPEAKS
    IF(DATA(I).GT.DATA(MAX)) MAX=I
    IF(DATA(I).LT.DATA(MIN)) MIN=I
200 CONTINUE
    DO 201 I=1,NPEAKS
    DATA(I)=DATA(I)*SCALE/999.*7.26
201 CONTINUE
    RANGE=DATA(MAX)-DATA (MIN)
    SEG=RANGE/50.
    ISTART=MAX
```

```
      IF(ABS(DATA(MIN)).GT.ABS(DATA(MAX))) ISTART=MIN
      RUN=DATA(ISTART)
      SGN=RUN/ABS(RUN)
      DO 202 I=1,51
      AV (I)=SGN
  202 CONTINUE
      SIG=SEG*AKF/2.
      DO 500 I=1,51
      STR=FICAT(I)*SEG/2.
      IF(I.EQ.51) STR=ABS(RUN)
      ANEUB=(STR*AKF)**2
  501 Y1=SIG**2+SIG*(SIG/CSC)**(1./SHE)*E-ANEUB
      Y2=2.*SIG+(1./SHE+1)*(SIG/CSC)**(1./SHE)*E
      SIG=SIG-Y1/Y2
      IF(AES(Y1).GT.0.01*ANEUB) GO TO 501
      STRESS(I+1)=SIG
      STRAIN(I)=SIG/E+(SIG/CSC)**(1./SHE)
  500 CONTINUE
      STRESS(1)=0.
      STRS=STRESS(52)*SGN
      ISTART=ISTART+1
      IF(ISTART.GT.NPEAKS) ISTART=1
      ISTOP=NPEAKS
      DAMAGE=0.
  300 DO 400 I=ISTART,ISTOP
      J=0
      SGN=1
      IF(DATA(I).LT.RUN) SGN=-1.
  301 J=J+1
      RUN=RUN+SEG*SGN
      STRS=STRS+(STRESS(J+1)-STRESS(J))*SGN*2.
      AV(J)=SGN
      IF(AV(J+1).NE.SGN) GO TO 350
      SM=STRS-STRESS(J+1)*SGN
      EST=(STRAIN(J)*E/(SF-SM))**(1./B)
  302 Y1=STRAIN(J)-EF*EST**C-(SF-SM))**(1./B)
      Y2=-C*EF*EST**(C-1.)-B*(SF-SM)/E*EST**(B-1.)
      EST=EST-Y1/Y2
      IF(ABS(Y1).GT.0.C1*STRAIN(J)) GO TO 302
      DAMAGE=DAMAGE+2./EST
  303 J=J+1
      IF(J.GE.51) GO TO 400
      IF(AV(J+1).EQ.SGN) GO TO 303
  350 IF(DATA(I)-RUN)*SGN+0.01*SEG.GE.0.) GO TO 301
  400 CONTINUE
      IF(ISTART.EQ.1) GO TO 401
      ISTOP=ISTART-1
      ISTART=1
      GO TO 300
```

```
401 DAMAGE=1./DAMAGE
    TYPE 104,DAMAGE
104 FORMAT(' BLOCKS TO FAILURE ',E)
    STOP
    END
```

REFERENCES

1. Brose, W.R., Dowling, N.E. and Morrow, JoDean, "Effect of Periodic Large Strain Cycles on the Fatigue Behavior of Steels", Paper 740221 presented at SAE Automotive Engineering Congress, Detroit, MI, February 1974.

2. Dowling, N.E., "Fatigue Failure Predictions for Complicated Stress-Strain Histories", J. of Materials, JMSLA, Vol. 7, No. 1, March 1972, pp 71-87.

3. Martin, J.F., Topper, T.H. and Sinclair, G.M., "Computer Based Simulation of Cyclic Stress-Strain Behavior with Applications to Fatigue", Materials Research and Standards, MTRSA, Vol. 11, No. 2, February 1971, pp 23-9.

4. Wetzel, R.M., "A Method of Fatigue Damage Analysis", Ph.D. Thesis, Department of Civil Engineering, University of Waterloo, Ontario, Canada, 1971

5. Richards, F.D., LaPointe, N.R. and Wetzel, R.M., "A Cycle Counting Algorithm for Fatigue Damage Analysis", Paper 740278 presented at SAE Automotive Engineering Congress, Detroit, MI, February 1974.

6. Landgraf, R.W., Richards, F.D. and LaPointe, N.R., "Fatigue Life Predictions for a Notched Member under Complex Load Histories", Paper 750040 presented at SAE Automotive Engineering Congress, Detroit, MI, February 1975.

7. Topper, T.H., Wetzel, R.M. and Morrow, JoDean, "Neuber's Rule Applied to Fatigue of Notched Specimens", J. of Materials, JMSLA, Vol. 4, No. 4, March 1969, pp 200-09.

8. Stadnick, S.J. and Morrow, JoDean, "Techniques for Smooth Specimen Simulation of the Fatigue Behavior of Notched Members", Testing for Prediction of Material Performance in Structures and Components, ASTM STP 515, 1972, pp 229-52.

9. Socie, D.F., "Fatigue Life Prediction Using Local Stress-Strain Concepts", Paper presented at 1975 SESA Spring Meeting, Chicago, IL, Mary 1975.

10. Tucker, L. and Bussa, S., "The SAE Cumulative Fatigue Damage Test Program", Paper 750038 presented at SAE Automotive Engineering Congress, Detroit, MI, February 1975.

11. Peterson, R.E., Stress Concentration Factors, John Wiley and Sons, 1974.

12. Mitchell, M.R., "A Unified Predictive Technique for the Fatigue Resistance of Cast Ferrous-Based Metals and High Hardness Wrought Steels", FCP Report No. 23, September 1976.

13. Higashida, Y. and Lawrence, F.V., "Strain Controlled Fatigue Behavior of Weld Metal and Heat-Affected Base Metal in A36 and A514 Steels Weld", FCP Report No. 22, August 1976.

CHAPTER 9

A NOTE ON FATIGUE SCATTER AND LIFE PREDICTIONS

Volker Weiss and Albert Kuo

Syracuse University

Syracuse, New York

One of the major problems in predicting fatigue life for service components from laboratory data is that the scatter of laboratory data often differs from that of actual service components. In this note a method to estimate fatigue lives from crack propagation models is presented. It will be shown that the result is a scatter and life prediction which well represents those desired for typical manufactured parts.

Fatigue crack growth can be reasonably well characterized by

$$\frac{da}{dN} = c(\Delta K)^m \tag{1}$$

where a is the crack length, N the number of cycles, c a constant, K the stress intensity factor, and m an exponent, usually about 4. From notch analysis of fracture [1] a similar relation is obtained, namely

$$\frac{da}{dN} = a\left(\frac{\sigma}{\sigma_o}\right)^{m'} \qquad m' \simeq 4 \tag{2}$$

where σ is the maximum cycle stress and σ_o a material constant. Mean stress and threshold effects are also readily incorporated in these formulae [2]. For the present estimate the second equation is used. After integration one obtains

$$\ln \frac{a}{a_o} = \left(\frac{\sigma}{\sigma_o}\right)^4 N \tag{3}$$

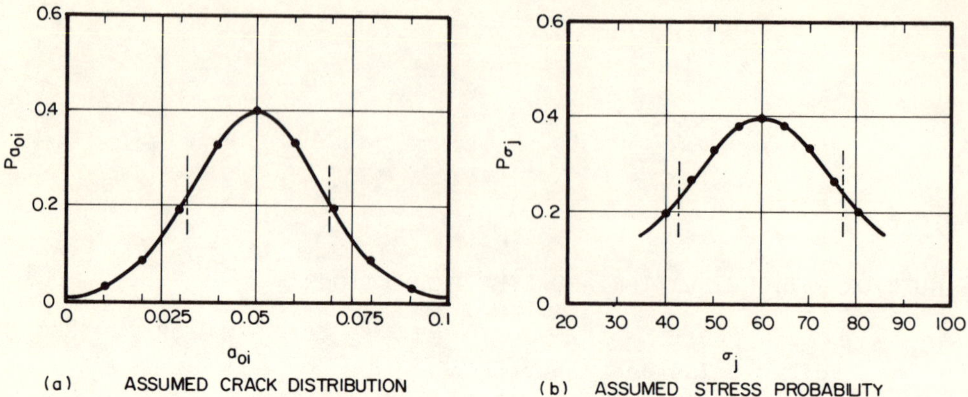

Figure 1. Assumed crack distribution and stress probability for hypothetical manufactured parts.

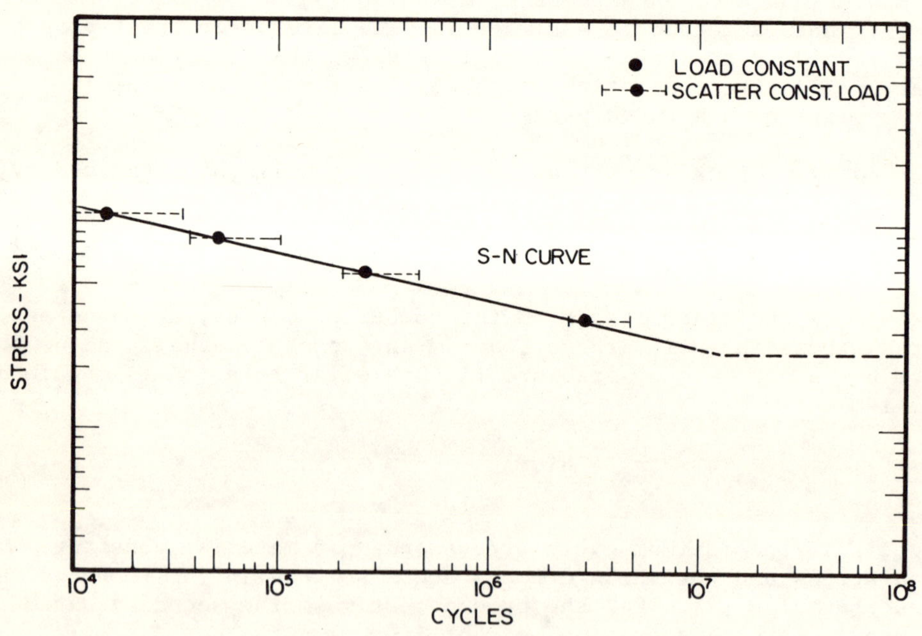

Figure 2. Assumed S-N curve for material with scatter bars resulting from the assumed initial crack distribution.

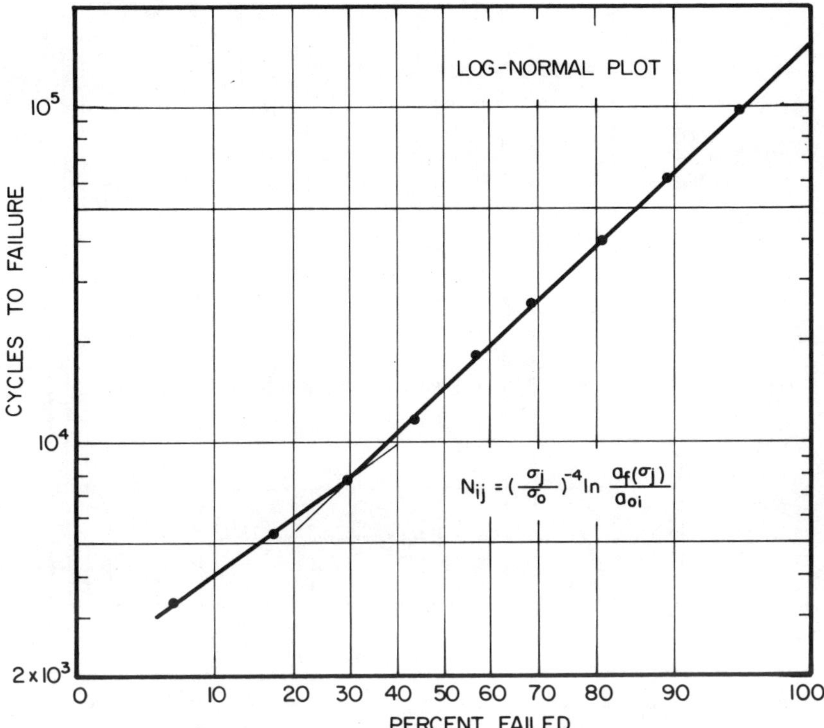

Figure 3. Life prediction for hypothetical manufactured lot, log-normal plot.

or, for varying stress ranges

$$a_i = a_o \exp \{\Sigma \frac{\sigma_i}{\sigma_o} N_i\} \tag{4}$$

This result indicates that the cumulative damage (or crack length) is independent of the loading sequence (obviously an oversimplification that needs to be taken into account in future treatments). It further follows from Equation (3) that a normal distribution of initial crack sizes, a_o, will lead to a log-normal distribution for the lives.

A large number of manufactured parts containing cracks of various lengths, scattering around 0.050 in., and subjected to various loading spectra is considered. The assumed Gaussian crack distribution and stress (load) probability are presented in Figure 1. The assumed S-N curve and the scatter resulting from the distribution of initial crack sizes are indicated in Figure 2. The resulting life distribution for the assumed initial crack and stress distribution is given in Figure 3 as a log-normal plot and

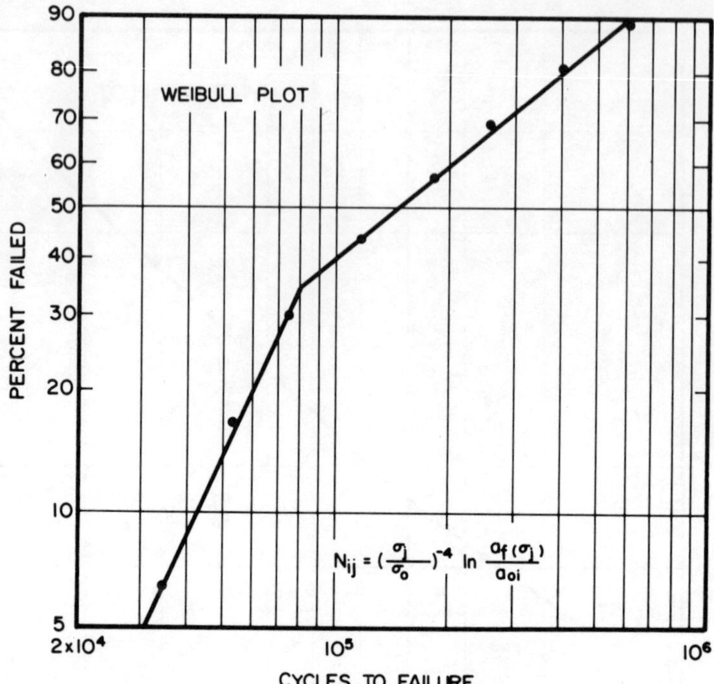

Figure 4. Life prediction for hypothetical manufactured lot, Weibull plot.

in Figure 4 as a Weibull plot. The double slope Weibull plot, frequently observed for actual components and often ascribed to the action of two different failure mechanisms, is a direct consequence of these assumptions, with only one failure mechanism active.

Often the hazard function

$$H = \frac{f(N)}{1-F(N)} \tag{5}$$

that is the failure rate of the survivors, where f(N) is the rate of failure in a given time period and F(N) is the accumulated fraction of failed parts, is of interest to the manufacturer. In practice the data for a part often follow the so-called bathtub curve. From statistics it is well known that no single distribution function f(N) can define a bathtub curve. However, the results of our calculations clearly define a bathtub curve, as encountered in real life, and as indicated in Figure 5. This lends considerable confidence in the proposed approach. In

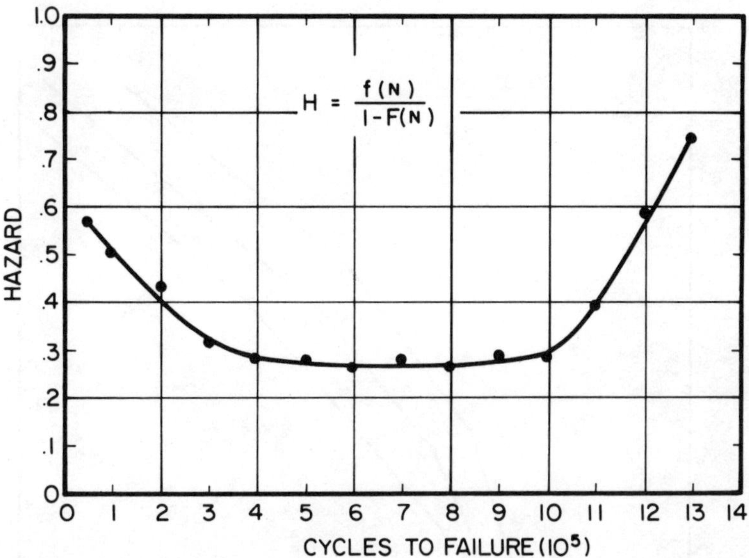

Figure 5. Hazard function (bathtub curve) for hypothetical lot.

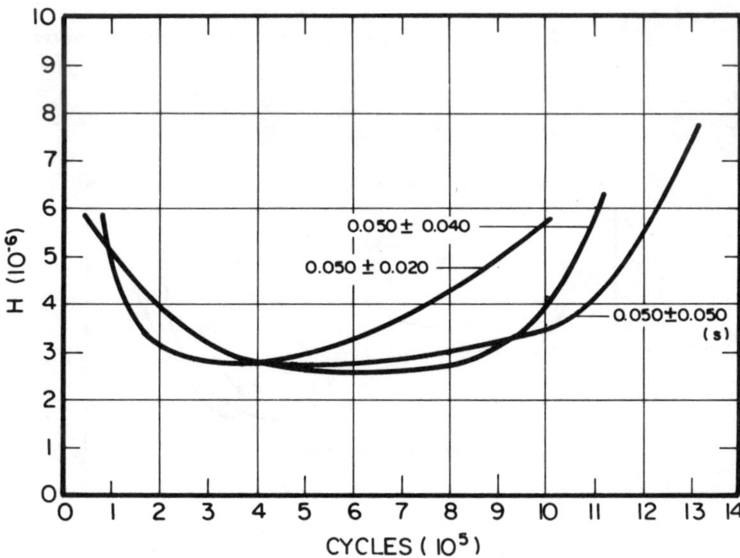

Figure 6. Effect of crack size variability on hazard function (bathtub curve).

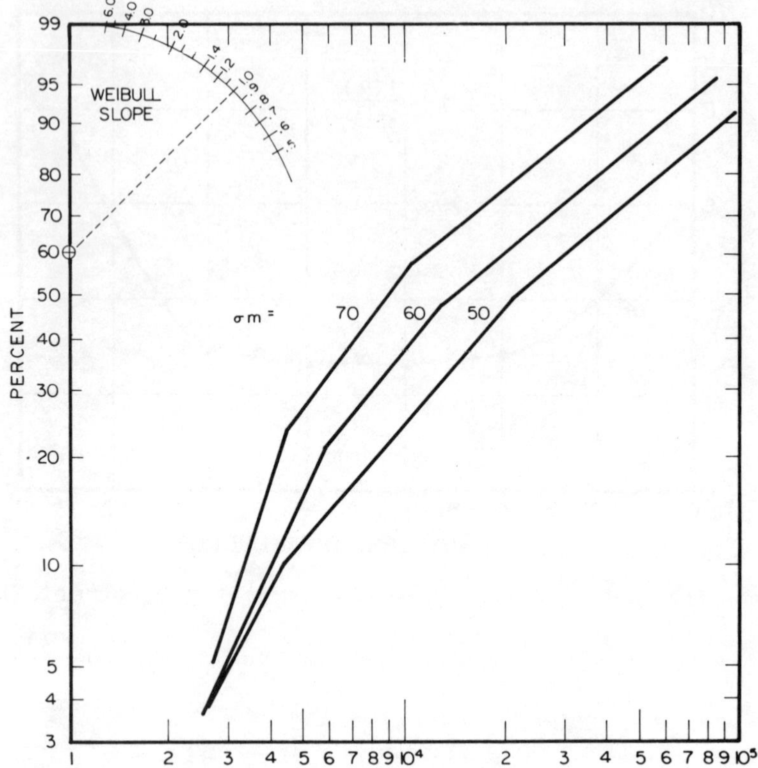

Figure 7. Effect of mean stress on Weibull plot.

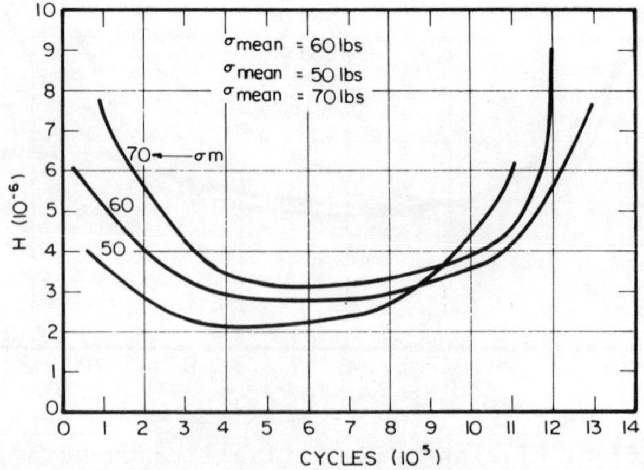

Figure 8. Effect of mean stress on hazard function (bathtub curve).

additional calculations the effects of average stress and of variations of the initial crack size scatter were also briefly examined. It was surprising to note that variations of the scatter of the initial crack size around a mean of 0.050 in. has little effect on the Weibull plot. The hazard function plots do show an effect, however, particularly in the second half beyond the mean life (Figure 6). Changes in the average stress are reflected, as expected, in the Weibull plot (Figure 7), although they appear difficult to notice for accumulated failures below 5% of the total production. However, from the hazard function plots, i.e., failures per year or monthly, the effect of different average stresses is quite evident (Figure 8).

REFERENCES

1. Weiss, V., "Notch Analysis of Fracture", in book Fracture-An Advanced Treatise in Seven Volumes, H. Liebowitz (ed.), Academic Press, III, 1971, p. 34.

2. Weiss, V., "Recent Advances in Notch Analysis of Fracture and Fatigue", Ingenieur-Archiv, 45 Springer-Verlag, 1976, pp. 281-9.

CHAPTER 10

ENVIRONMENTALLY ASSISTED FAILURES IN ORDNANCE COMPONENTS

P. A. Thornton and V. J. Colangelo

Watervliet Arsenal

Watervliet, New York

ABSTRACT

The production of ordnance equipment in common with private industry utilizes a wide variety of manufacturing processes including: forging, heat treatment, electroplating, welding, machining, etc. Each of these processes has the propensity to cause material problems which may lead to premature failure. The failure analyst must, therefore, be familiar with the processing history of a component as well as details of a failure incident and the service environment.

To demonstrate the interaction between failure analyses and detrimental environments, both in manufacturing and in service, we have restricted this chapter to a few examples involving environmentally associated failures. Case histories are presented dealing with the following type failures in steel: liquid metal embrittlement, hydrogen embrittlement, pitting corrosion and stress corrosion, which have occurred in weapon components. These cases are reviewed in some detail to convey the techniques utilized in arriving at the reason(s) for failure. Implementation of the recommendations subsequent to an analysis, demonstrates how failures, although unfortunate, can serve to improve a product and refine a design or process.

INTRODUCTION

The production of ordnance components involves a variety of engineering materials and corresponding processing. Among the various processes performed on these materials are: hot forging, heat treatments, machining, electroplating, welding, etc. Consequently, we encounter, as do other industries, a diversified range of problems often synergistically related to material, processing, and environment.

The object of this chapter is to review a rather select number of environmentally assisted failures associated with material and related processing. These examples will serve to demonstrate the relationship between manufacture of ordnance equipment and a broad spectrum of private industry. Furthermore, we will attempt to show how implementation of failure analysis recommendations can improve the reliability and performance fo a product. Occasionally, valuable information available from failures is not recognized or utilized and the investigator's findings or recommendations are overlooked. Unfortunately, under these circumstances the best possible failure analysis "wins the battle, but loses the war".

LIQUID METAL EMBRITTLEMENT

The mechanisms of liquid metal embrittlement (LME) have been extensively studied in recent years in a multitude of metals and alloys. It is generally agreed that this type of failure if promoted in steels by the following conditions [1,2]:

1. A stress in excess of the yield strength
2. Melting of the embrittling species
3. Wetting of the steel by the liquid metal

Investigation

During the manufacture of certain low alloy steel tube forgings, the process of swage autofrettage employs lead (pb) as a lubricant. The bore of the tube is electroplated with a thin film of Pb to facilitate passage of the swaging mandrel. This operation imposes a residual compressive stress, in the hoop direction, at the bore of the tube. However, a corresponding tensile stress of lesser magnitude develops in the longitudinal direction or long axis of the tube. Subsequently, these forgings are given a stress relief treatment at 343°C (650°F) for 5 hours.

After utilizing this process for approximately ten years without incident, a series of catastrophic failures occurred, "in house", following this operation. The subject tubes cracked virtually through the wall thickness and occasionally fractured into two halves during a later pressing sequence as shown in Figure 1. These failures displayed certain characteristic features, viz., (1) the cracks were oriented normal to the long axis of the forging, (2) the fractures appeared brittle, exhibiting practically no deformation

Figure 1. Example of transerve fracture in 105mm M68 gun tube. Arrow denotes fracture surface. Scale in inches.

and were coated with Pb and Pb oxide, (3) the cracks penetrated roughly 90% of the well thickness, leaving a thin annular rim of material intact at the OD. This problem eventually involved 38 forging with some individual components containing as many as 60 transverse cracks.

The mechanical property results from longitudinal (L-R orientation - ASTM E399) tensile and Charpy impact specimens exhibited acceptable levels of yield strength, ductility and impact energy as given in the following table:

0.1% YS		UTS		% RA	-40° CHARPY IMPACT	
MPa	(ksi)	MPa	(ksi)		J	(ft-lb)
1104-1173	(160-170)	1242-1276	(180-185)	40-60	28-46	(20-30)

Consequently, insufficient mechanical properties were not considered a factor in the cracking problem. Also, chemical analysis of the steel showed the material to be within specifications and dissolved gas contents (H_2, O_2, N_2) to be well below the embrittlement range. Metallographic examination of the transverse crack profiles disclosed a completely interfranular fracture (Figure 2) indicating the cracks selectively followed prior austenitic grain boundaries. Therefore, a grain boundary fracture mechanism such as hydrogen

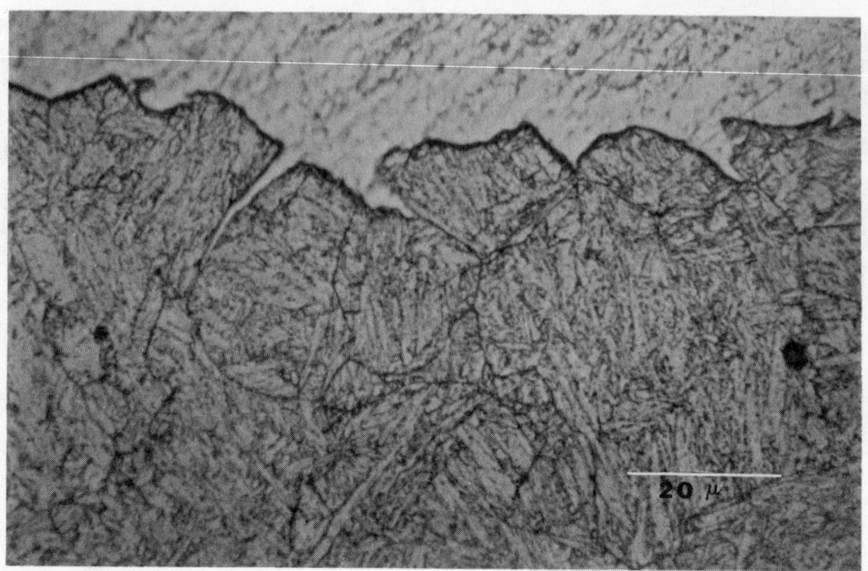

Figure 2. Photomicrograph of transverse crack showing intergranular fracture mode. 2% Nital.

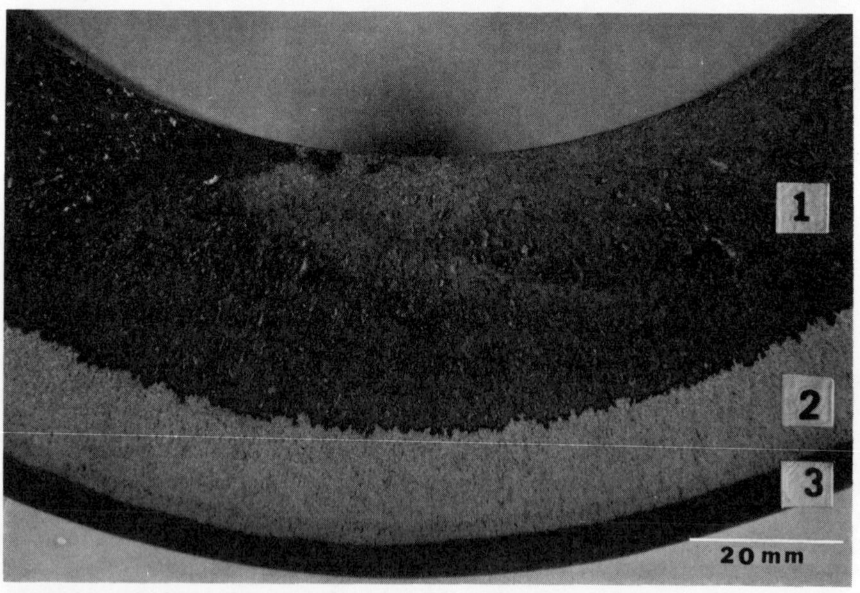

Figure 3. Close-up of fracture surface showing various zones. (1) lead oxide, (2) lead, and (3) shear lip on O.D.

embrittlement, stress corrosion or liquid metal embrittlement was suggested. The possibility of interaction between these mechanisms was also considered.

The aspect of LME was intensively explored because prima facie evidence indicated that the Pb lubricant from swaging was involved. This was implied by the prominant coating of Pb and Pb oxide on the fractures as illustrated in Figure 3. A laboratory experiment was conducted to analyze the role of the Pb lubricant in the failures, while extensive examinations of the manufacturing process, namely, electroplating, swaging, and stress relief treatment continued. One primary factor identified in the overall operation was the involvement of "repaired" anodes in the plating process. These anodes were pulled from service after a specific amount of depletion, and restored to original dimensions by a weld repair. The Pb wire utilized, contained approximately 4% Antimony (Sb). Antimony, when added in dilute amounts to Pb, increases the embrittlement effects produced by the lead in high strength steels [3,4]. The presence of certain substances that prevent the formation of hydrogen gas molecules at the cathode during plating increases the concentration of atomic hydrogen that can be absorbed by the steel; thereby increasing the potential for hydrogen embrittlement [5]. These substances, called cathodic poisons, include Sb. Conspicuously, when the repaired anodes were returned to service, transverse cracking occurred.

Correspondingly, transverse cracks were produced in the laboratory on scaled-down cylinders processed similarly to the full size tubes. The particular cylinders that cracked did so when Sb was present in the anode. This observation, along with fracture features identical to the tube forgings, indicated that a two-stage mechanism was likely operating during the production of the cracks. First, hydrogen embrittlement resulting from plating and swaging initiates transverse microcracks or fissures in the bore surface. Then, these fissures, shown in Figure 4, act as the initial defect (stress concentrator) and promote liquid Pb embrittlement (cracking) when the LME criteria are satisfied during the subsequent thermal soak.

Conclusions and Recommendations

The embrittlement mechanism which describes the transverse cracking takes into consideration the evidence gathered in the shop operations and data from the corresponding laboratory experiments. This information includes (1) a longitudinal, residual tensile stress in the tube from swaging, (2) a cathodic poison (Sb) in the plating operation, (3) sensitivity to time lapse between plating and swaging for hydrogen diffusion, (4) availability of liquid Pb for embrittlement (crack extension) during the post swage thermal soak.

Figure 4. SEM micrograph showing transverse fissures on bore surface (white arrow). Dark arrow denotes fissure depth on fracture surface.

Based on this analysis, it was recommended that a lubricant other lead be utilized for the swaging process. However, if it is necessary to use Pb, then antimony or any other contaminants, viz., zinc, arsenic, phosphorous, sulphur, etc., must be eliminated from the plating operation. In essence, a pure Pb anode must be used.

Subsequently, an investigation was conducted for alternate swaging lubricants. Candidates included: water-soluble wax emulsions used in drawing operations, synthetic and hydrocarbon oil base drawing compounds containing high pressure agents, high pressure greases, and sodium-stearate compounds. Laboratory testing showed the above type of lubricants would not cause embrittlement in this steel, under the respective manufacturing conditions. Presently, a "high pressure grease" is being employed. This compound contains ester, animal fat, fatty acid soap and hydrocarbon oil. Substitution of this grease has eliminated both the transverse cracking problem and the toxicity hazards of working with lead.

HYDROGEN EMBRITTLEMENT

Cracking or failure resulting from hydrogen embrittlement has been a problem in high strength steels for many years [6-8]. The damage may occur in the form of flakes (fish eyes), surface cracks, or a single major crack which results in premature failure. The general characteristics of hydrogen embrittled materials are:

1. A loss of tensile strength and ductility which diminishes with low test temperature and high strain rates.
2. Increased hydrogen contents which generally result in a loss of mechanical properties.
3. Increasing the tensile strength of the material causes increased susceptibility, as manifested by a decrease in failure time at a particular stress level.
4. Increased notch acuity lowers the applied stress required to cause failures.

Hydrogen primarily enters steels in two ways: (1) during the melting and pouring stages of production, and (2) during chemical or electrochemical treatments to the semi-finished steel. Although both methods can result in embrittlement and cracking, the latter will be examined because it is involved in the preponderance of hydrogen embrittlement problems in high strength steels.

Investigation

A vivid example of hydrogen embrittlement associated with electroplating was encountered in a large number of 1/4"-28 UNF, detent screws from the 152mm M81 gun. These fastening devices are self-locking, cylindrical head, cap screws, manufactured from alloy steel (SAE 4037) according to Military Standard MS21262(AGS) and cadmium (Cd) plated as per specification QQ-P-416. Two screws illustrating this problem are shown in Figure 5. Although these particular parts were cracked just above the nylon locking insert, this was not always the case. Other screws exhibited cracks further up the shank and occasionally in the head. The failure investigation consisted of tensile and hardness testing, metallography, and SEM of the fracture surface.

The mechanical tests showed the shank portion of the screws to have an average breaking load of 32500 N (7300 lb) in tension, as compared to the ultimate load requirement of 25900 N (5820 lb) for this part. The hardness measurements revealed a level of 42-43 Rc.

The microstructure was tempered martensite, containing very little non metallic included matter. However, a feature quite obvious during the examination was considerable intergranular

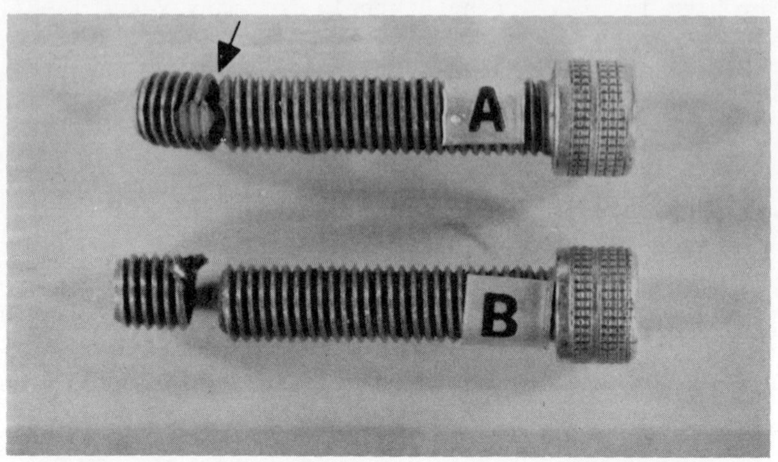

Figure 5. Cracked "detent" screws. Arrow denotes crack. Scale in inches.

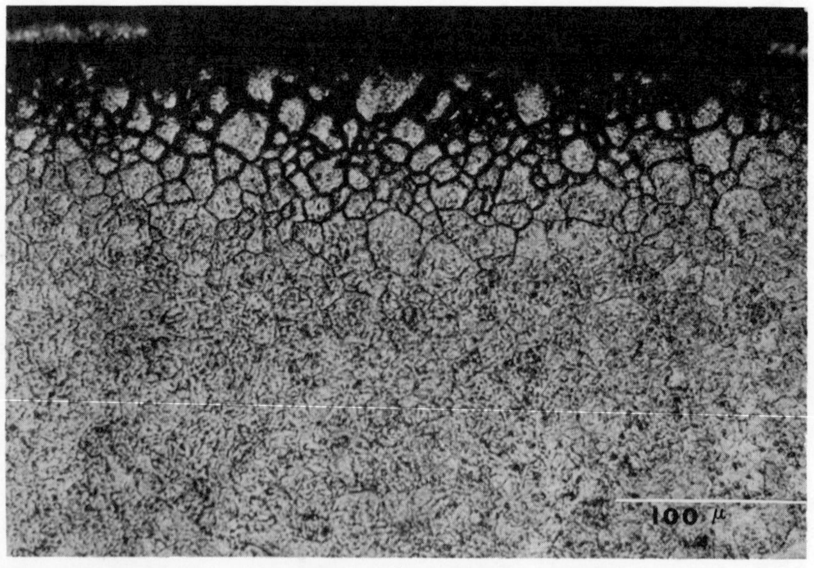

Figure 6. Photomicrograph showing extensive grain boundary degredation at surface of shank portion.

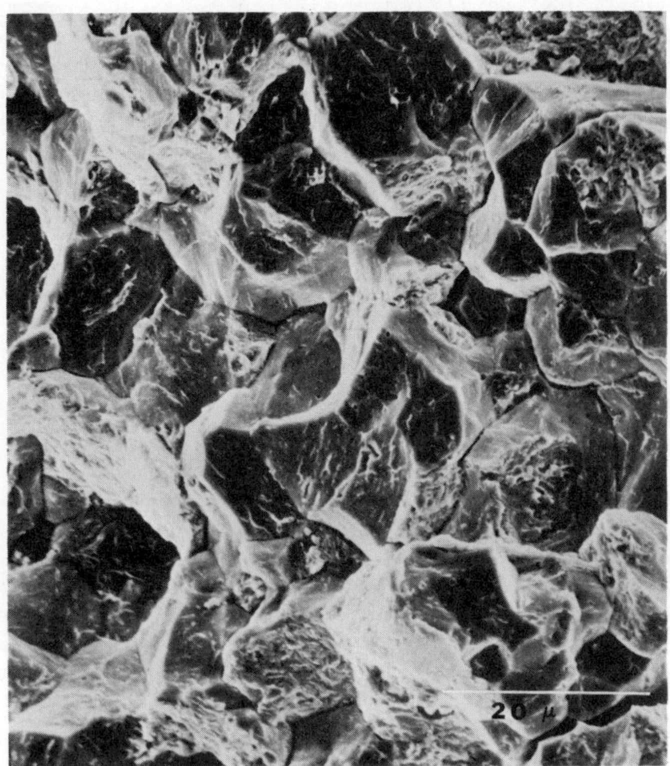

Figure 7. SEM micrograph showing intergranular fracture and secondary cracking.

separation or cracking along the surface of the screw. This "crazing", illustrated in Figure 6, was adjacent to the Cd plating.

Fractography by SEM verified the optical microscopic observations. The fracture surfacr prominently displayed a faceted, "rock candy" appearance characteristic of failure along prior austenitic grain boundaries (Figure 7). This failure mode is typical of hydrogen embrittlement wherein the hydrogen evolves during the plating process and is absorbed by the steel. Subsequently the hydrogen diffuses to the austenitic grain boundaries which act as sinks for these atoms. When a tensile stress is present, such as produced by tightening the screws, the grain boundaries are high susceptible to failure.

Conclusions and Recommendations

The evidence collected during this investigation demonstrated that hydrogen embrittlement was responsible for premature failure of the screws. The embrittled condition resulted from the Cd electroplating process and was not alleviated by an adequate thermal treatment to "bake-out" the hydrogen. Plating specification QQ-P-41 calls for a post plating bake out treatment for hardness levels of Rc 40 and above. For high strength steel it is standard practice in plating operations to thermally soak the material in the neighborhood of 204°C (400°F) to remove hydrogen [9.10]. It is likely that no hydrogen relief treatment was given these screws. In any event, it was recommended that that Cd plated cap screws, regardless of their hardness level, receive a bake-out at 190°C - 204°C for 4-6 hours. This action was sufficient to eliminate any reoccurrence of the cracking.

PITTING CORROSION

Corrosion in a very broad sense has been defined as the degredation of a material by reaction with its environment [11]. Pitting is a form of localized corrosion in which the attack is confined to numerous small cavities in the metal surface. The resultant cavities may cause only slight metal losses during their growth; hence pitting is a very dangerous form of environmental attack because perforation or failure may occur quite rapidly and without warning. Furthermore, pits can act as stress concentrators and contribute to premature failures in highly stressed components.

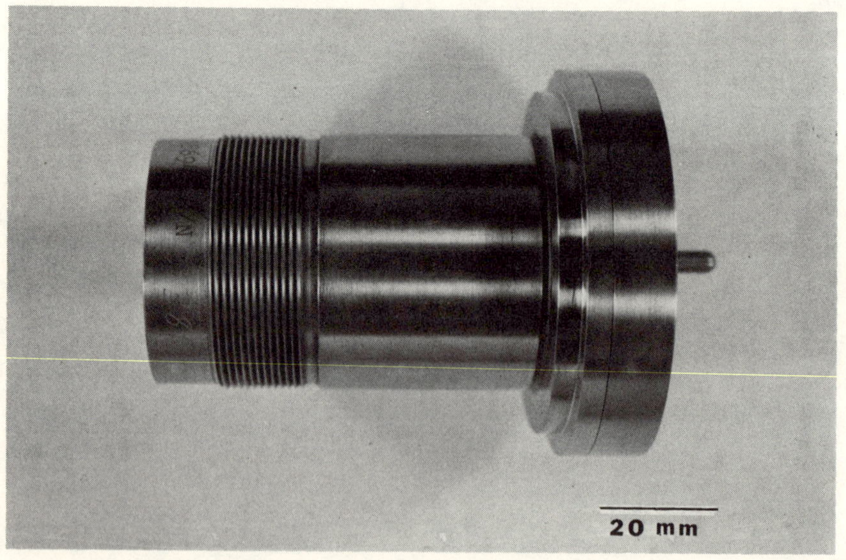

Figure 8. Firing probe component from 152mm gun.

Figure 9. Head area of firing probe showing surface attack. Scale in inches.

Investigation

During routine field inspections of a firing probe component (Figure 8) employed in the breech mechanism of the 152mm gun, several areas of severe surface attack were discovered. This damage, which interferes with the gas seal function of the probe, is illustrated in Figure 9. Sixteen components fabricated from 416 stainless steel were involved and rendered unserviceable by this condition. Only 10 rounds had been fired on this particular probe, thereby minimizing the probability that erosion or gas wash was responsible for the degredation. In fact, some probes involved were never subjected to firing.

The appearance of the cavities (Figure 10) strongly suggested pitting corrosion, where surface and subsurface concentrations cells are produced sporadically in the part due to a reaction with its environment. Pitting is intermediate between general corrosion (extensive surface attack) and complete immunity [11]. Unfortunately, the so-called "corrosion resistant" stainless steels are not immune to localized attack in certain aqueous environments, especially those containing chlorides. Furthermore, the geometry and design considerations of this component allow potential corrosives, v.z, condensed water vapor, firing residues, etc., to collect in the annular space between the probe and the tube chamber.

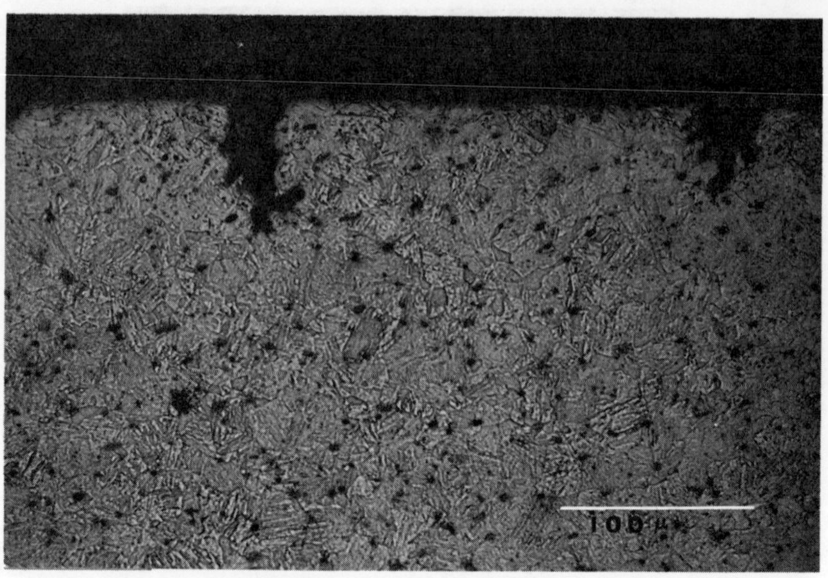

Figure 10. Photomicrograph through damaged region illustrating the pit-like appearance of the damage.

Metallography revealed the microstructure of the probe to be tempered martensite containing a relatively large amount of sulphide stringers. The sulphur is added to improve machinability; however, this addition decreases the alloy's resistance to pitting [11]. Pitting associated with the sulphide stringers is shown in Figure 11. These inclusions very likely provide excellent pit nucleation sites where they intersect the machined surface and may also contribute to the electrochemical processes within a pit during its progression.

Conclusions and Recommendation

Pitting corrosion is responsible for the surface degredation of the stainless steel firing probe. In addition to atmospheric conditions, propellant residues from firing can contribute to the initiation and growth of the pits. Moreover, the sulphur additions to the material result in stringers which also contribute to the pitting attack.

A design change to the probe, minimizing or eliminating contact with potential corrosives was deemed inappropriate. Therefore a material change to one with greater pitting resistance and protective coating on the present material were considered. Since there were a considerable number of components already in existence it was

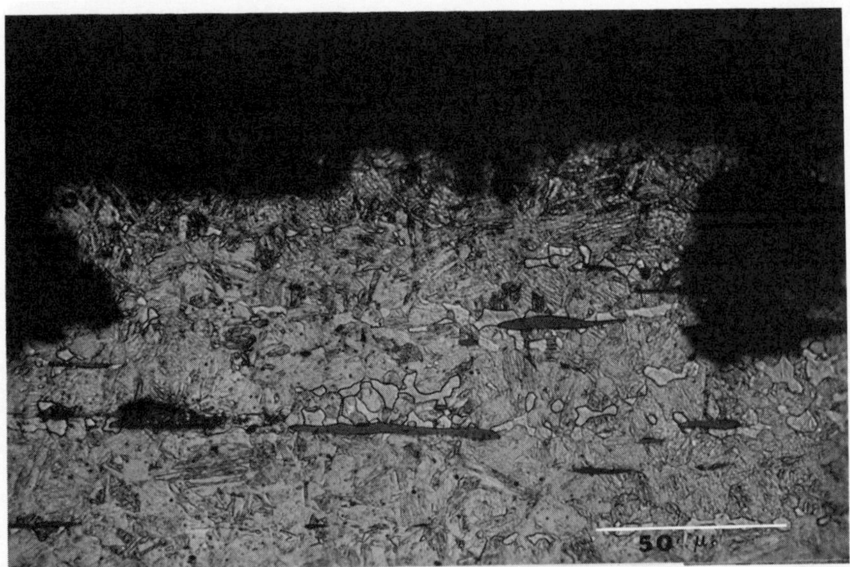

Figure 11. Photomicrograph showing the pitting attack associated with the sulphide stringers in the matrix.

Figure 12. 152mm firing probe with dry film lubricant applied to head portion (arrow).

decided to concentrate our efforts on protective coatings. Thus a solution would also be applicable to those components.

The ensuing study [12] included the following protective coatin as applied to the 416 stainless steel:

Electroplated lead
Electroplated chrome
Electroplated nickel
Dry film lubricant (thermal curing)
Fe-phosphate + Dry film lubricant (thermal cure)
Dry film lubricant (air cure)
Fe-phosphate + Dry film (air cure)

This "comparative" experiment utilized a 10% solution of ferric chloride ($FeCl_3$). $FeCl_3$ is an extremely agressive pitting environment for stainless steels and was selected in order to accelerate the testing [13].

The results of this investigation demonstrated that the dry fil lubricants (especially the thermal cure type applied over Fe-phosphate) provided better pitting resistance than the electroplated coatings (Cr and Ni). Admittedly, this determination is influenced by the experimental conditions rather than the exact corrosive conditions in the field. But the fact that the tests were accelerated, utilizing an unusually aggressive medium, provides a measure of confidence that the dry film lubricant (w/Fe-P) will adequately protect the firing probe against pitting. Currently the probes are dry filmed as shown in Figure 12 and performing successfully in service.

STRESS CORROSION

Stress corrosion generally implies cracking in a metal due to a combination of tensile stress (applied and/or residual) and a corrosive medium [14-16]. During stress corrosion cracking the materia is virtually unscathed over most of its surface, while fine cracks propagate through it. These cracks may be transgranular or intergranular depending on the alloy and the corrodents involved. Also, as a general rule, the stress corrosion cracks exhibit "branching" or secondary cracks accompanying the main crack. Furthermore, it has been shown that stress corrosion cracking can be produced in almost any metal or alloy in certain environments and at a corrosion potential specific to the metal-environment combination [16].

Investigation

During a routine training mission at the Hone, Germany, firing range, a 175mm M113 gun tube experienced a premature catastrophic

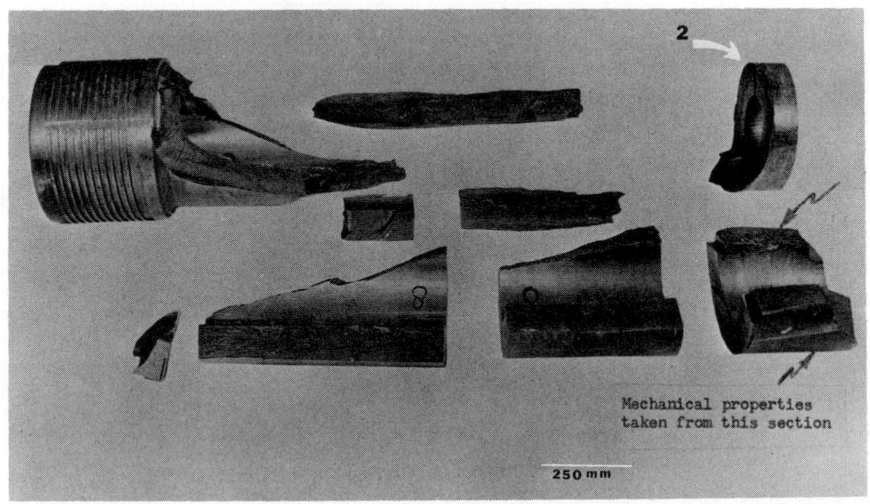

Figure 13. Fragments recovered from the camber area of 175mm M113 tube SN 1144.

failure. This incident resulted in fragmentation of the chamber area as illustrated in Figure 13. Despite an extensive search only a third of the failed portion was recovered, indicating the brittle nature of the fracture. Prior to the failure the tube had been fired on 21 occasions for a total of 163 rounds.

The ensuing examination consisted of a comprehensive, chemical, mechanical and metallurgical analysis. The chemical analysis showed the steel to have a typical gunsteel chemistry (4337 modified with V) and gas contents within acceptable limits. For example: H_2 - 0.21 ppm, O_2 - 46 ppm, and N_2 - 72 ppm.

Transverse tensile and Charpy specimens (C-R orientation as per ASTM E399) near the failed region produced the following results:

.1%YS		UTS		%RA	-40° Charpy Impact	
MPa	(ksi)	MPa	(ksi)		J	(ft-lb)
1313-1368	(190-198)	1416-1465	(205-212)	12-28	7-8	(5-6)

This data shows the steel to have a relatively high yield strength level with correspondingly low Charpy impact values. In addition, longitudinal (L-R orientation) Charpy specimens were tested to establish a fracture transition temperature for this particular material and for specimens retempered at 554°C (1030°F) for one hour. These results are displayed in Figure 14 and demonstrate that the ductile to brittle transition temperature of this material

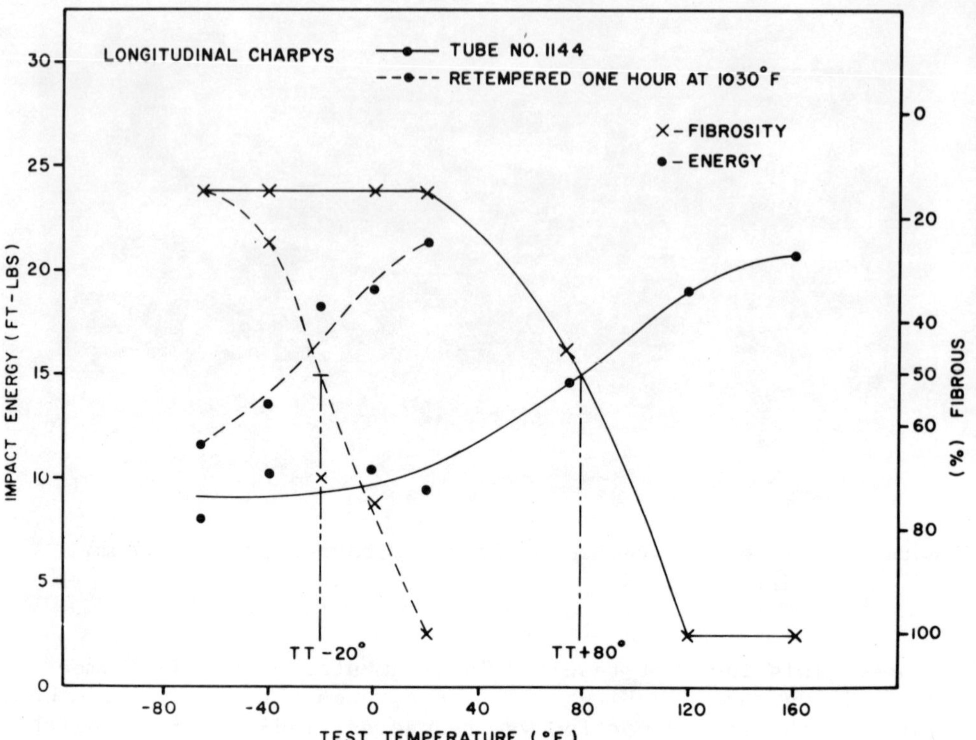

Figure 14. Plot showing impact energy behavior and FATT for tube 1144 and the retempered material.

undergoes a dramatic improvement for the retempered condition, as measured by both impact strength and percent fibrosity. The transition temperature decreased approximately 100°F (+80°F to -20°F), indicating that this tube was tempered improperly and that temper embrittlement resulted.

The metallographic specimens containing the fracture profiles exhibited an intergranular failure mode (Figure 15), accompanied by considerable secondary cracking (Figure 16). The fracture surfaces were so obscurred by ocrrosion, however, that characterization of the failure mode by electron fractography was impractical.

In a subsequent study [17] it was determined that this tube, tempered at 490°C (915°F), together with others tempered in the temper embrittlement range were susceptable to stress corrosion cracking. This vulnerability is graphically demonstrated in Figure 17. It was postulated that a hydrogen assisted growth mechanism was operating with the hydrogen being generated by the corrosion reaction at the crack tip [18].

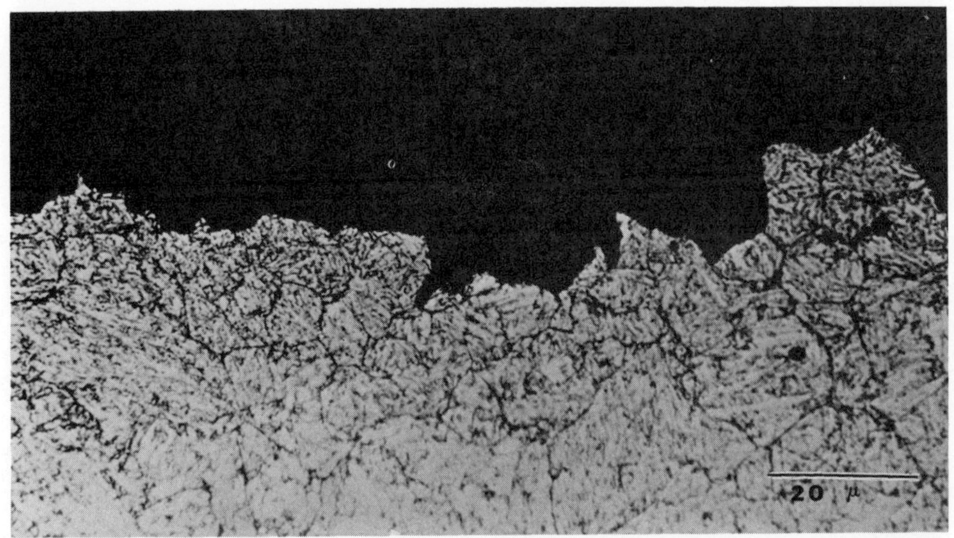

Figure 15. Photomicrograph displaying intergranular failure mode along the fracture surface at area 2 in Figure 12.

Figure 16. Photomicrograph showing secondary cracking and branching associated with the fracture.

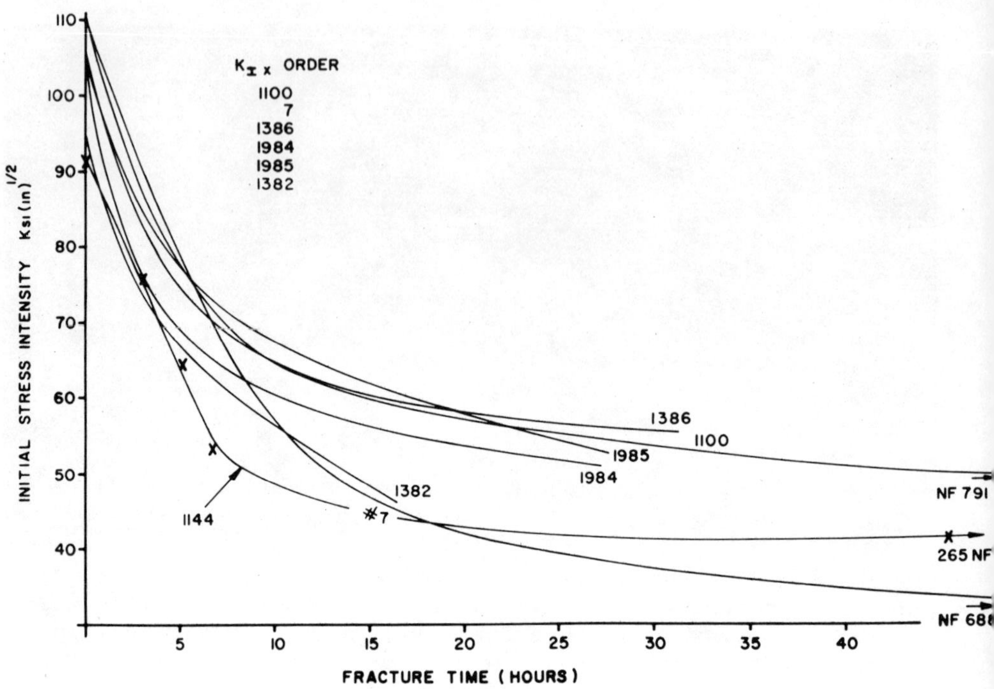

Figure 17. Plot illustrating the susceptibility of gunsteel to stress corrosion cracking. Note relative position of tube 1144. Stress corrosion susceptibility of gunsteel; fracture in oxygenated 3.5% sodium chloride.

Conclusions and Recommendations

The mechanical property data showed the steel to have relatively low impact toughness. Although this is not necessarily the cause of the failure per se, it played a contributory role because as the toughness of a steel decreases its tendency for brittle cracking increases. The fracture appearance transition temperature determinations demonstrated that the tube material was temper embrittled. Temper embrittlement in steel manifests itself as a loss of fracture toughness or impact strength with relatively little or no change in tensile strength and ductility [19,20]. When low alloy steels are held for tempering within their embrittling temperature range (427°C-566°C) or cooled too slowly through this interval, an intrinsic brittle material condition results. This condition, in conjunction with stress concentrators such as heat checking (bore crazing from firing) and a corrosive environment likely resulting from water vapor and firing residues, acted synergistically to produce the

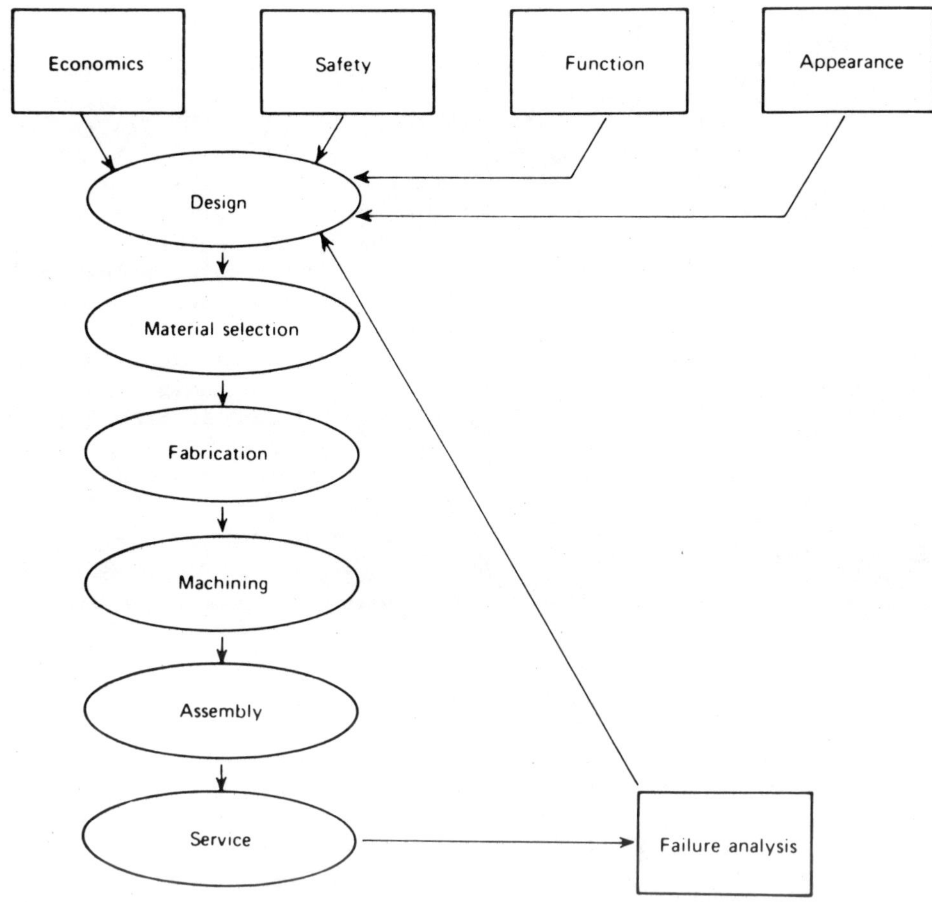

Figure 18. Schematic showing the relationship of failure analysis to the production chain.

stress corrosion cracking during service. Whe this cracking reached a critical stage the tube failed catastrophically.

Based on this analysis and the corresponding studies of temper embrittlement/stress corrosion cracking, previously referred to, a minimum tempering temperature of 552°C (1025°F) was established for heat treatment of gun tubes. Although the precise limites for temper embrittlement are chemistry dependent, this criteria has avoided any further embrittlement problem from tempering in gun tubes.

SUMMARY

The aspects of a few specific environmentally assisted failures have been discussed with emphasis on the metallurgical techniques used during the analysis and the implementation of the results in avoiding re-occurrence of the failures. Failures, although unfortunate, are virtually unavoidable. Thsi is due in part to the large number of interacting factors involved in the production and utilization of engineering or structural components. For example, material defects and variation, design, processing, workmanship, maintenance, adverse or unexpected service conditions, abuse and negligence. Although unpleasant, the information possible from a failure is very valuable and must not be wasted. This data is perhaps the last link on the product chain and should be fed back to the design/engineering stage. This procedure is schematically illustrated in Figure 18.

Judicious implementation of failure analysis recommendations can result in beneficial changes to a component, viz., its material, processing or design. In short, this represents Product Improvement, with the attendant increases in performance, reliability and profitability.

One final comment regarding the implementation of failure analysis results must deal with the subject of "Technology Transfer". This term refers to the communication of scientific or laboratory data to the shop personnel (workers actually performing the manufacturing operations). Frequently, this information is complex and presented in a manner which is not readily understood, except perhaps by the investigators themselves. Again we must emphasize that effective utilization of the failure may not occur unless the recommendations are understandable and acceptable to personnel at all levels of the production chain. This is especially true at the shop level where the fabrication and processing take place.

REFERENCES

1. Rostoker, W., McCaughey, J.M. and Markus, H., Embrittlement by Liquid Metals, New York: Reinhold Publishing Corp., 1960.
2. Kamdar, M.H., "Embrittlement by Liquid Metals", Progress in Materials Science, Vol. 15, No. 4, 1973, p. 289.
3. Breyer, N.N. and Gordon, P., Third International Conference on the Strength of Metals and Alloys, Conference Proceedings, Cambridge, England: The Institute of Metals, London, 1973, p. 493.
4. Breyer, N.N. and Johnson, K.L., J. of Testing and Evaluation, Vol. 2, No. 6, 1974, p. 471-7.
5. "Hydrogen Damage Failures", in Metals Handbook, Vol. 10, Failure Analysis and Prevention, 8th Ed., Metals Park, OH: American Society for Metals, 1975, p. 230
6. Colangelo, V.J. and Heiser, F.A., Analysis of Metallurgical Failures, New York: John Wiley and Sons, 1974.
7. Croucher, T.R., "Delayed Static Failure", in Source Book in Failure Analysis, Metals Park, OH: American Society for Metals, 1974, p. 20.
8. "Hydrogen Damage Failures", in Metals Handbook, Vol. 10, Failure Analysis and Prevention, 8th Ed., Metals Park, OH: American Society for Metals, 1975, p. 230.
9. Read, H.J., Hydrogen Embrittlement in Metal Finishing, New York: Reinhold Publishing Corp., 1961.
10. Daniels, W., "A Study of Thermal Relief After Chromium Plating", Watervliet Arsenal, NY, Report No. WVT-6735, Nov. 1967.
11. Fontana, M.G. and Greene, N.D., Corrosion Engineering, New York: McGraw-Hill, 1967.
12. Tauscher, S.G. and Thornton, P.A., "Pitting Corrosion of the 152mm Firing Probe", to be published as Watervliet Arsenal Technical Report.
13. Uhlig, H.H., Corrosion and Corrosion Control, New York: John Wiley and Sons, 1963.
14. Fontana, M.G. and Greene, N.D., Corrosion Engineering, New York: McGraw-Hill, 1967.
15. Uhlig, H.H., Corrosion and Corrosion Control, New York: John Wiley and Sons, 1963.
16. "Stress Corrosion Cracking", in Metals Handbook, Vol. 10, Failure Analysis and Prevention, 8th Ed., Metals Park, OH: American Society for Metals, 1975, p. 205.

17. Colangelo, V.J. and Ferguson, M.S., "Susceptibility of Gun Steels to Stress Corrosion Cracking", Watervliet Arsenal, NY, Report No. WVT-7012, Nov. 1970.

18. Colangelo, V.J. and Ferguson, M.S., "The Role of the Strain Handening Exponent in Stress Corrosion Cracking of a High Strength Steel", Corrosion, 25, 1969, pp. 509-14.

19. Heiser, F.A. and DeFries, R.S., "Temper Embrittlement in Steel for Thick-Walled Gun Tubes", J. Metals, 27, 1975, p. 8.

20. Holloman, J.H., "Temper Brittleness", Trans. ASM, 36, 1946, pp 473-542.

CHAPTER 11

HIGH TEMPERATURE ENVIRONMENTAL EFFECTS ON METALS

S. J. Grisaffe, C. E. Lowell, and C. A. Stearns

NASA-Lewis Research Center

Cleveland, Ohio

INTRODUCTION

Propulsion and power generation systems require components that will function for extended periods of time under high temperatures and stresses in hostile environments. These components are designed primarily on the basis of mechanical properties such as strength, fatigue life and creep resistance [1]. With this type of design approach, for systems where the operating parameters of temperature, stress, etc., are well characterized, component life is often found to be controlled by environmental attack. The factors involved in environmental attack are not readily included in component design criteria because they are not sufficiently well defined or understood. This situation has made it difficult and risky to predict long time service behavior on the basis of results obtained in accelerated, well controlled laboratory tests. Furthermore, little reliable and systematic data are available on service failures and operational histories of components so there is no good base for making life prediction correlations.

This chapter is an overview of present understanding and ability to predict high temperature environmental attack of metals. The gas turbine engine is used as an example but most of the technology applies equally to other systems.

ENVIRONMENTAL ATTACK IN A GAS TURBINE ENGINE

Environmental attack in a gas turbine engine derives from (1) high temperatures, (2) combustion products of the air and fuel

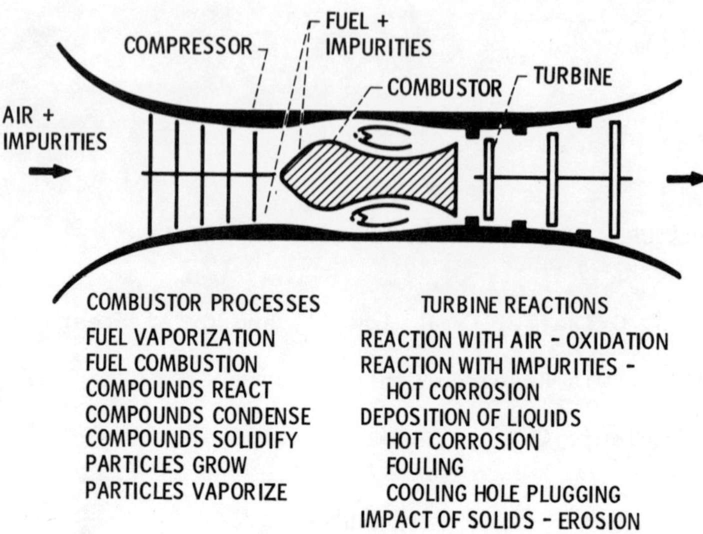

Figure 1. Origins of environmental attack in the hot section of a gas turbine engine.

burned, and (3) impurities. The impurities enter the system with both the fuel and air as represented in Figure 1. In the combustor where the fuel is burned, other physical and chemical processes also occur. The impurities can oxidize, vaporize, and react to form chemical compounds [2,3]. Thus the hot combustion gases, which exit the combustor and enter the turbine, can contain a variety of potentially harmful constituents. Oxidation attack of hot-gas-path metallic components results from reaction with oxygen and oxides of carbon. Liquid and gaseous species can cause an accelerated type of oxidation attack called hot corrosion [4]. Airfoil cooling holes can be reduced by fouling as a result of deposition of materials from the combustion product as flow [5]. Solid particles can impact components and produce erosion which both destroys optimum airfoil configurations and reduces load-bearing cross-sections [6]. All of these processes limit component life over a range of service temperatures as indicated very schematically in Figure 2. Of course, it must be empahsized that the domains of the environmental attack life limits are variable depending on environmental conditions. Thus we see from Figure 2 that while components are designed to rather well-defined fatigue and creep limits, environmental attack can result in very early failures.

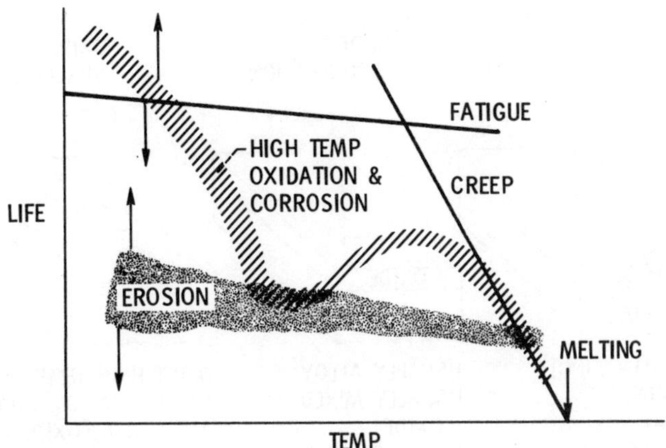

Figure 2. Schematic representation of turbine components life-limiting factors.

High Temperature Oxidation

Of all the modes of environmental attack, oxidation has been the most studied and is probably the best understood. Oxidation attack of metallic surfaces manifests itself in a number of different ways, some of which are depicted in Figure 3. For a one component metal system, the oxidation reaction converts the surface to a single oxide scale which grows rapidly at first but slows to a low growth rate. The amount of oxide formed is a function of the particular metal, the temperature, and oxygen pressure. This type of oxidation behavior is characterized by a parabolic growth rate curve and involves only increases in weight resulting from oxygen pickup. Bulk diffusion through the oxide is the controlling mechanism in this type of oxidation: the process has been studied extensively and is rather well understood [7].

When the oxidation attack involves alloys (multicomponent metal systems) and includes the simultaneous processes of vaporization and spallation, the situation becomes very complex. Unfortunately, this is usually the case encountered in practice for alloys developed to possess specific high temperature mechanical properties. If the oxide formed has a sufficiently high vapor pressure or if the oxide further oxidizes to form a volatile species, metal can be transported from the oxidizing component by the volatile oxide species. This type of attack generally manifests itself as a weight loss and for some systems it has been studied extensively [8]. In addition to oxide growht and vaporization, spallation can contribute to overall oxidation attack. Spallation also results in a net weight loss of the component, but the mechanisms of this type of

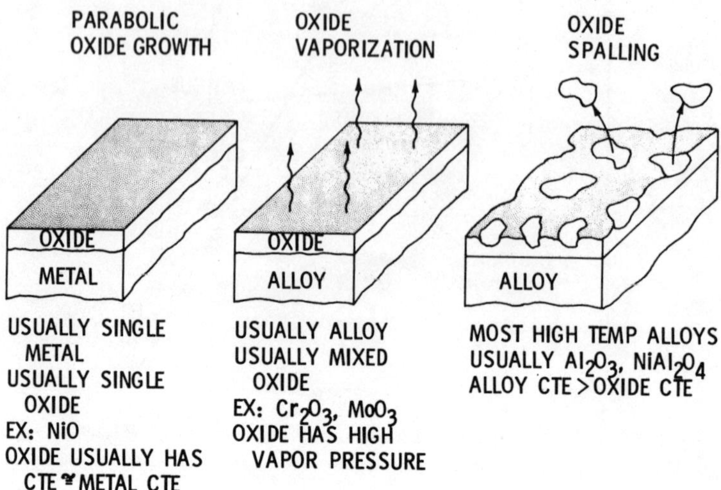

Figure 3. Schematic representation of modes of oxidation attack (CTE-coefficients of thermal expansion).

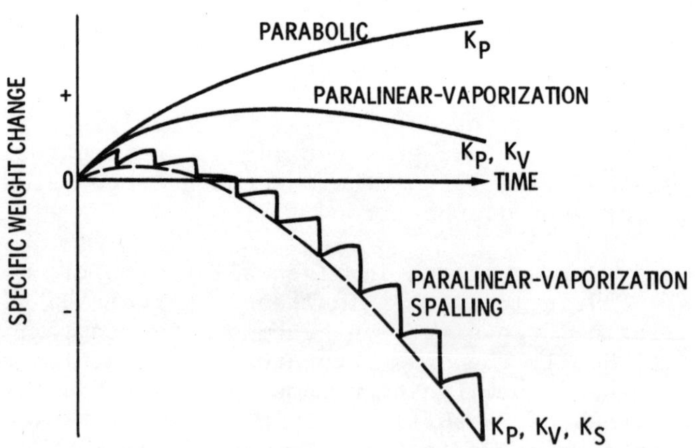

Figure 4. Specific weight changes with time for major modes of oxidation attack.

attack are not well established. Because spallation is often associated with thermal cycling, it is viewed as the result of a mismatch in the coefficients of thermal expansion of the metal and the oxide [9,10].

The specific weight changes characteristic of some modes of oxidation attack are represented in Figure 4. Parabolic oxide growth rate behavior is characterized by the rate constant K_p. The combination of parabolic oxide growth and linear vaporization oxide loss results in a net paralinear behavior which can be characterized by a combination of K_p and the vaporization rate constant K_v. Finally, when spalling is also included, still another rate constant, K_s, must be added to characterize the behavior. The spallation rate constant, K_s, represents the linear type of spallation weight loss approached at long times. The dashed line shown in Figure 4 is the resultant actually observed in laboratory cyclic oxidation tests [9]. The stepped curve represents schematically what happens on each heating-cooling cycle. During each heating cycle, new oxide forms and possibly simultaneously vaporizes to some extent; on cooling, weight is lost as oxide spalls.

Actual weight change curves measured for a number of advanced gas turbine engine airfoil materials are shown in Figure 5. These advanced materials were cyclically tested at 1100°C in a Mach 0.3 burner rig burning Jet A fuel. The large weight losses observed for TD-NiCr and MA-953 result from vaporization. These two alloys

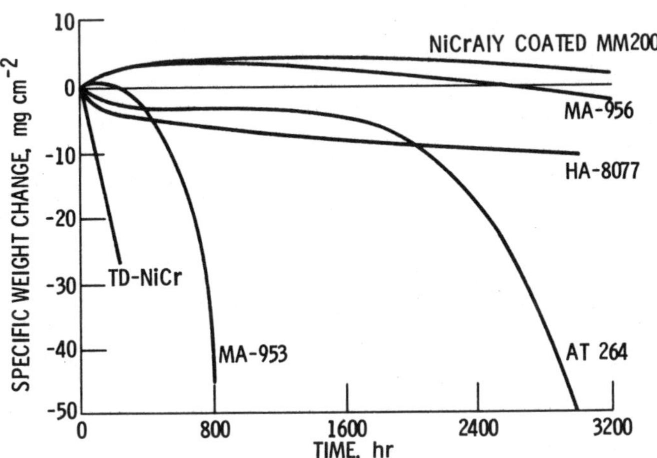

Figure 5. Cyclic oxidation results obtained in Mach 0.3 burner rig. Metal temperature was 1100°C; one hour at temperature per cycle.

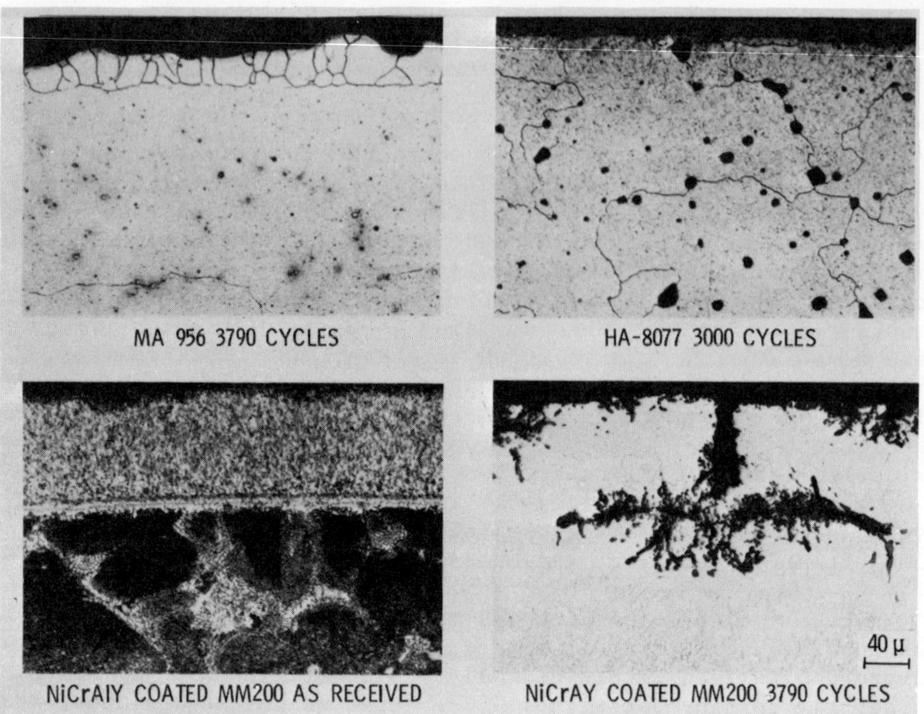

Figure 6. Microstructures of materials tested in cyclic oxidation. Metal temperature was 1100°C; one hour at temperature per cycle.

form a chromia (Cr_2O_3) scale which further oxidizes to CrO_3. At high temperatures, CrO_3 readily vaporizes in a high velocity gas stream. Chromium is thus transported away from the alloy to produce a rapid weight loss and ultimately the integrity of the alloy is compromised. The NiCrAlY-coated MM-200 forms a predominately aluminum oxide (Al_2O_3) scale which does not vaporize and the observed slight weight loss is due to spalling. The iron-chromium-aluminum alloy, MA-956, likewise shows good oxidation resistance while the other two materials undergo considerable spallation and show significant weight loss at intermediate times. Metallographic examination of these oxidized materials show microstructures that reflect the levels of their respective weight losses (Figure 6). The NiCrAlY-coated MM-200 and MA-956 show generally minor surface attack with only isolated regions showing more severe attack in the former. On the other hand, the HA-8077 shows considerable attack with many voids widespread throughout the material [11].

From the above examples, one can see that it is far more desirable to have some ability to predict oxidation attack rather than to perform tests that require thousands of hours. Except for some single metals which follow simple oxide growth rate kinetics, there have been relatively few attempts to predict long term oxidation attack. At the NASA-Lewis Research Center, we have been addressing this problem. The parabolic oxidation attack equation has been extended to a more generalized attack prediction scheme [12,13]. This approach relates specific weight change to mechanisms of attack by the equation

$$\frac{\Delta W}{A} = (K_p t)^{1/2} - K_v t - K_s t \pm \sigma$$

where ΔW is weight change, A is surface area, K_i is the respective rate constant, t is time and σ is an error parameter. Data from short time laboratory tests are fitted to this equation by regression analysis. If there is a satisfactory fit of the data (of 33 oxidation resistant alloys, only two were found to be anomalous [14]), then the data are entered into a computer program called COREST 9 [11] to analyze and extrapolate the oxidation behavior. COREST is based on a material balance approach where information on the oxide scale composition is used to calculate the average effective thickness of metal lost. Figure 7 is a comparison of COREST predicted thickness losses with measured values for numerous alloys tested cyclically at 1150°C. It is to be noted that the measured values are those for maximum attack while the COREST values are estimates of average attack. The predicted values are seen to be within at least a factor of three of measured values and this demonstrates that our ability to predict oxidation attack is progressing.

We have also used the above technique in a program aimed at developing more oxidation resistant alloys. The results presented in Figure 8 show iso-attack contours on ternary phase diagrams for a series of Ni-Cr-Al alloys [13]. The values indicated on the contour lines are a measure of the degree of oxidation attack and lower numbers indicate increased oxidation resistance. The contours, for materials melted in ZrO_2 crucilbes, indicate that at 1200°C best cyclic oxidation weight change resistance is obtained for the compositions Ni45Al and Ni35Cr15Al. The first composition was expected because it is in the β-phase region typical of the widely used aluminide coatings used for superalloy protection. The second composition, however, was only identified by the present approach. Another interesting feature is shown by comparison of the two diagrams in Figure 8. For the alloys where Zr was added by melting in ZrO_2 crucibles, spalling resistance was increased. Apparently the Zr promotes and increase in the oxide-metal bond.

The foregoing indicates that our ability to predict oxidation attack continues to develop. Spallation, the most common form of weight loss, is now predicted by the NASA approach. While current

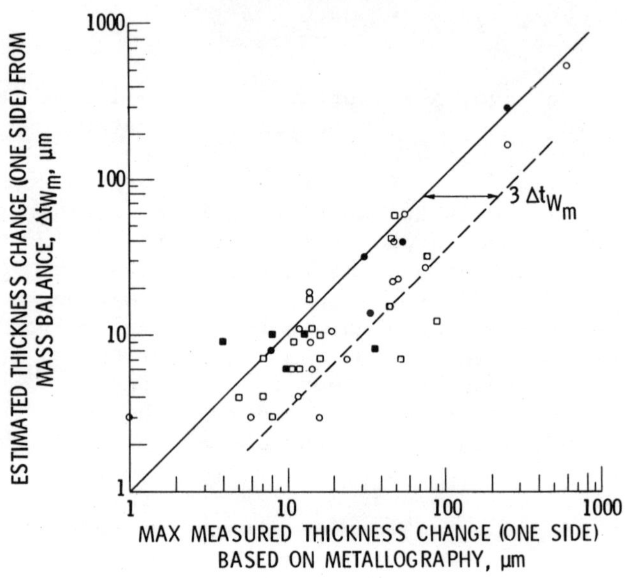

Figure 7. COREST estimated versus measured metal thickness loss for 100 hours cyclic oxidation at metal temperature of 1150°C and one-hour cycles.

predictions are not as accurate as desired, this methodology is maturing. Problems associated with alloy surface depletion and changing oxide scale composition are still to be solved before reliable long time predictions can be made with confidence.

Hot Corrosion and Other Impurity Effects

Impurities or contaminants which enter the high temperature system can cause a number of different kinds of environmental attack of components. In gas turbine engines, operating on quite pure fuels, it is not uncommon for the fuel to contain trace quantities of elements such as sulfur and vanadium as impurities. Air ingested into the engine can contain varying amounts of such

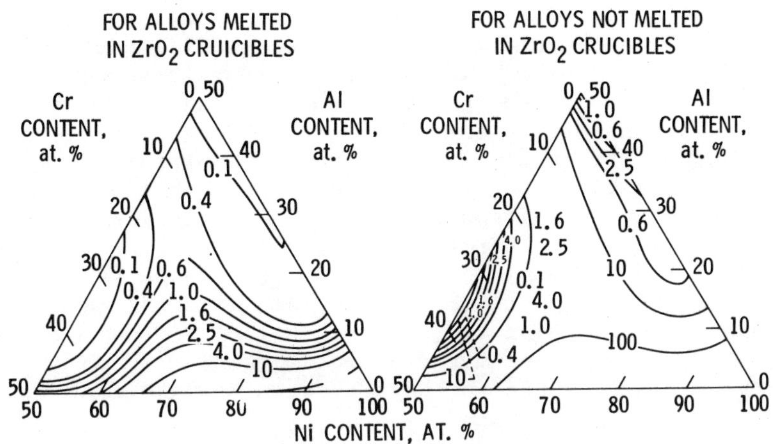

Figure 8. Attack contours for Ni-50Cr-50Al system at 1200°C.

airborne substances as sand and sea salt aerosol (composition approximately 69% NaCl, 15% $MgCl_2$, 11% Na_2SO_4, 3% $CaCl_2$, and 2% KCl). These various "impurities" can undergo chemical reaction in the combustion process to form other compounds which subsequently exit the combustor along with the fuel-air combustion products. Metal components at high temperatures which are subject to the flow of these materials from the combustor can undergo environmental attack beyond that of oxidation.

Some types of environmental attack associated with impurity effects are represented very schematically in Figure 9. Certain

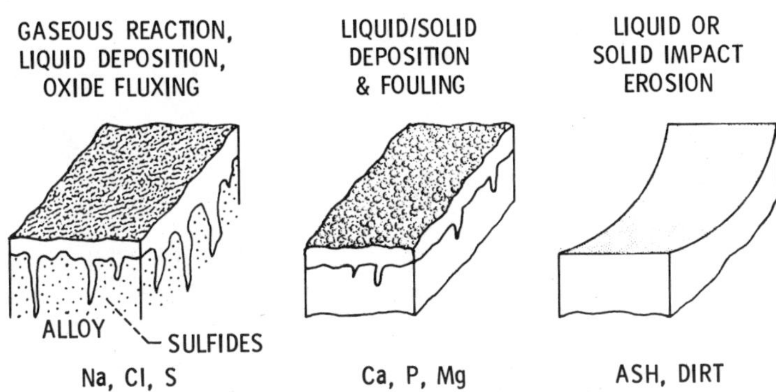

Figure 9. Schematic representations of environmental attack.

gaseous species can react with or otherwise compromise the integrity of the normally self-limiting (protective) oxide developed on alloy components [14,15]. Once the protective oxide is compromised, bare metal surface is exposed which can be further attacked. Certain compounds that deposit on components as liquids can also compromise the oxide and attack the substrate to produce hot corrosion and sulfidation attack [17-19]. Compounds containing the elements Na, V, S, and Cl have been identified as contributing to the above modes of attack. Deposition of other chemically less aggressive substances causes other harmful effects like cooling hole plugging and fouling of aerodynamic configureations. These harmful effects are also classified as attack and they are generally associated with compounds containing the elements Ca, P, Fe, and Mg [5]. Erosion attack results from the direct mechanical removal of component materials by impact of hard substances like ash, sand or dirt.

The weight change behaviors associated with these types of environmental attack are indicated in Figure 10. Of course in actual service, it is probable that several of these types of attack take place simultaneously. As expected, deposition/fouling is manifested as a weight gain. This attack can be sporadic and it is

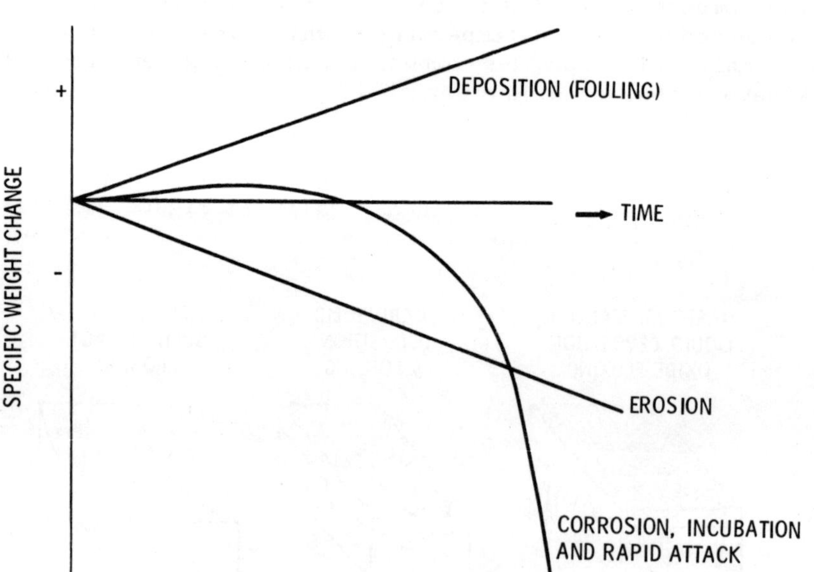

Figure 10. Specific weight changes for three modes of "impurity" associated environmental degradation.

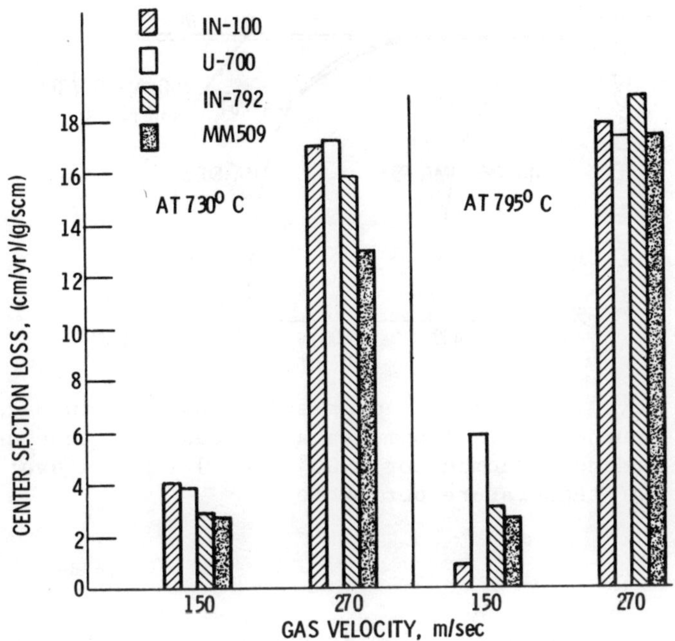

Figure 11. Erosion of turbine alloys in the effluent of a pressurized fluidized bed.

not catastrophic to components although it can reduce airflow through the gas turbine and cause engine stall. Mechanism-wise, deposition is expected to be related to the chemistry of formation of compounds in the combustion process, dew points, mass transport rates, and possibly nucleation. No systematic approach has yet been applied to developing a predictive capability for deposition/ fouling attack.

Erosion attack always results in a weight loss. The attack is generally quite linear in time although it may be sporadic. Mechanistically one would expect this mode of attack to be relatively uncomplicated, but little has been done to develop an ability to predict high temperature erosion of the type illustrated in Figure 11 [20].

As was indicated in Figure 10, hot corrosion attack is catastrophic in that once initiated, it proceeds very rapidly in time. Catastrophic attack is proceeded by an incubation period of unpredictable and varying duration during which very little weight change takes place. Of all the modes of attack associated with

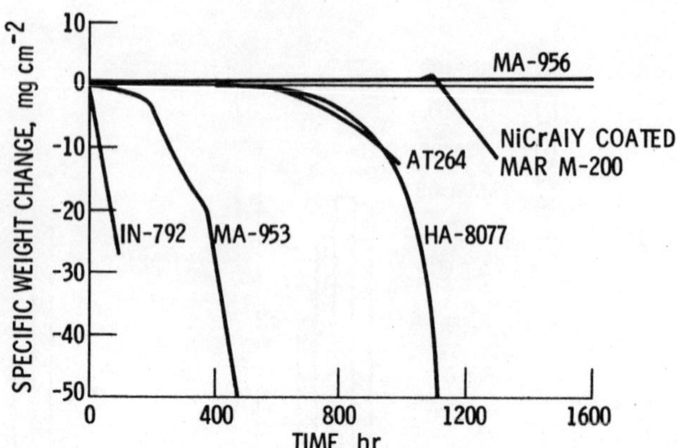

Figure 12. Cyclic hot corrosion results obtained in Mach 0.3 burner rig. 5 ppm synthetic sea salt aqueous solution added to combustor; 900°C metal temperature; one hour at temperature per cycle.

impurity effects, hot corrosion is unquestionably the most complicated mechanistically.

A few weight change examples reflecting hot corrosion attack are presented in Figure 12. These results were obtained in a burner rig hot corrosion test which is basically similar to a cyclic oxidation burner rig test except that a synthetic sea salt aqueous solution is continually added into the combustion chamber. At a metal test temperature of 900°C, the oxidation behavior of the materials in Figure 12 is known to be good, but the weight losses indicate that all except the MA-956 sustained an unexpectable amount of hot corrosion attack [11]. Post test metallographic examination of the test samples likewise indicated severe attack for all except the MA-956 which showed only minimal damage. Typically observed microstructures are shown in Figure 13.

The severity of hot corrosion attack has provided the impetus for much research in many laboratories in the United States and England [21]. At the NASA-Lewis Research Center there has been a continuing effort to (1) characterize the phenomena, (2) understand the mechanisms involved, and (3) alleviate the hot corrosion problem. Basic research has been directed at determining the chemistry and kinetics of corrosive salt formation in combustion systems. Doped laboratory flames have been investigated by high pressure mass spectrometric sampling techniques [3,22]. These studies have established that sodium sulfate, considered the

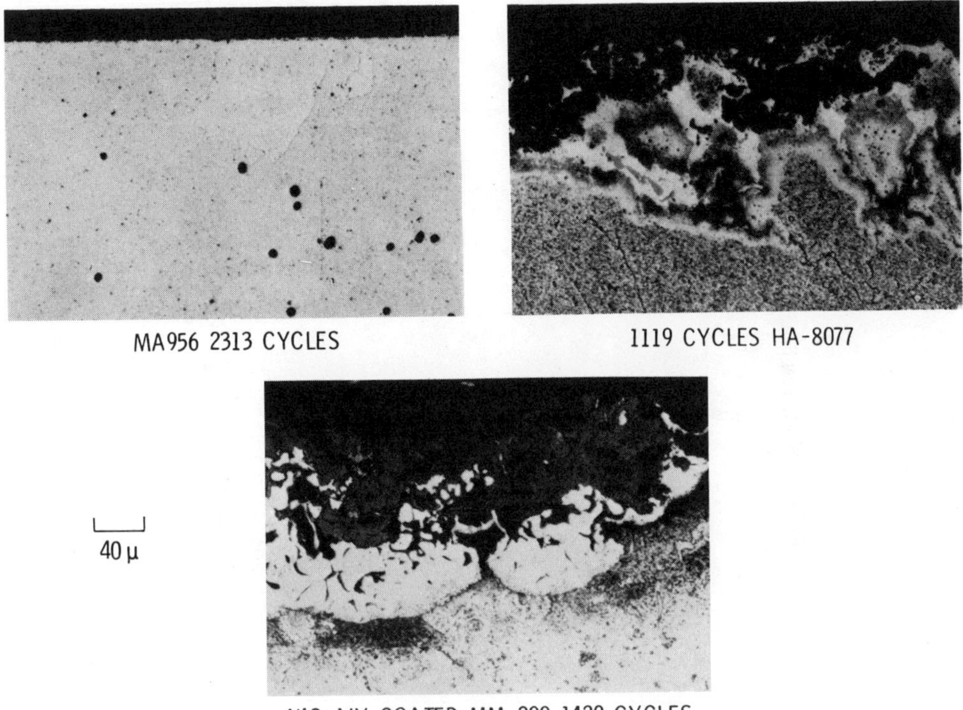

Figure 13. Microstructures of materials tested in cyclic hot corrosion. Metal temperature was 900°C; one hour at temperature per cycle; 5 ppm synthetic sea salt.

primary salt responsible for hot corrosion attack [4,17,18], is formed in flames doped with sodium-containing salts and sulfur compounds [23]. The time required for formation of the sulfate is very short (< 1 millisecond) compared to the residence time in gas turbine engines and therefore the observed flame chemistry is applicable to real engines. Typically measured composition profiles for a particular laboratory flame system are compared with equilibrium thermodynamically calculated composition in Figure 14. The agreement shown between measured and calculated compositions is satisfactory enough to estalbish that equilibrium calculations can be applied to other aspects of the hot corrosion problem.

Equilibirum calculations have been used to determine dew points for sodium sulfate and other compounds as a function of various burner rig and engine parameters [2]. Calculated condensation temperatures have been found to be in good agreement

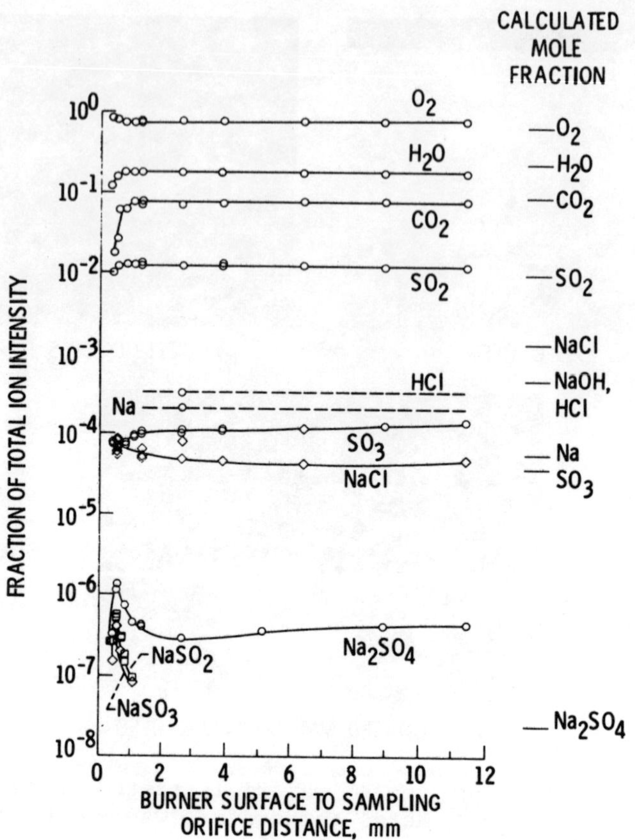

Figure 14. Measured composition profiles and calculated compositions for doped flame. Reactants (wt %): CH_4 = 4.7, O_2 = 89.4, H_2O = 3.5, SO_2 = 2.0, and NaCl = 0.35.

with measured values [24] and a transport theory is being developed to predict deposition rates. This is considered to be the first step toward the objective of ultimately being able to predict attack rates. The present state of development of our predictive approach is indicated in Figure 15. While calculated and measured dew points are seen to be in good agreement, theoretical rate predictions are not yet in satisfactory agreement with measurements. However, these initial results have provided valuable insights into the hot corrosion process and are considered encouraging enough to justify further work to refine the transport theory.

Solutions to the hot corrosion problem can be sought semi-empirically in (1) improved alloys or ceramics, (2) protective surface coatings, (3) use of additives to the engine environment,

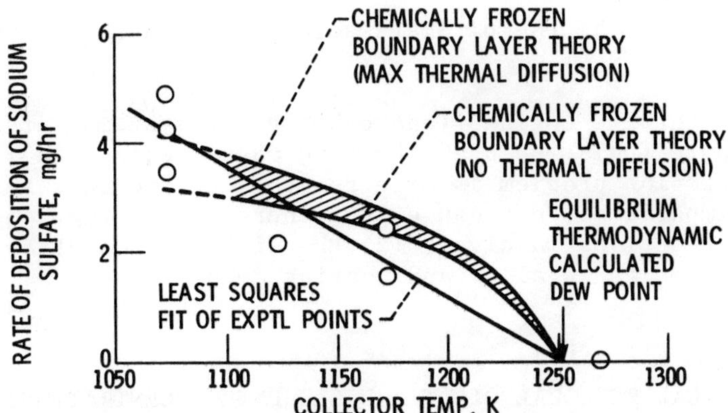

Figure 15. Comparison of experimentally measured and theoretically calculated deposition rates of Na_2SO_4 from burner rig (Mach 0.3) flame doped with 8 ppm NaCl.

and (4) air/fuel cleanup to eliminate harmful impurities. A number of these approaches are being pursued at the NASA-Lewis Research Center. For example, Figure 16 shows the reduction in hot corrosion attack that can be obtained by heavily doping a highly hot corrosive test environment with a chromium type fuel additive [25]. Other additives are also being evaluated, but because of considerations of price, logisitics, etc., additives

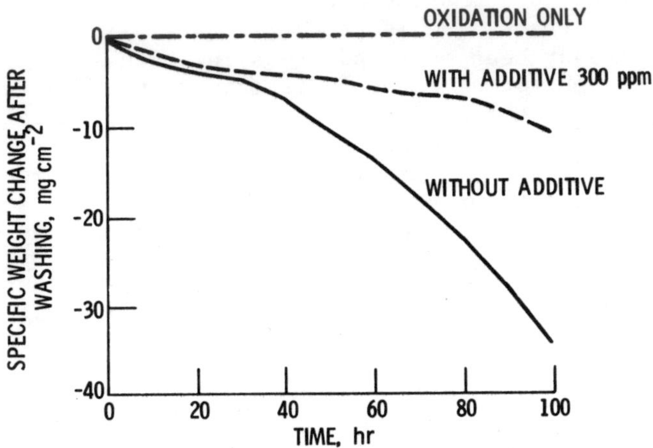

Figure 16. Effect of chromium additive on cyclic hot corrosion; Mach 0.3 burner rig; 900°C metal temperature; one hour at temperature per cycle; 5 ppm synthetic sea salt.

are not deemed to be the ultimate solution to the hot corrosion problem.

As gas turbine engines (and other power systems) are operated on more impure fuels, such as the type characterized in Table 1, the hot corrosion problem is expected to increase in severity. With more impurities, one can expect increases in the complexity of (1) the reactions involved, (2) the effects of additives, and (3) the trade-offs in alloy and coating compositions.

Table 1

TYPICAL RESIDUAL OIL TRACE ELEMENT CONCENTRATIONS

	ppm		ppm		ppm
S	12,000	Na	35	Pb	2
Si	300	Cl	2	Zr	2
V	180	Ba	4	Rb	2
Fe	150	Zn	4	Nd	1
Ni	120	P	4	Y	1
Ca	120	Cr	3	Se	1
K	80	Co	3		
Al	75	Mn	2.5		
Mg	75	Cu	2.5		

CONCLUDING REMARKS

In attempting to summarize the current status of understanding of environmental attack, one is faced with areas in which good progress has already been made and areas in which much more remains to be done. In teh former category is oxidation. The ability to understand and predict the effects of oxidation, especially cyclic oxidation, has grown rapidly in recent years. Hot corrosion in relatively "clean" fuel systems is father behind; the mechanisms are becoming clear but our predictive capabilities are still in their infancy. Erosion and deposition, while presumably more straightforward, are only beginning to be considered. Finally, our understandings of the effects of all the impurities found in "dirty" fuels, e.g., residual oils and some coal-derived liquids, is in the very earliest development stages and this area offers many challenges.

REFERENCES

1. Spera, D.A. and Grisaffe, S.J., NASA TM X-2664, 1973.

2. Kohl, F.J., Stearns, C.A. and Fryburg, G.C., in Metal-Slag Gas Reactions and Processes, Z.A. Foroulis and W.W. Smeltzer, Eds., The Electrochemical Society Soft Bound Series, 1975, p 649. Also, NASA TM X-71641, 1975.

3. Stearns, C.A., Miller, R.A., Kohl, F.J. and Fryburg, G.C., J. Electrochem. Soc., 124 (7), 1977, p 1145. Also NASA TM X-73600, 1977.

4. Stringer, J., "Hot corrosion in Gas Turbines, MCIC Rpt. 72-08, 1972.

5. Deadmore, D.L. and Lowell, C.E., NASA TM X-73661, 1971.

6. Committee on Conservation of Materials Through Reduction of Erosion, Erosion Control in Energy Systems, NMAB-334, 1977.

7. Hauffe, K., Oxidation of Metals, Plenum Press, NY, 1965.

8. Tedmon, C.S., J. Electrochem. Soc., 113 (8), 1966, p 766.

9. Barrett, C.A. and Lowell, C.E., Oxidation of Metals, 9 (4), 1975, p 307.

10. Deadmore, D.L. and Lowell, C.E., Oxidation of Metals, 11 (2), 1977, p 95.

11. Lowell, C.E. and Deadmore, D.L., NASA TM X-73656, 1977.

12. Barrett, C.A. and Pressler, A.F., NASA TN D-8132, 1976.

13. Barrett, C.A. and Lowell, C.E., Oxidation of Metals, 11 (4), 1977.

14. Barrett, C.A., Proceedings of Conference on Environmental Degradation of Engineering Materials, Oct. 10-12, 1977, College of Engineering, Virginia Polytechnical Institue and State University, Blacksburg, VA, 1977, pp 319-27.

15. Stearns, C.A., Kohl, F.J. and Fryburg, G.C., NASA TM X-73467, 1976.

16. Fryburg, G.C., Miller, R.A., Kohn F.J. and Stearns, C.A., NASA TM X-73599, 1977. Also, J. Electrochem. Soc., 124 (11), 1977, p 1738.

17. Goebel, J.A., Pettit, F.S. and Goward, G.W., Met. Trans., 4 (1), 1973, p 261.

18. Bornstein, N.S., DeCrescente, M.A. and Roth, H.A., Met. Trans., 4 (8), 1973, p. 1799.

19. Stearns, C.A., Kohl, F.J. and Fryburg, G.C., NASA TN D-8461, 1977.

20. Zellars, G.R., Rowe, A.P. and Lowell, C.E., "Corrosion/ Erosion of Turbine Blade Materials in the High-Velocity Effluent of a Pressurized Fluidized Coal Combustor", Fifth International Conference fo Fluidized-Bed Combusion, Dec. 12-14, 1977, Washington, DC.

21. Proceedings of Third Conference on Gas Turbine Materials in a Marine Environment, U. of Bath, England, Sept. 20-23, 1976.

22. Stearns, C.A., Kohl, F.J., Fryburg, G.C. and Miller, R.A., NASA TM-73720, 1977.

23. Fryburg, G.C., Miller, R.A., Stearns, C.A. and Koh, F.J., NASA TM-73794, 1977.

24. Kohl, F.J., Santoro, G.J., Stearns, C.A., Fryburg, G.C. and Rosner, D.E., NASA TM X-73683, 1977.

25. Lowell, C.E. and Deadmore, D.L., NASA TM X-73465, 1976.

CHAPTER 12

EFFECT OF WEAR ON PERFORMANCE AND RELIABILITY

Nam P. Suh and Nannaji Saka

Massachusetts Institute of Technology

Cambridge, Massachusetts

ABSTRACT

Many systems consisting of an assembly of mechanical parts in relative motion fail due to the damage done to the sliding and rolling surfaces by wear. Failure of a system may be precipitated by just the dimensional loss caused by wear, or wear-induced fatigue and fracture of such machine elements as bearings, gears, and splines. The wear rate of mechanical systems rapidly increases as the wear debris accumulates, and as the lubricant breaks down due to both the increase in surface traction and the activation of new wear mechanisms. The sudden failure of machines may be prevented through on-line monitoring of the wear process, which may be accomplished by checking the density and the nature of the wear particles and the deterioration of the lubricant. To determine the cause of failure, the worn surface and the subsurface should be analyzed for damages. Reliability of machines can be increased by proper design of sliding surfaces (geometry and topography), proper selection and quality control of wear resistant materials, adequate provision for lubrication, proper sealing of the sliding components from the environment, and the filtration of wear debris.

INTRODUCTION

One of the common causes of failure of machines, vehicles, and even electrical and electronic equipment is wear of mechanical parts in relative motion. Even the failures attributed to other causes can sometimes be traced to wear, having initiated a chain of events which leads to the final failure. The problems associated with wear are many: need for maintenance, loss and replacement of parts, loss of productivity and the risk to life.

The economic impact of wear is subtle but significant. The annual cost to the U.S. caused by tribological problems is estimated to be at least 16 billion dollars [1]. The strategic significance of the problem is illustrated by the following facts [2]:

1. Approximately 40 percent extra aircraft are needed to maintain a viable military posture since at any time a similar percentage is undergoing maintenance.
2. Procurement of military items costs several times more than similar civilian items due to stringent military specifications. These specifications are in part a reflection of the lack of understanding of the precise failure mechanisms.
3. The U.S. Navy spends as much for the maintenance of its aircraft as for the fuel.
4. Even the most advanced engineering systems such as inertial guidance devices are limited by tribological considerations.

Although many of the failures caused by wear are gradual, there are some catastrophic failures precipitated by seizure and gouging of sliding surfaces. Wear is usually accelerated in corrosive environments unless the corrosion product forms a continuous and coherent protective layer.

A good deal of progress has been made over the last quarter-century in understanding the basic modes and mechanisms of wear. However, the wear rates of materials cannot yet be predicted from the first principles. Therefore, actual tests of the wear life become indispensable. Since there is always an element of uncertainty about the reliability of sliding surfaces, on-line monitoring and failure analysis appear to be promising means of reducing maintenance costs.

The purpose of this chapter is to review the basic wear mechanisms, the factors that control the wear rate of machine elements under various conditions, methods of on-line monitoring and control, and the techniques of failure analysis. An effort is made to emphasize the fact that wear can be caused by a number of different mechanisms, and that wear problems cannot be solved adequately without understanding the specific causes and basic mechanisms underlying the failure.

BRIEF REVIEW OF WEAR MECHANISMS AND IMPLICATIONS

Wear many be defined as the removal of material from solid surfaces as a result of mechanical and chemical interactions. The wear rate is a complex function of several variables such as the load, speed of sliding, temperature, environment, materials, etc. In most cases, the wear rate is dominated by a single mechanism although more than one mechanism may be present in a specific situation and one form of wear may lead to the other modes of wear.

Some important wear mechanisms observed in various wear processes are:

1. Cumulative subsurface damage due to plastic deformation, crack nucleation and crack propagation observed in sliding wear.
2. Cutting of the surface layer which is partially responsible in abrasive wear.
3. Large deofrmation and crack propagation due to solid particle impingement leading to erosive wear.
4. Formation of new surface layers (through chemical reactions) which flake off due to mechanical interaction between sliding surfaces such as observed in corrosive wear.
5. Crack nucleation and propagation with little plastic deformation as in wear of rolling contact elements.

The term fatigue wear is often used in the literature, but it is not descriptive enough to be meaningful, since most wear processes are essentially a form of fatigue wear involving a cumulative damage process under repeated loading (e.g., the deformation of subsurface layer, crack nucleation in the subsurface material due to large plastic deformation, crack propagation under the influence of the surface traction [3]). The wear rate for sliding and abrasive wear is usually expressed in terms of the wear coefficient K which is defined as

$$K = \frac{3VH}{LN} \tag{1}$$

where V is the wear volume, H the hardness of the material, L the distance slid and N the normal load. A physical interpretation of the wear coefficient for sliding wear is given as [4]

$$K = \frac{\text{Volume of material worn}}{\text{Volume of plastically deformed material}} \tag{2}$$

For most metals and plastics, K is about 10^{-4} to 10^{-3}, indicating that a very large amount of plastic work is expended to remove a small amount of material. The wear coefficient of a given set of materials varies over a wide range. Further, it is also a sensitive function of the coefficient of friction. This type of wear occurs in cams, gears, splines, etc.

Abrasive wear is caused when hard particles plow and cut the softer material. The cutting and plowing are accompanied by large subsurface plastic deformation which is the primary energy consuming mechanism in typical abrasive wear [5]. This conclusion is supported by the typical wear coefficients observed in abrasive wear. The wear coefficient defined in Equation (1) may be rewritten as [6]

$$K = \frac{3\mu V u}{\mu L N} \quad \frac{\text{Work done to remove the material by cutting}}{\text{Total external work done}}$$

where u is the specific energy to remove a unit volume of material by cutting, and μ is the coefficient of friction. In the case of cutting without subsurface deformation such as in cutting of metals with sharp tools, the specific energy for cutting, u, is equal to the material hardness, H [6]. Therefore, the wear coefficient can be interpreted as the fraction of the external work consumed in removing the metal by a cutting mechanism. In typical abrasive wear, the wear coefficient is about 10^{-3} to 10^{-2} [7,8], indicating that only a small fraction of the external work done is consumed in generating the chip. The physical reason for such a small wear coefficient is that many abrasive particles do not cut, but rather simply slide along the surface deforming the surface layer plastically.

Erosive wear due to solid particle impingement occurs either when hard abrasive particles cut the surface at a glancing incident angle or when the surface is plastically deformed due to the mechanical impact between the particle and the surface [9-12]. In recent years, the mechanisms of erosive wear aroused renewed interest because of the severe wear and corrosion problems encountered in coal utilization. The erosive wear mechanisms depend on the impingement angle, particle velocity, particle size, and the hardness of particles relative to the specimen. When the particle impinges on a surface at low speeds and at low impingement angles, the wear phenomenon is similar to abrasive wear, involving cutting and subsurface deformation. However, at high impingement angles not much cutting is observed and wear is primarily caused by plastic deformation, subsurface crack nucleation and crack propagation processes [13]. Even at very high speeds, wear is still caused by subsurface deformation processes although fracture and melting of the surface may play a significant role.

Erosive wear of metals by hot gases is a problem in guns. Although no definitive work is done to date on the subject matter, a correlation exists between the erosion rate and the surface temperature rise. Thermally-activated softening of the surface layer through recovery, annealing, and tempering as well as such thermally-activated processes as oxidation and chemical reactions may all play a role. Since the wear rate depends on the transient temperature rise at the surface, any means of lowering the temperature rise will minimize the erosion rate.

Corrosive wear is governed by the concomitant action of electrochemical and mechanical effects on the surface pit formation and subsurface damage process [14-16]. When the corrosion product is a hard particle, it can subsequently cause abrasive wear. When the corrosion rate is lower than the surface damage rate by plastic deformation, crack nucleation and crack propagation, delamination wear may dominate. In this case, the major effect of corrosion is to alter the surface traction which in turn controls the delamination wear rate [17].

The failure of hard rolling contact elements is caused by either the crack initiated at the surface or the damage created by the sliding action caused by the deformation of the contact points [18]. For hard metals, crack nucleation is favored at the trailing edge of the contact point. Since crack nucleation is accelerated by the presence of wear particles, the bearing life is very sensitive to cleanliness of the lubricating oil.

The foregoing discussion shows that wear can be caused by different mechanisms; and, therefore, any assessment of the risk and any attempt to failure analysis will be meaningful only when the basic mechanisms of wear are clearly understood.

FAILURE ANALYSIS

Failure analysis of worn parts consists of several interrelated steps. Preliminary analysis of the circumstance of failure, estimation of approximate values of load, and friction coefficients, observation and alanysis of worn surface and subsurfaces and, in some cases, testing to simulate failure are the necessary steps of failure analysis. In the following paragraphs, a variety of approaches and experimental techniques currently used are described. Although no rigorous rules can be laid down, a logical approach to the problem should yield satisfactory results in most cases.

Preliminary Examination

In the majority of failures, it is not possible to isolate a single cause for the failure. Often the cause is obscured by the overwhelming damage it creates, and wear failures are no exceptions. Nevertheless, it is necessary to gather as much information about the failure as possible. It is preferable to obtain information about the design, materials, manufacturing and assembly procedures, service conditions, etc. An on-site inspection of failure is desirable. Only then is it possible to infer whether the failure is due to wear or due to other modes of failure. An examination of existing working machines will provide, by comparison, additional clues to the cause of failure.

Before conducting any rigorous failure analysis, it is necessar to make some simple calculations of the friction coefficient, wear rate, and wear coefficients, etc. By measuring the dimensions of the wear scar (or volume of material worn) and with some information about approximate loads and duration of operation and hardness of the starting materials, it is possible to calculate the wear coefficient. As discussed earlier, wear coefficients of the order of 10^{-3} to 10^{-2} indicate abrasive wear, whereas 10^{-4} to 10^{-3} indicate sliding wear.

Although the numerical values of wear and friction coefficients will indicate the probable wear mechanisms, it is desirable to confirm the diagnosis with a microscopic examination of surface and subsurface and the nature and shape of wear particles. It should be noted that the mode of wear need not be the same throughout the part life; it may change from sliding to abrasive in which case the wear coefficient thus calculated is an average of several modes.

Surface Examination

Once the preliminary analysis indicates the probable wear mechanism (from the approximate value of wear coefficient), it is necessary to conduct a more rigorous examination of the wear track and subsurface. However, these procedures require sample preparation which may distort the information already available. Therefore the next step is to conduct, if possible, nondestructive testing of the surface. Many nondestructive techniques are presently used for the examination of worn surfaces. The surface examinations consists of topographing and evaluation of the chemical nature of the surface physical structure, and even measurement of some of the properties such as hardness.

Surface topography is obtained by mechanical methods (using a stylus) very easily, but a quantitative characterization may require many readings and becomes a tedious job. The simplest method of observing the surface is optical microscopy or scanning electron microscopy. Replica transmission microscopy will provide better information. When the surfaces are relatively smooth, optical interferometric techniques will reveal excellent information abou the surfaces. It should be kept in mind, however, that it is useful to start with simple techniques; and if enough information is not obtained with simple techniques, more advanced techniques should be used. This not only saves time, labor and cost, but also leads to less confusion. Observation of surfaces at very high magnification and refined techniques may distort the overall picture as attention is focused on minor details.

Surface composition and physical structure are examined by electron microprobe and X-ray diffraction techniques. The former gives the average composition which can also be obtained by the scanning electron microscopes which are equipped with energy dispersive X-ray analyzers. The objective of this analysis is to examine whether any chemical reaction of the constituents with the environment, and any transformations of the metallurgical structure due to stress, frictional heating, etc., took place. A comparison with original surfaces gives some idea whether such transformations have taken place and the approximate temperatures attained in service. This information may provide additional clues about the causes of wear.

The shiny and smooth region of the surface does not mean that the surface is in good conditionl It simply indicates that the subsurface damage process is not completed yet. When the subsurface crack reaches the surface, wear sheets and flat pits will again form. The pit density is approximately proportional to the wear rate. Scanning electron micrographs of surface and subsurfaces for delamination, abrasive, corrosive and erosive wear are shown in Figures 1-4. In all cases, it is clear that the subsurface deforms quite extensively, and crack nucleation and growth processes take place independent of the mode of wear.

Other interesting features of the surface appearance are [19]:

1. When two identical metals are slid against each other, the surface will look rough with plowing grooves. The roughness is greater for softer metals.

2. When the applied load and the frictional force are large, there are more craters. This type of wear is sometimes referred to as "severe wear".

3. When one of the sliding surfaces is much harder than the other, the hard asperities will gradually be removed creating an extremely smooth surface on the hard surface. Once the hard surface becomes smooth, the softer surface will also become smooth.

4. While subsurface damage proceeds, the surface may look smooth and polished with regions where surface craters are present. The number of surface craters indicates the rate of subsurface damage process. The delamination wear sheets about to come off can be identified easily since they lift off in the direction opposite to the sliding direction.

Figure 1(a). Wear sheet on the wear track of Fe-Mo specimen, tested under a normal load of 2.5 Kg and a sliding distance of 90 m.

5. The cracks appearing at the surface may be due to the subsurface cracks reaching the surface or due to the crack initiated at the surface. Very hard surfaces which experience rolling contacts may fail by the crack initiated at the surface perpendicular to it, whereas metals which can undergo plastic deformation may show cracks at the surface due to the subsurface crack reaching it.

Subsurface Examination

Subsurface examination of worn parts gives a wealth of information about the failure. It also provides information about the entire history of the wear mechanism as various points at various depths are at different stages of the wear process. Indeed, failure to observe the subsurface damage may lead to an incorrect hypothesis about the wear mechanism. This is especially true for the sliding wear of metals. The subsurface examination yields information about the microstructure of materials and how that structure is affected by normal and tangential loads. For example, large

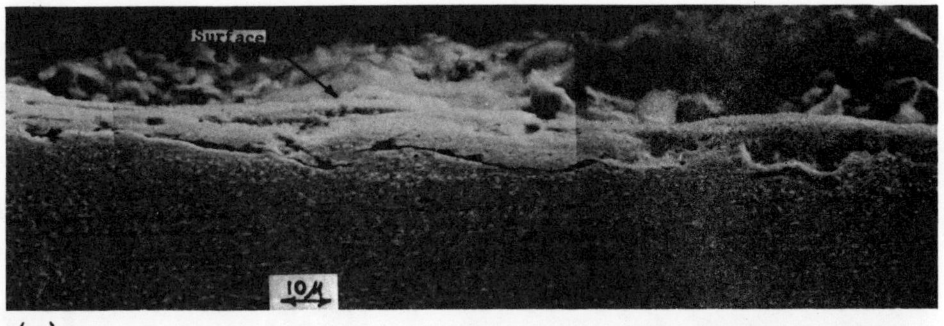

(a)

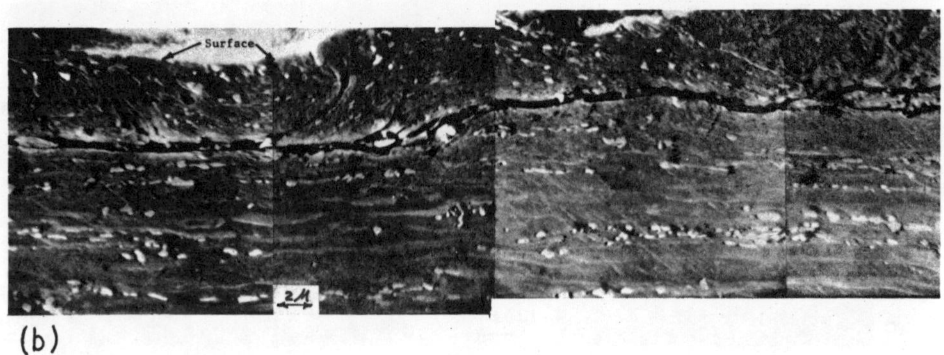

(b)

Figure 1(b). Subsurface crack formation in (a) OFHC copper, tested under a normal load of 1.8 Kg and a sliding distance of 15 m; (b) AISI 1020 steel, tested under a normal load of 1.8 Kg and a sliding distance of 45 m.

scale subsurface deformation, nucleation and propagation of cracks from inclusions and second phases indicate that the delamination type wear is the dominant mode of wear. Dendritic structure indicates that surface melting took place (if the starting material did have such structure).

The subsurface examination may also indicate the damage introduced during the surface preparation. When the etched cross section of a hardened steel is white near the surface (with cracks), it may indicate the formation of martensite which can form when the temperature at the surface is raised high during high speed sliding (or grinding) followed by rapid quenching. In some cases, the surface layer may be overtempered during the grinding operation, which can be detected by microhardness indentation tests.

The presence of flat pits, such as those shown in Figure 1, and subsurface cracks parallel to the surface make it possible to

Figure 2(a). Surfaces of worn OFHC copper specimen for different grits: (a) 60, (b) 180, (c) 600, (d) 4/0. The normal load was 4 Kg and the sliding distance was 4 m.

delineate the failure by the delamination wear. The details of the fracture pattern in the pit differ a great deal from material to material. Ductile metals show dimples while brittle metals show smooth pit bottom.

All the techniques that are used for the surface examination, except the stylus method, can be used for the subsurface examination also. In general, hardness measurements near the worn surface yield some additional information about the microstructural transformation.

Examination of the Lubricant

The majority of the sliding and rolling surfaces are lubricated to reduce friction and wear. In many cases, deterioration of the quality of lubricant is responsible for increased wear and

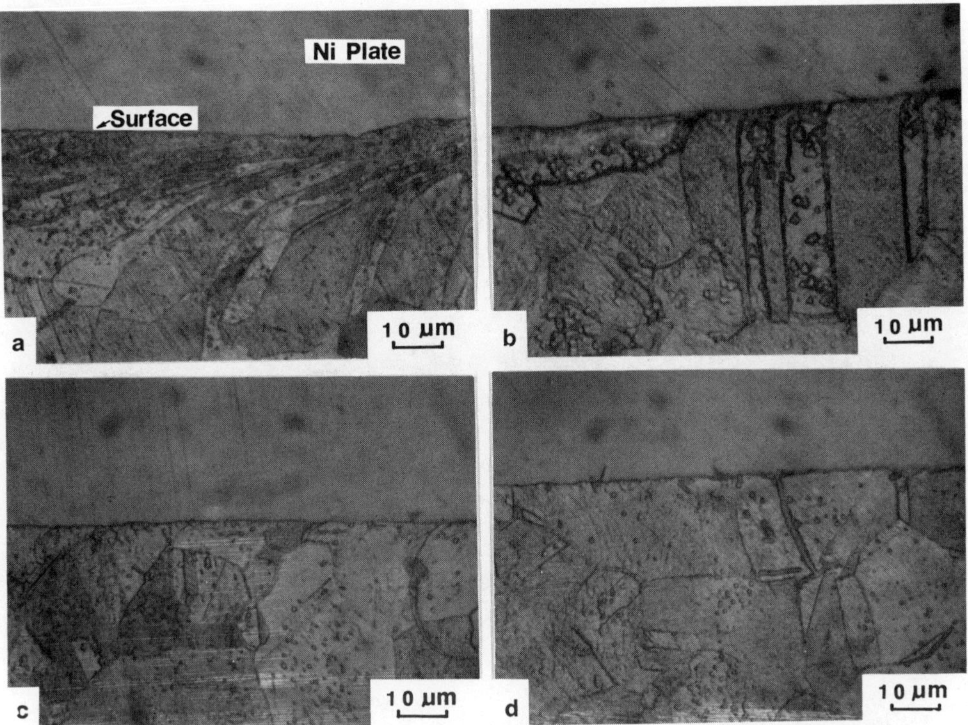

Figure 2(b). Optical micrographs of the OFHC copper subsurface for different grits: (a) 60, (b) 180, (c) 320, (d) 600 grit. The normal load was 4 Kg and the sliding distance was 4 m.

ultimate failure of sliding and rolling components. Therefore, an examination of the lubricant, either during or after failure, provides additional reasons for failure. It is often necessary to compare the used and virgin lubricants to assess the changes brought about by contamination and thermal degradation of the lubricants. Chemical and spectroscopic analyses are available for such an investigation. It is also necessary in certain cases to conduct simple tests to evaluate the pH, viscosity, etc., of the used lubricant to detect a variation in physical and chemical properties of the lubricant due to service.

Although a good deal of research has been done to detect various chemical species present at the surface due to the use of boundary lubricants, the effort has not illuminated our understanding of the failure mechanisms primarily because the identification of those species was equivalent to identifying symptoms rather than the causes of failure.

Figure 3. Wear particle formation by conjoint actions of surface pits and subsurface cracks. The wear test was in 0.001 M NaCl at pH 6.

Examination of the Wear Debris

As the surfaces wear off, loose wear particles are formed; and they contaminate the lubricant in the case of lubricated systems. The size, shape and composition of the wear particles provide some information about the mode of wear. Therefore, monitoring wear particles by lubricant analysis is a very important aspect of wear analysis. The presence of chip-like wear particles is an indication that abrasive wear is occurring, while the wear particles produced by the delamination process are flat and those due to plowing tend to be small. Attempts have been made to correlate the wear rate with the formation of various means.

One of the major contributions to wear particle analysis is the introduction of Ferrography [20-23]. It is a technique of entrapping wear particles in oil using a magnetic field and observing the trapped wear particles under an optical microscope. By such an observation, the particles created by cutting can be separated from the flat wear particles created by delamination wear (see Figures 5 and 6).

Simulated Testing

More frequently than not, it is impossible to predict the mode and causes of failure by failure analysis alone. It is necessary to conduct tests to simulate the failure. There are

Figure 4(a). Erosion surface of 2024 aluminum impacted at different temperatures by SiC particles at impingement angle of 90°: (a) 17°F, 280 m/sec; (b) 200°F, 150 m/sec.

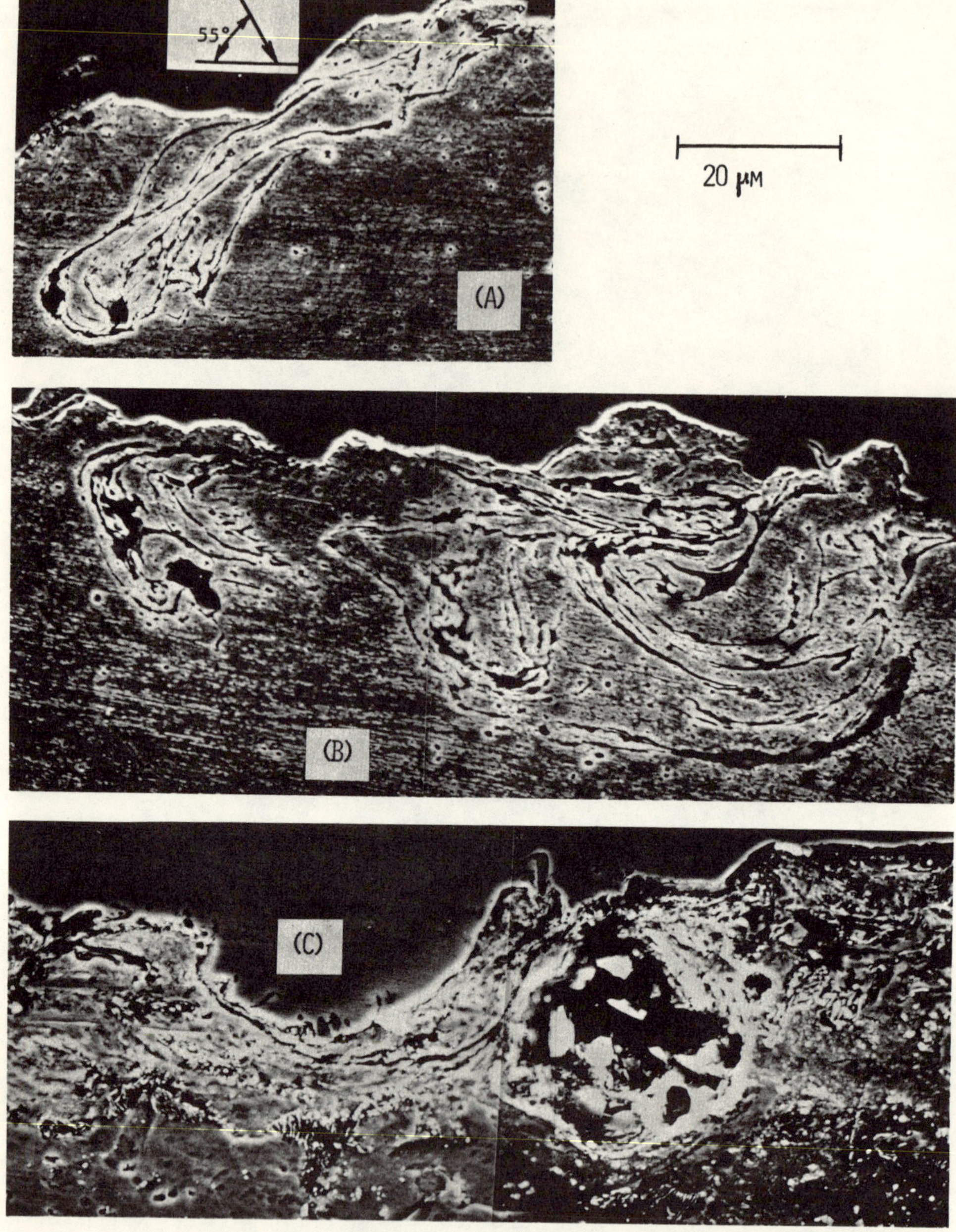

Figure 4(b). Subsurface damage in erosive wear: (a) copper, 254 μm SiC particles, impacted at 55°, 105 m/sec; (b) same as (a) except the impact angle is 90°; (c) AISI 1020 steel, same conditions as (b).

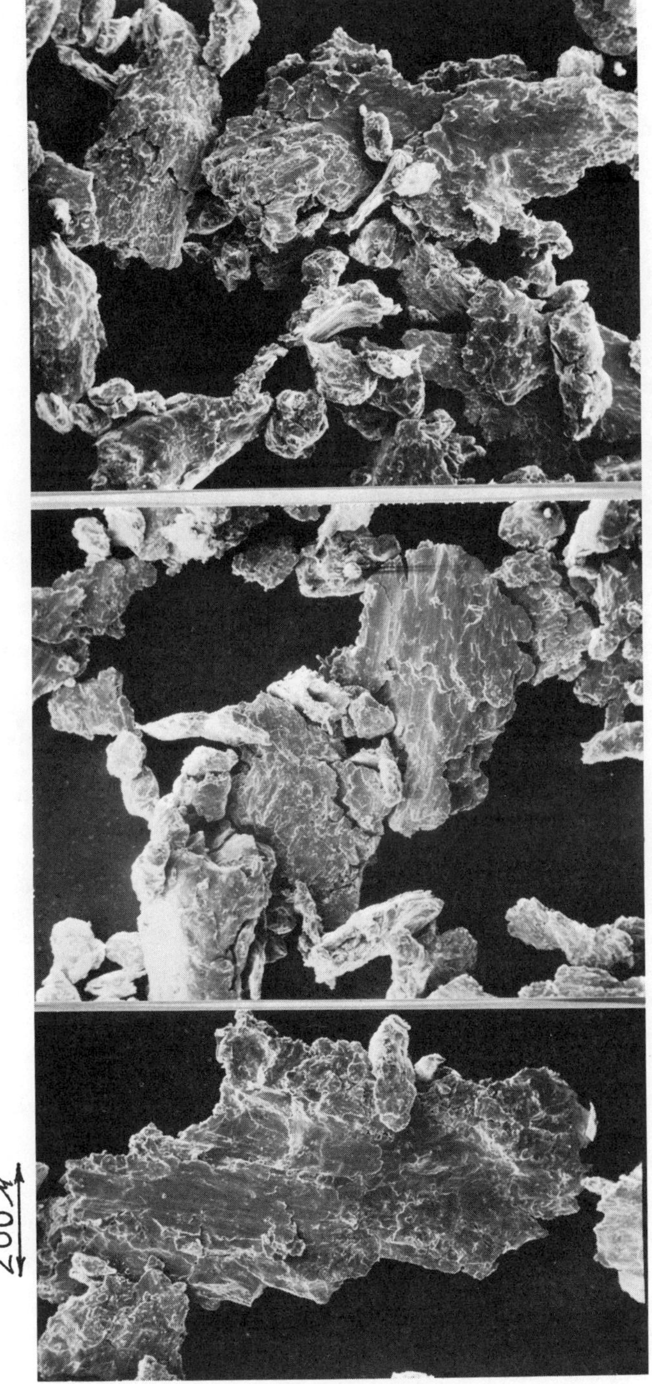

Figure 5. Wear particles collected after tests in argon; AISI 1020 steel under a normal load of 2.25 Kg.

several advantages of this test. First, as various parameters are under control during the test, it will be easier to identify the exact reasons for failure. Second, such tests will prove or disprove the statistical significance of the failure itself. Finally, it is possible to alter certain experimental parameters and provide a quick solution to the failure problem.

WEAR PREVENTION

The best way of minimizing the risk associated with wear, of course, is to prevent wear itself. A widely used means of preventing wear is lubrication. Both boundary lubrication and hydrodynamic lubrication may be used to lower the frictional force which in turn decreases the wear rate. However, there are other techniques which can be used to minimize wear. These will be discussed here briefly.

Sliding wear of metals can be decreased by coating the surface with a very thin layer of soft metal [24,25]. The reason for this is that a very thin layer of soft metal cannot work-harden indefifintely due to the dislocation instability in such a thin layer, and it can deform continuously. Since the frictional force is decreased due to the lower flow stress of the softer metal, the substrate does not undergo the delamination process as rapidly as it would in the absence of the softer metal on the surface. The wear rate has been decreased by more than three orders of magnitude by applying a 0.1 μm thick nickel layer on a steel surface [25]. This technique is currently used to prolong the life of splines and door hinges for the U.S. Navy at M.I.T.

Metals can also be prevented from wearing by making the surface layer harder. Nitriding, carburizing, and carbonitriding are some of the traditional techniques. In addition to these techniques, deposition of hard materials such as carbides, nitrides, borides and oxides by physical or chemical vapor deposition can decrease the wear rate [26].

Since the second phase particles or inclusions act as crack nucleation sites, any attempt to harden the surface without introducing them will be desirable. In this sense, laser treatment of the surface shows promise. Another approach is to eliminate inclusions by using special refining techniques. It has been shown that even soft metals can last much longer than hard metals when the crack nucleation sites are eliminated [27].

Elimination of wear particles from lubricants and from the sliding surface is an important step in minimizing the wear rate. The importance of good filtration of wear particle debris from oil

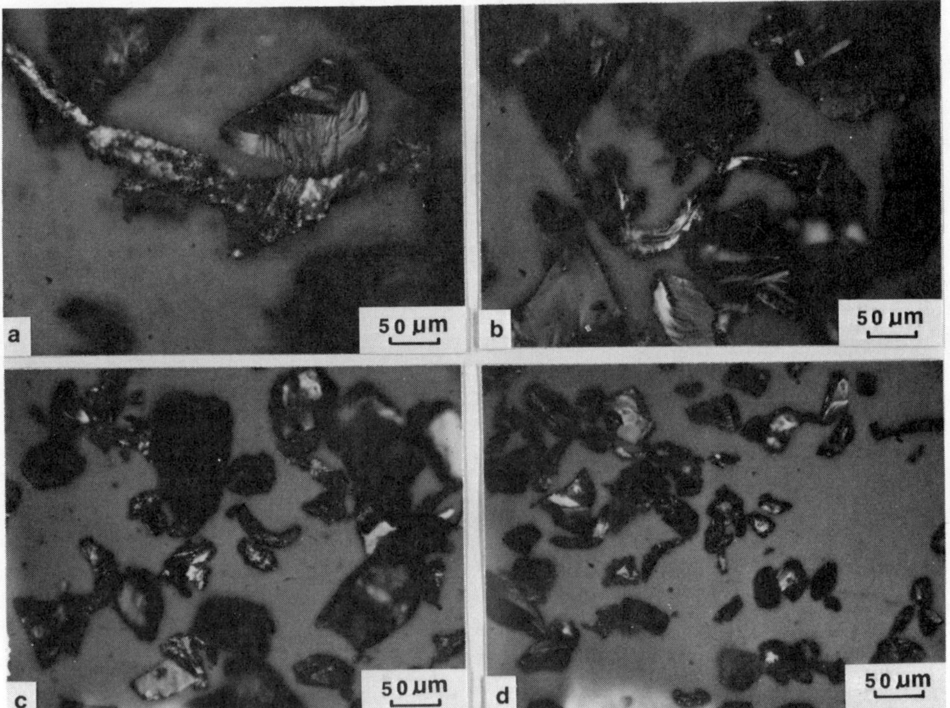

Figure 6. Abrasive wear particles of OFHC copper for different grit sizes: (a) 60, (b) 120, (c) 180, (d) 320 grit. The normal load was 4 Kg and the sliding distance was 4 m.

cannot be overemphasized. The oil should also be free of corrosive constituents other than those which form a stable hard surface layer through its reaction with the constituents of the surface.

CONCLUSIONS

1. Wear is often the first step in the failure of machines and vehicles, and the cost associated with wear is high.
2. Progress is being made to diagnose the wear process through inspection of wear particles in lubricants, surface topography, subsurface damages, and analyses of wear rates and friction coefficients. Economic on-line monitoring of wear is not yet possible.
3. The basic wear mechanisms should be better understood. The design of wear-resistant materials and the analysis of failure cannot be done without that understanding.

ACKNOWLEDGEMENTS

The authors are indebted to Drs. Arden L. Bement, Edward Van Reuth of the Defense Advanced Research Projects Agency, and Dr. Richard S. Miller, Commander Harold P. Martin and Lt. Commander Kirk Petrovic of the Office of Naval Research for their support and encouragement.

REFERENCES

1. Jost, H.P., "Economic Impact of Tribology", Mechanical Engineering, Vol. 97, No. 8, 1975, pp 26-33.

2. Peterson, M.B., "Wear Prevention", International Conference on Fundamental of Tribology, M.I.T., Cambridge, MA, June 1978.

3. Suh, N.P. and Co-workers, Delamination Theory of Wear, Elsevier Sequoia, S.A., Lausanne, 1977.

4. Shaw, M.C., "Dimensional Analysis for Wear Systems", Wear, Vol. 43, 1977, pp. 263-6.

5. Suh, N.P., Sin, H-C. and Saka, N., "Fundamental Aspects of Abrasive Wear", International Conference on Fundamentals of Tribology, M.I.T., Cambridge, MA, June 1978.

6. Sin, H-C., Saka, N. and Suh, N.P., "Abrasive Wear Mechanisms and the Grit Size Effect", to be published in Wear.

7. Finkin, E.F., "Abrasive Wear", Evaluation of Wear Testing, ASTM STP 446, ASTM, 1969, pp 50-90.

8. Moore, M.A., "A Review of Two-Body Abrasive Wear", Wear, Vol. 27, 1974, pp 1-17.

9. Finnie, I., "Erosion of Surfaces by Solid Particles", Wear, Vol. 3, 1960, pp 87-103.

10. Tilly, G.P., "Erosion Caused by Airborne Particles", Wear, Vol. 14, 1969, pp 63-79.

11. Bitter, J.G.A., "A Study of Erosion Phenomena - Part I", Wear, Vol. 6, 1963, pp 5-21.

12. Neilson, J.H. and Gilchrist, A., "Erosion by a Stream of Solid Particles", Wear, Vol. 11, 1968, pp 111-22.

13. Sherman, C.J., "Wear of Metallic Surfaces by Particle Impingement", S.M. Thesis, Department of Mechanical Engineering, M.I.T., 1971.

14. Tao, F.F., "A Study of Oxidative Phenomena in Corrosive Wear", ASLE Trans., Vol. 12, 1969, pp 97-105.

15. Quinn, T.F.J., "Oxidational Wear", Wear, Vol. 18, 1971, pp 413-9.

16. Rabinowicz, E., "Lubrication of Metal Surfaces of Oxide Films", ASLE Trans., Vol. 10, 1967, pp 400-07.

17. Tse, M-K. and Suh, N.P., "Chemical Effects in Corrosive Wear of Aluminum Alloys", Wear, Vol. 44, 1977, pp 145-62.

18. Suh, N.P., "Wear Mechanisms: An Assessment of the State of Understanding", International Conference on Fundamentals of Tribology, M.I.T., Cambridge, MA, June 1978.

19. Tohkai, M., Saka, N. and Suh, N.P., "Microstructural Aspects of Friction", to be presented at ASLE-ASME Lubrication Conference, October 16-18, 1979, Dayton, OH.

20. Scott, D., Seifert, W.W. and Westcott, V.C., "The Particles of Wear", Scientific American, Vol. 230, No. 5, May 1974, pp 88-97.

21. Westcott, V.C., "Ferrographic Analysis - Recovering Wear Particles from Machines to Humans", Naval Research Reviews, March 1977, pp 1-18.

22. Barwell, F.T., Bowen E.R., Bowen, J.P. and Westcott, V.C., "The Use of Temper Colors in Ferrography", Wear, Vol. 44, 1977, pp 163-71.

23. Westcott, V.C., "Monitoring of Wear", International Conference on Fundamental of Tribology, M.I.T., Cambridge, MA, June 1978.

24. Jahanmir, S., Suh, N.P. and Abrahamson, E.P. II, "The Delamination Tehory and the Wear of a Composite Surface", Wear, Vol. 32, 1975, pp. 33-49.

25. Jahanmir, S., Abrahamson, E.P. II, and Suh, N.P., "Sliding Wear Resistance of Metallic Coated Surfaces", Wear, Vol. 40, 1976, pp 75-84.

26. Suh, N.P., "Coated Carbides - Part, Present and Future", The Carbide Journal, Vol. 9, No. 1, 1977, p 1.

27. Jahanmir, S., Abrahamson, E.P. II and Suh, N.P., "The Effect of Second Phase Particles on Wear", Proc. 3rd North American Metal Working Research Conference, Carnegie Press, Pittsburgh, PA, 1975, pp 854-64.

CHAPTER 13

CORROSION FATIGUE BEHAVIOR OF COATED 4340 STEEL FOR BLADE RETENTION BOLTS OF THE AH-1 HELICOPTER

M. Levy and C. E. Swindlehurst, Jr.

Army Materials and Mechanics Research Center
U. S. Army Air Mobility Research and Development Laboratory

INTRODUCTION

The main rotor blade bolt for the 540 roto system (AH-1G, UH-1C, UH-1M helicopter) is proposed to be improved by the substitution of plasma-sprayed tungsten carbide coating on the outer shank for the present cadmium or chromium plate. The WC coating was originally suggested to reduce the costly machining and plating operations required to recondition blade retention bolts for the AH-1G helicopter. Fretting-induced corrosion resulted in signifi-cnet surface pitting on the bolts after less than 500 hours of service. Based on prior results with a similar configuration bld for the 240 rotor system, a WC-coated bolt is expected to last over 3000 hours (the approximately design life of the airframe) as compared to about 300 hours for current Cd-plated new production bolts and about 500 hours for Cr-plated reworked bolts. Thus a significant cost savings has been projected if bolt reconditioning and replacement is eliminated.

Although WC-coated bolts for the 240 rotor system has not experienced any fatigue problems, the Army Aviation Systems Command decided that coating effects on fatigue life should be investigated prior to approval of the WC coating for the following reasons: (1) the blade retention mechanism is subjected to fatigue conditions due to steady centrifugal loading; (2) stress analysis of the blade retention bolt provided no assurance that the part was not fatigue critical; (3) available data on the effect of the WC coating on the fatigue strength of 4340 steel was inadequate [1,2]; (4) the need for utilizing the Coricone sealer in conjunction with the WC coating for enhanced corrosion resistance had not been confirmed.

This study was undertaken to determine the effects of these coating systems (WC, Cd, Cr) on the fatigue behavior of 4340 steel in environments likely to be encountered in service. In addition, the efficacy of Coricone 1700 sealant (an organic film for enhanced corrosion resistance) in combination with the WC and the solid film lubricant was also determined. Full-scale axial fatigue tests of the coated blade retention bolts were carried out by the Army Air Mobility Research and Development Laboratory (AMRDL), Langley Research Center, Virginia.

MATERIALS

The substrate alloy (bolt material) was VAR 4340 steel. Axial tension-tension and rotating beam fatigue specimens were rough machined (turned) and heat treated according to the following schedule: normalized at 1650 F (899 C) for 1 hour; air cooled; austenitized at 1525 F (829 C) for 1 hour; oil quenched (130 to 170 F, 54 to 77 C); tempered at 900 F (482 C) for 4 hours directly from the oil quench before reaching room temperature. The ultimate tensile strength of the alloy was 194 ksi (1338 MN/m^2). The fatigue specimens were subsequently finish machined (turned and polished to 8-16 rms finish), shot peened, and the following coating systems applied:

a. Cd plating plus chromate treatment;
b. Cr plating plus solid film lubricant (SFL);
c. plasma-sprayed WC plus solid film lubricant with Coricone; and
d. plasma-sprayed WC plus solid film lubricant without Coricone.

Detailed coating procedures are shown in Table 1. Coating thickness requirements are contained in Table 2. The coated test specimens were fatigue tested (both axial tension-tension and rotating bending) in air and 3.5% sodium chloride solution (to simulate marine atmosphere). Fatigue testing of the bare 4340 alloy was also carried out to obtain baseline data.

EXPERIMENTAL PROCEDURES

Fatigue Tests

Rotating bending fatigue (intended as a screening test). Stress versus cycles-to-failure studies of smooth fatigue specimens were carried out using a Krouse rotating bending fatigue machines which applied a cyclic stress at a frequency of 50 hertz and a stress ratio R = -1. The standard Krouse fatigue machine was modified by the addition of a lucite environmental chamber for

Table 1. Coating procedures.

a.	Cadmium Plating	1.	Degrease and rinse in flowing water
		2.	Cyanide dip 30 to 60 sec at RT
		3.	Cd plate at 20 A/ft^2 (215 A/m^2)
		4.	Rinse in water
		5.	Bake at 385 F (196 C), minimum 23 hours (within 1 hour of plating)
		6.	Degrease
		7.	Cyanide reactivation - 5 to 10 sec at RT
		8.	Rinse in water
		9.	15 to 20 sec in chromate conversion bath
		10.	Rinse in water and dry
b.	Chromium Plating	1.	Degrease and rinse in flowing water
		2.	Reverse etch in chromium plating bath (131 F, 55 C)
		3.	Cr plate (131 F, 55 C)
		4.	Rinse
		5.	Bake at 375 ± 25 F (191 ± 4 C) for minimum of 3 hours
		6.	Dip plated samples into solid film lubricant (MIL-L-46010)
		7.	Air dry for 30 minutes at RT
		8.	Cure at 400 F (204 C) for minimum of 1 hour
c.	Plasma-Sprayed Tungsten Carbide	1.	Grit blast fatigue specimens according to MIL-A-21380B
		2.	Plasma spray WC-Co (METCO 72 FNS) 7 to 9 mil per side (178 - 229μ)
		3.	Surface grind to 3.25 to 4.0 mil per side (82.6 - 116μ)
		4.	Apply Coricone 1700 sealant by spraying (specimens were prepared with and without Coricone)
		5.	Dry at RT
		6.	Cure at 350 F for at least 20 minutes
		7.	Dip specimens into SFL
		8.	Air dry for 30 minutes at RT
		9.	Cure at 400 F (204 C) for minimum of 1 hour

Table 2. Thickness of coatings.

Cadmium	0.3 - 0.5 mil/side (7.62 - 12.7μ)
Chromium	2.0 - 2.5 mil/side (50.8 - 63.5μ)
Tungsten Carbide (After Finish Grind)	3.25 - 4.0 mil/side (82.6 - 101.6μ)
Coricone Sealer	0.2 mil/side (5.1μ)
Dry Film Lubricant	0.35 mil/side (8.9μ)

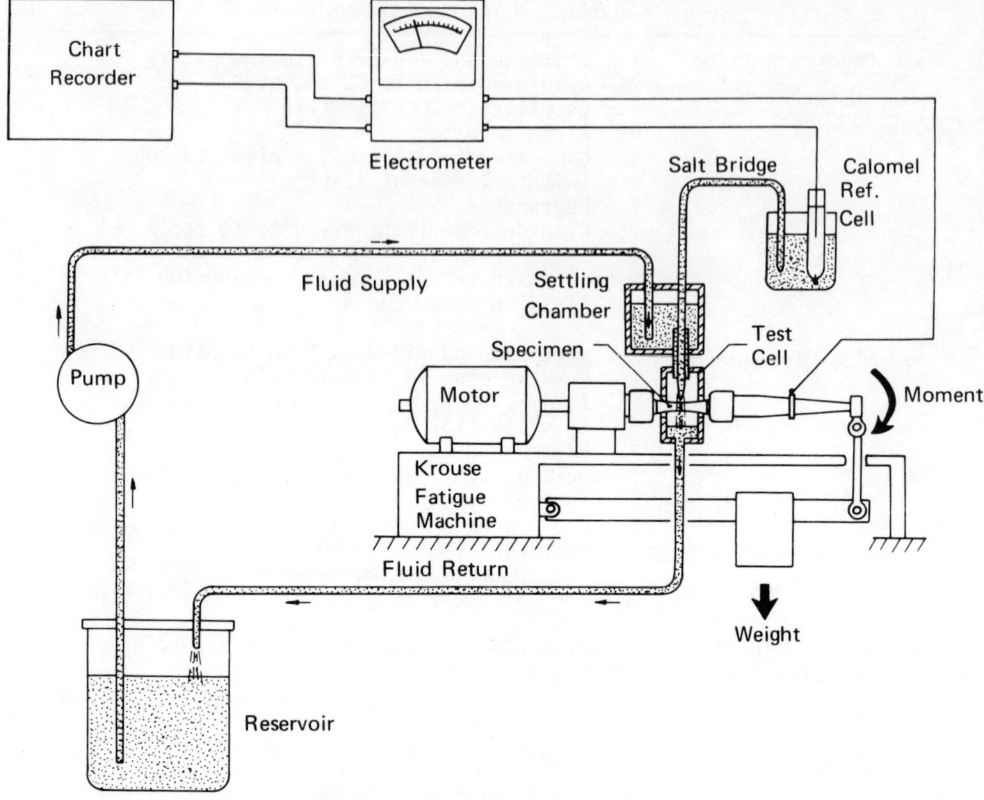

Figure 1. Schematic for corrosion fatigue apparatus.

corrosion fatigue studies, as shown schematically in Figure 1. A constant displacement tubing pump delivered the test fluid through PVC tubing from a two-liter reservoir at the rate of 15 liters per hour. The pulsations of the pumping system were removed in a settling chamber, so that a steady stream of fluid was then returned to the reservoir by gravity flow. At speeds up to 3000 rpm, the fluid maintained good contact with the rotating specimen as it flowed over and round the test section.

Rotating bending fatigue specimens measured 1/2" (12.7 mm) diameter by 4" long (101.6 mm), with a 2-1/4" (63.5 mm) radius reduced section giving a minimum cross section of 1/4" (6.5 mm) diameter at the specimen center as shown in Figure 2(a).

Axial tension-tension fatigue test (intended to simulate the high mean centrifugal loads on the blade retention bolt). Fatigue testing of axial tension-tension smooth fatigue specimens was

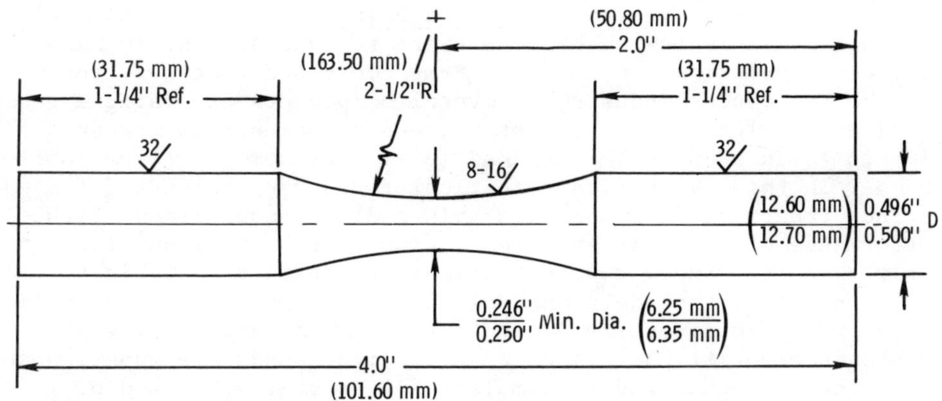

Figure 2. Fatigue specimen geometry. (a) Rotating bending fatigue specimen.

carried out using the Instron Dynamic Cycler Model 1211 system and sinusoidal loading. The apparatus was operated at a cyclic frequency of 33 hertz and a stress ratio R = 0.8. For tests in 3.5% NaCl solution, a plastic cell containing this environment was attached to completely surround the gage length of the specimen. Unlike the corrosion cell of the rotating bending fatigue apparatus, the solution was quiescent. Axial tension fatigue specimens are described in Figure 2(b).

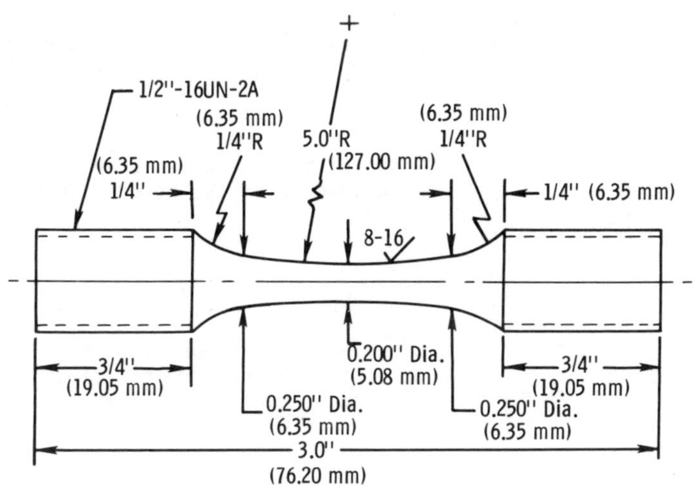

Figure 2(b). Axial tension-tension fatigue specimen.

Full-scale AH-1G blade retention bolt tests (AMRDL-Langley). The full-scale retention bolt test program was similar to many component qualification test programs conducted routinely throughout the helicopter industry. Several components are fatigue tested at different load levels to establish a relationship between the time to failure and the test load level with some level of confidence. In this evaluation, identical tests were conducted for each surface treatment in order to obtain a direct comparison of the performance of the bolt specimens. The test plan is outlined in Table 3 and the three test conditions are defined in Table 4. All test conditions include ground-air-ground cycles, which for helicopter rotor components represents the centrifugal loading which occurs once per flight. Accelerated flight loads are superimposed on the centrifugal load to simulate flight-by-flight loading conditions on a rotor, as illustrated in Figure 3.

A test fixture was designed to apply loads on the retention bolts in essentially the same manner as on the actual hardware. Two significant changes were necessary to maintain compatibility with available test equipment. Only axial loads could be applied effectively. This meant that flatwise bending loads had to be

Table 3. Test plan for AH-1G blade retention bolts.

Specimen ID		Test	
Test No.	Bolt No.	Cond.	Bolt Descriptions and Remarks
1	CD-1 CD-2	I	Test 1 was preceded by the instrumented static proof test.
4	CD-3 CD-4	II	CD series bolts - Cadmium-plated, as received from the airframe manufacturer.
7	CD-5 CD-6	III	
2	CR-1 CR-2	I	CR series bolts - Chromium-plated, reconditioned at the rework facility.
5	CR-3 CR-4	II	
8	CR-5 CR-6	III	
3	WC-1 WC-2	I	WC series bolts - Tungsten carbide coated per tentative process specification proposed by rework facility.
6	WC-3 WC-4	II	
9	WC-5 WC-6	III	Bolt WC-3 was retested as bolt WC-5.

Table 4. Test conditions for full-scale fatigue test.
Gag Cycle: 5-105,000 lb

Test Condition	Flight Loads, lb	Flight Loads per Flight	Inspection Interval, Flights
I	±15,000	325	1,400
II	±25,000	80	1,800
III	±35,000	16	4,500

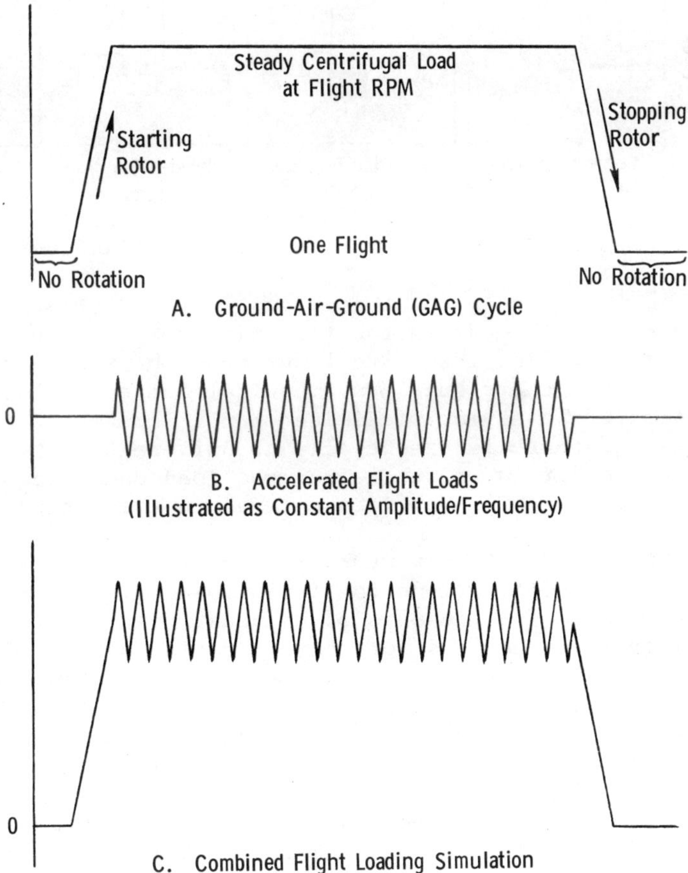

Figure 3. Synthesis of loading conditions for AH-1G blade retention bolts fatigue tests.

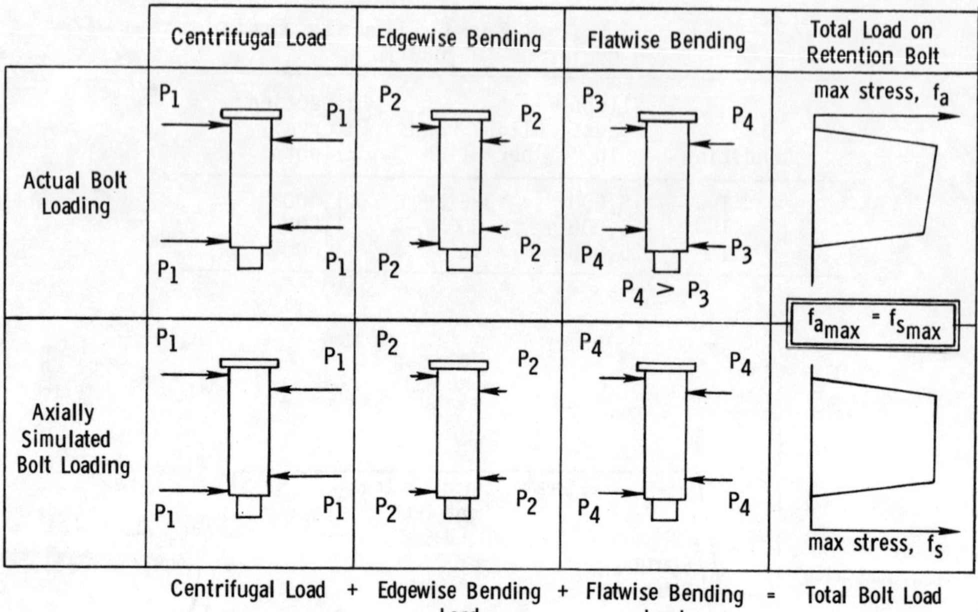

Figure 4. Rationale for axial simulation of blade bending loads.

simulated by axial loads on the test fixture. As shown in Figure 4, the bending loads are lower than the steady centrifugal loads and merely contribute a symmetrical increment for the axial test instead of an asymmetrical increment. In terms of critical bolt stresses, there is essentially no difference between the actual load condition and the simulated load condition. As a result, there is a high confidence level in the fatigue test data.

A sketch of the test fixture in Figure 5 illustrates the results of the modifications for axial loading. The materials used in the various components of the test fixture are equivalent to the materials specified for the production parts. Critical geometric aspects of the rotor grip and blade assembly relative to the loads on the retention bolts are also maintained in the test fixture design. Photographs of the test fixture, Figures 6 and 7, illustrate the installed retention bolt specimens.

All tests were conducted on a closed-loop hydraulic, servo-controlled fatigue test machine which was built to NASA-Langley specifications. The capacity of the test machine is 400,000 lb with up to 30 Hz cyclic loading rates. A digitally programmed control feature was used to provide the flight-by-flight loading sequence. The control console for the fatigue test machine with

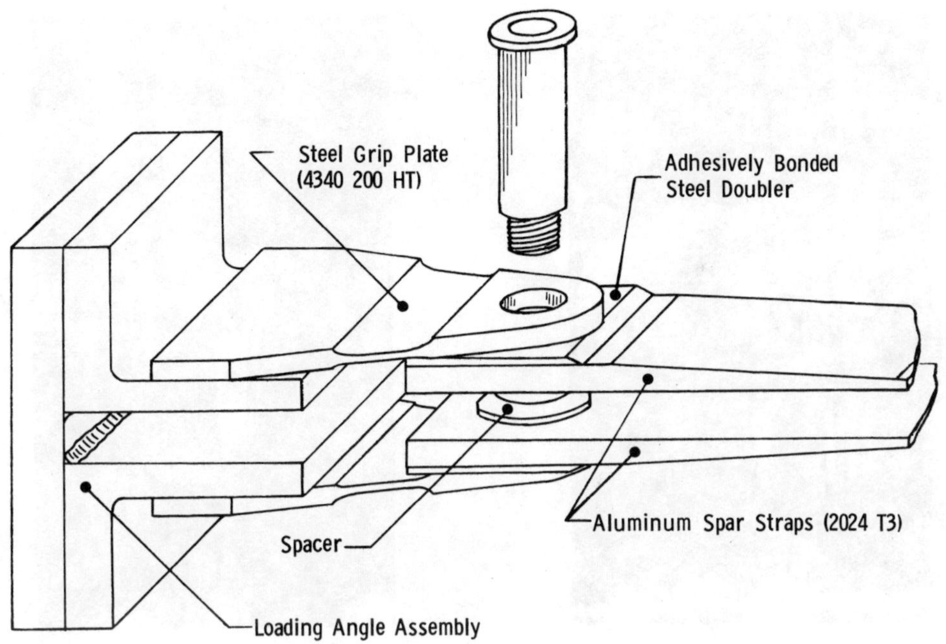

Figure 5. Schematic for AH-1G blade retention bolt test fixture.

accessory strain recording equipment for the static proof test is shown in Figure 8.

The static proof test was necessary to assure that the test fixture would function under the various fatigue-loading conditions. Text fixture instrumentation allowed an assessment of stress levels throughout the assembly, as well as detection of any load eccentricities which may have been introduced. In addition, instrumentation of the interior surfaces of the retention bolt specimens for the static proof test indicated stress levels which could be anticipated during the fatigue tests.

Since it was quite possible that the fatigue tests would produce no failures, some measure of the relative performance fo the bolt specimens was required. This was provided by periodically interrupting the fatigue tests to inspect the bolt specimens. As indicated by Table 4, the lengths of the testing segments between inspections varied for the different load conditions. The bolt specimens were physically removed and visual inspections of the surfaces were conducted to assess wear. The appearance of the bolts was documented by photographs and supplementary notes when

Figure 6. Test fixture installed in 400,000 lb fatigue testing machine.

Figure 7. Installed retention bolt instrumented for static tests.

Figure 8. Cantrolls for full-scale retention bolt fatigue tests.

required. Chemical spot tests were also employed to detect penetration of the protective coatings. Dilute nitric acid was used as the indicator. Doping with sodium thiocyanate was attempted to enhance the positive indication, but the extreme sensitivity of the doped nitric acid led to false indications from iron impurities in the tungsten carbide coating itself. The nitric acid also reacted with the cadmium plating material, so no spot tests of the production bolt specimens were possible.

RESULTS

Rotating Bending Fatigue

Figure 9(a) shows the deleterious effect of NaCl solution on the fatigue life of bare 4340 steel. The fatigue strength (value at 10^7 cycles) of the bare specimen decreases from 105 to 20 ksi (724 to 138 MN/m^2), a reduction of 81%. Figure 9(b) contains S-N curves for the 4340 steel coated with electroplated cadmium plus chromate treatment. The air value of the bare material is identical to that of the coated material (105 ksi, 724 MN/m^2). Parallel to data for steels in general, the fatigue limit of the bare alloy in air is approximately one half the tensile strength. The NaCl solution reduced the fatigue strength of the coated alloy by 24% (from 105 ksi to 80 ski, 724 to 552 MN/m^2). If we compare the fatigue strength of the coated alloy in NaCl solution with that of the bare alloy in the same environment, it is apparent that the cadmium plus chromate treatment significantly improves the fatigue strength of the alloy in NaCl solution (from 20 to 80 ksi, 138 to 552 MN/m^2). Figure 9(c) shows that 3.5% NaCl solution caused a 5.3% reduction in the fatigue strength of 4340 steel coated with electroplated chromium plus solid film lubricant (from 95 to 90 ksi, 655 to 621 MN/m^2). This coating system produced a 9.5% degradation in the fatigue life of the bare steel [Figures 9(a), (c), Table 5] in air. The fatigue strength of the alloy coated with plasma-sprayed tungsten carbide plus solid film lubrican with or without the Coricone sealant was 90 ksi (621 MN/m^2) regardless of the environment [Figure 9(d), Table 5], i.e., there was neither a deleterious effect of environment nor a beneficial effect of the Coricone sealant. A fatigue reduction of 14.3% was attributed to the coating system.

Axial Tension-Tension Fatigue

Figuer 10, in conjunction with Table 5, contains S-N curves and data obtained in air for the alloy, bare and coated. Fatigue data at 10^7 cycles showed that the cadmium and chromium electroplates, particularly the chromium, improved the fatigue strength

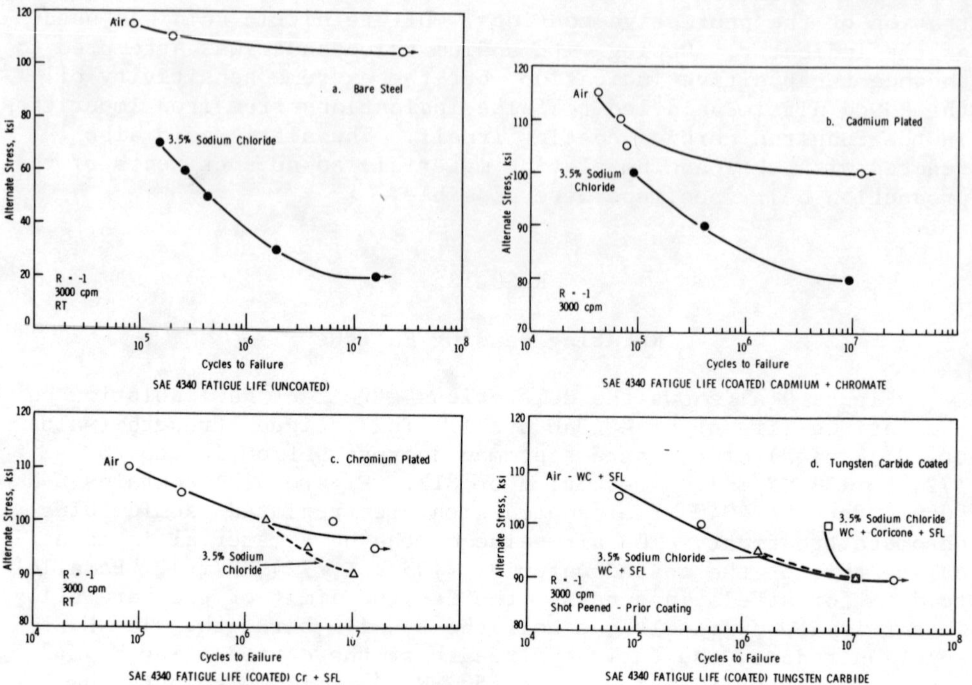

Figure 9. S-N curves (rotating bending) of bare and coated 4340 steel showing effect of environment.

of the bar alloy. The tungsten carbide coating reduced fatigue strength from 160 to 140 ksi (1103 to 965 MN/m^2) which represented a fatigue reduction of 12.5%. The Coricone sealant has no effect on fatigue strength of the alloy in air. Table 5 contains fatigue data at 10^7 cycles for the uncoated and coated alloys in 3.5% NaCl solution. This environment caused a 31.2% reduction in the fatigue strength of the bare alloy and a 45% to 60% reduction in the fatigue strength of the WC-coated alloy and the Cr-plated alloy. The Coricone again provided no beneficial effect on the corrosion fatigue resistance of the WC-coated alloy. The Cd-plated alloy was unaffected by the NaCl solution. It is evident that the axial tension fatigue data differs from the rotating bending fatigue data in the following manner: (a) degradation of the fatigue strength of the bare alloy due to NaCl solution is considerably less in axial tension. Gould [3] found that the attack of steel subjected to fatigue tests in sea water was more severe in bending than in direct tension due to a stretching of all the anodic areas in bending tests, whereas only some of these areas were stretched in direct tension tests; (b) based on air values only, fatigue strength reductions due to the WC and Cr coatings were quite

Table 5. Effects of coatings and environment on the fatigue strength of 4340 steel.

Test	Condition	Air Stress ksi	Air Stress MN/m²	Air Change, %	3.5% NaCl Stress ksi	3.5% NaCl Stress MN/m²	3.5% NaCl Change, %
Rotating Bending R = -1	Bare	105	724	-	20	138	-81
	Cd + Chromate	105	724	0	80	552	-24
	Cr + Dry Film*	95	655	-9.5	90	621	-14.3
	WC + Dry Film*	90	621	-14.3	90	621	-14.3
	WC + Coricone + Dry Film*	90	621	-14.3	90	621	-14.3
Axial Tension R = 0.8	Bare	160	1103	-	110	758	-31.2
	Cd + Chromate	165	1138	+3.1	165	1138	0
	Cr + SFL*	175	1207	+9.4	90	621	-43.8† -48.6‡
	WC + SFL*	140	965	-12.5	60	414	-62.5† -57‡
	WC + Coricone + SFL*	140	965	-12.5	60	4.4	-62.5† -57‡

*Shot peened
†Compared to bare alloy air value
‡Compared to coated alloy air value

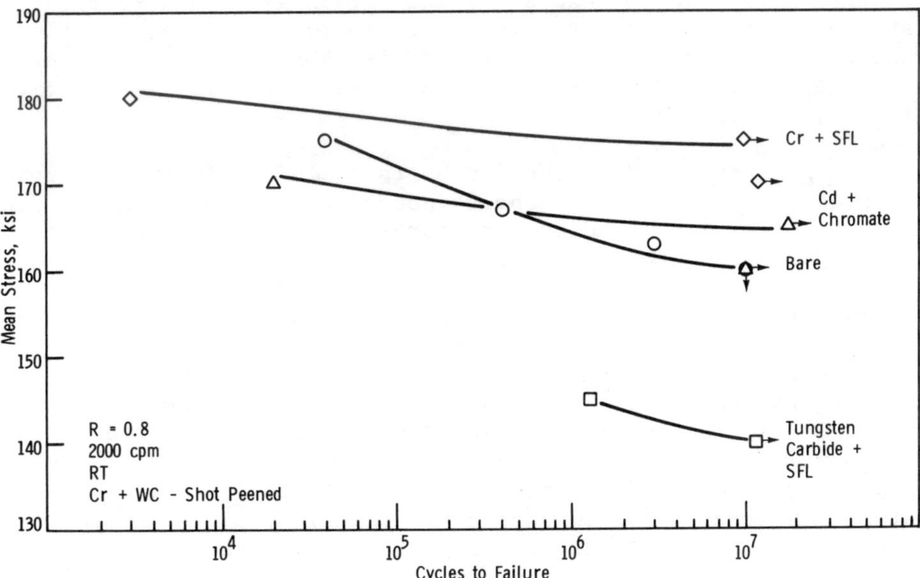

Figure 10. S-N curves (axial tension) of bare and coated 4340 steel in air environment.

similar in both rotating bending and axial tension fatigue tests. But in NaCl solution, significantly greater reductions in axial fatigue strength of the coated alloys were observed due to environmental effects which remains to be elucidated. Since the Cr and WC hard (brittle) coatings have a relatively low intrinsic fatigue strength in comparison with the steel, they will become discontinuous at a relatively low stress level owing to the development of fatigue cracks. (The Cr normally contains internal cracks.) These cracks will permit access of the corrosive NaCl solution to the steel base at the root of the fatigue crack. In the case of the axial tension test (high steady tensile oad), it may be easier for the environment to reach the crack tip.

Full-Scale Retention Bolt Tests

Static proof test. The test fixture was found to be suitable for fatigue testing. Bolt specimen instrumentation detected significant strain levels on interior surfaces. But no stress levels above endurance limits were indicated at a static load greater than scheduled fatigue loads. Excessive compliance of the test fixture due to rotation of the loading angles was noted. This would have increased the hydraulic fluid pumping requirements and severely limited cyclic loading frequency in the fatigue tests. The angles were subsequently restrained by welding them to heavy base plates and mounting the welded assembly on the test machine platens. Load eccentricities were not excessive and strain gage data from the instrumentation on the retention bolts indicated linear elastic properties but a very complex stress distribution. Significant strain levels that suggested that the retention bolts were fatigue critical were recorded.

Fatigue tests. No retention bolts failed during the fatigue test program. All retention bolts sustained the equivalent of four lifetimes of fatigue loading, or approximately 14,000 simulated hours of flight, without failure.

The fretting resistances of the tungsten carbide coating and the chrome plating were superior to that of the cadmium plating. This result confirmed previously developed data. However, since the test arrangement did not duplicate the wear mechanism of the hardware, quantitative improvements could not be estimated. No deterioration of the tungsten carbide coating (cracking, debonding, chipping) was observed during the fatigue tests. Chemical spot tests which were conducted during periodic inspections of the bolts detected no penetration or cracking of the tungsten carbide coating.

There was evidence of deposited cadmium and iron-based fretting products from adjacent steel test fixture components on the bolt surfaces, but these materials could be removed by polishing with fine steel wool and did not degrade the tungsten carbide coating. The abrasive action of the tunsten carbide-coated bolts on the steel grip plates, which produced the detectable fretting products, was insignificant in the laboratory environment. No hole elongations and very little corrosion were produced after twelve lifetimes of fatigue loading.

Metallography

The WC-coated specimens were examined metallographically for an assessment of coating integrity, bonding, and the sealing capability of the solid film lubricant (SFL) and Coricone. Figure 11 contains micrographs of the cross-sectional area of the coated specimens. Figure 11(a) demonstrates the good bonding between plasma-sprayed WC and the substrate 4340 steel. Note that some pososity is present and that the pores are discontinuous. Figure 11(b) shows good bonding between the solid film lubricant and the WC coating. Also shown is the capability of the SFL to fill surface pores present in the WC coating. Figure 11(c) demonstrates the good bonding that can be obtained between SFL and Coricone and between Coricone and WC coatings. Note that the Coricone has filled any surface pores present in the WC coating. Since the pores in the WC coating are discontinuous, neither the SFL nor Coricone has infiltrated below the surface pores.

SUMMARY

1. NaCl solution significantly degrades fatigue strength of the bare 4340 steel. The degradation is much more severe under conditions of rotating bending fatigue.

2. Although the fatigue strength of WC-plus-SFL-coated 4340 steel is unaffacted by NaCl solution in rotating bending fatigue, it is significantly reduced by this environment in axial tension fatigue testing, which more closely simulated service operating conditions. Regardless of the method of testing, the Coricone sealant does not impart additional resistance to corrosion fatigue and the WC coating reduces the air fatigue strength of the 4340 steel by about 14%.

3. NaCl solution further degrades the rotating bending fatigue strength of both Cd- and Cr-plated 4340 steel, but in axial tension fatigue testing, further degradation is limited to the Cr-plated alloy.

4. Regardless of the method of test, the sprayed WC and the plated Cr coatings exhibited comparable fatigue behavior in NaCl solution.

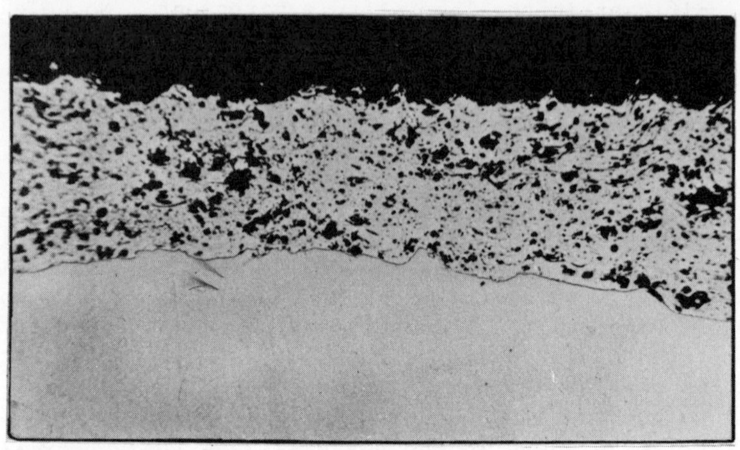

Figure 11. Micrographs of cross-sectional area of plasma-sprayed tungsten carbide coating interfaces. (a) Tungsten carbide coating/substrate interface. 150X

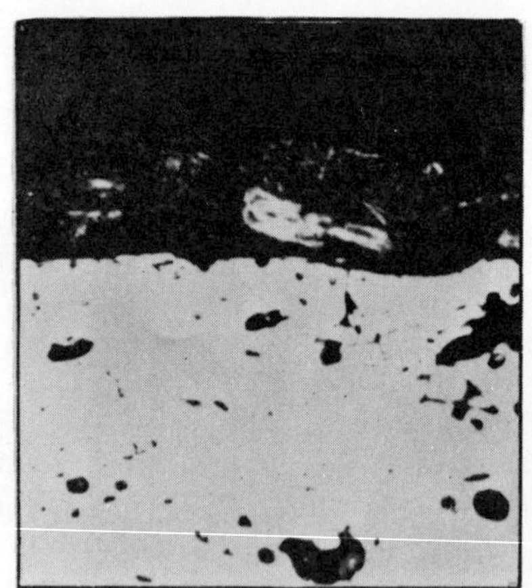

Figure 11(b). Solid film lubricant/tungsten carbide coating interface. 750X

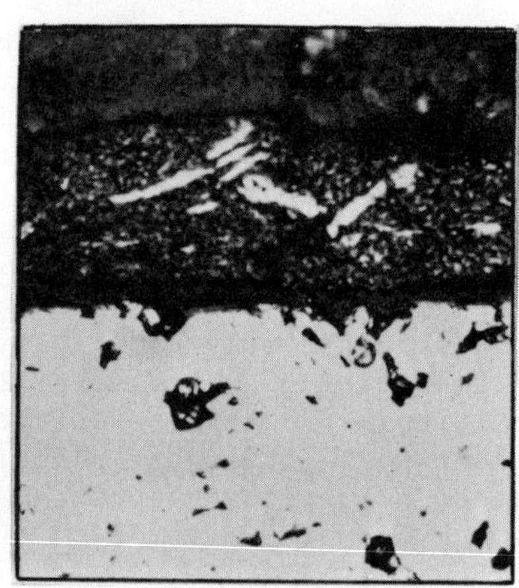

Figure 11(c). Solid film lubricant/Coricone/tungsten carbide interface. 750X

5. Despite the reduced fatigue strength of the WC-coated steel test specimens, full-scale fatigue tests of the blade retention bolts produced no failures through the equivalent of four lifetimes (approximately 14,00 flight hours of loading). Note that the full-scale component fatigue tests were carried out in laboratory air.

CONCLUSIONS

Prior experience in the field indicates a six to ten times improvement in life of the coated bolt can be achieved with the WC coating. Both laboratory and full-scale component fatigue tests indicate that the strength of the WC-coated bolt is adequate in a laboratory air environment. Although axial tension fatigue tests of WC-coated 4340 steel showed a significant degradation of fatigue strength in NaCl solution, no service fatigue failures of WC-coated retention bolts have been experienced. It appears, therefore, that the WC coating process for AH-1G blade retention bolts should be satisfactory for further production and rework aircraft as a cost-effective measure. But periodic visual inspection of the blade retention bolts should be conducted at a time interval within the 200 to 500 hours range until an adequate service data base is established. Any evidence of corrosion should be sufficient grounds for replacing the bolt since field experience indicates that corrosion can be detected prior to catastrophic failure.

ACKNOWLEDGEMENT

This work was supported by the U. S. Army Aviation Systems Command, St. Louis, Missouri.

REFERENCES

1. Viglione, J., Jankowsky, E.J. and Ketchan, S.J., "Effects of Metallic Coatings on the Fatigue Properties of High Strength Steels", Materials Protection and Performance, Vol. 11, March 1972, p 31-6.

2. Levy, M. and Morrossi, J., "Erosion and Fatigue Behavior of Coated Titanium Alloys for Gas Turbine Engine Compressor Applications", Army Materials and Mechanics Research Center, AMMRC TR 76-4, February 1976.

3. Gould, A.J., "Corrosion-Fatigue of Steel Under Asymmetric Stress in Sea Water", J. Iron and Steel Inst., Vol. 161, 1949, p 11-16.

CHAPTER 14

MICROCIRCUIT RELIABILITY CHARACTERIZATION

Joseph J. Naresky

Rome Air Development Center

Griffiss AFB, Rome, New York

ABSTRACT

With component densities doubling each year, semiconductor microcircuits present a particularly difficult challenge, in terms of characterizing their reliability. Conventional reliability analysis procedures are inadequate, and new test methods and instrumentation must be developed. The reliability physics approach to reliability characterization is discussed; it is concerned with developing a thorough understanding of the basic failure mechanisms in these devices, and how they proceed with time and stress. The mathematical models for this approach are reviewed. Also described is the use of the models in developing screening tests to "screen out" a particular failure mechanism, and tests to accelerate a particular mechanism to cause device failure. Typical microcircuit failure mechanisms are described as well as use of instruments, such as the scanning electron microscope and mass spectrometer, to develop a thorough understanding of the failure mechanisms, in order to perform a complete reliability characterization of the device. The evolution of microcircuit technology, and the need for increasingly complex and sophisticated failure analysis tools and procedures to characterize the reliability of microprocessors and large scale integrated circuits are reviewed. New trends and requirements for failure analysis are discussed in terms of where we are and where we have to go. Finally, the concept of "reverse engineering" is presented as the appraoch of the 1980s to microcircuit reliability characterization.

INTRODUCTION

In the conventional scheme of reliability, the failure of equipment derives from the failure of basic elements or parts of the equipment, the failure rates being additive under certain assumptions. The assessment of equipment reliability and its design can therefore be aided by parts reliability information which, supposedly, can be obtained independently from the particular equipment at hand, from previous experience. Although there are problems in actually implementing this reliability system, the fundamental premise is so clear and comfortable that it has become sacred - the law of equipment failure from part failure.

Several developments have put a stress on this concept over the years - the emergence of solid state devices with, it first seemed, no wear out; the dramatic increase in functional complexity of equipments with rapidly increasing parts counts demanding lower and lower failure rates; and the emergence of microcircuitry effectively redefining the basic part into something now often functionally more complex than whole equipments were just a decade ago. With increased part complexity causing increased part multiplicity - instead of resistors, capacitors, transistors and diodes, there are a plethora of digital circuit types classified by function, logic type, manufacturing processing, material, etc. Fortunately, of course, they have very low failure rates, particularly when attention is paid to all we have learned about proper design and application, testing and screening, etc. - so low, in fact, that actual measurements of failure rate, by counting failures tested in use environments simulated in the laboratory, is practically impossible. Yet because of the large numbers of parts used, these failure rate data are important and are needed.

In recent years, there has been a gradual awareness that there is considerable information about parts that may be applicable to microcircuit reliability and that this heretofore wasted information can be collected and brought to bear on the part problem. Certainly, information about details of semiconductor manufacturing processes, and relationships to product yield, and information on the assembly and packagaing processes, and product dropout at these stages of manufacturing relate to final product quality and reliability. Surely, experience with problems with certain elements of microcircuit parts, e.g., metallization, passivation, bonding, packaging, etc., is applicable to similarly manufactured devices ofdistinctly different function. As of this point in time, no formal method of utilizing all potentially applicable information which may be available on microelectronics is generally accepted; yet, short of full interpretation of these data with failure rate as an output is a growing freestyle discipline called reliability characterization, making best use of whatever valid information exists, for general part reliability applications.

RELIABILITY CHARACTERIZATION

Figure 1 schematically lays out the elements of part reliability characterization and assurance. Conventionally available inputs are field failure analysis, field failure statistics, applications experience, data from screening and qualification testing, and, albeit rarely available, use condition reliability tests. These conventional inputs to part reliability characterization result in reliability assurance technology, the general and detail specifications for functional application as well as reliability, including reliability prediction, electrical test methods, techniques for part screening, requirements for qualification, applications guidance, and design and process requirements.

Electrical characterization provides baseline knowledge of electrical function, comparison of requirements with what is actually available and potentially available, and defines the circuit implementation by logic diagram and schematic. This provides inputs to both reliability assurance (specifications) and to reliability physics characterization - the new part reliability synthesis.

Reliability physics is the discipline attempting the apply nonvolatile physics and chemistry information to needed part reliability requirements. It began with post mortem analysis when it was realized that parts fail because of mechanisms frequently inherent in the design or processes of parts. Failure mechanisms could be described by laws of physics or chemistry and failure phenomena could be treated by scientific experimental procedure with more generally applicable results than conventional reliability. Reliability physics characterization includes physical part baselining or product analysis, laboratory accelerated testing and subsequent failure analysis of resulting failures. To sustain these efforts as technology evolves, continuous upgrading of analysis techniques for fault detection, failure analysis, product evaluation, and accelerated testing are needed.

Full reliability characterization, including information about materials, design and processing, with applications to reliability prediction, reliability assessment, and finally reliability assurance through specifications must be performed on each new technology fo electronics, hopefully, before it is employed. This requires some gambling on future technologies - not every highly touted laboratory invention makes it into equipment. The status of device reliability characterization as of 1977 is presented in Table 1; some recent outputs of RADC characterization are shown in Table 2.

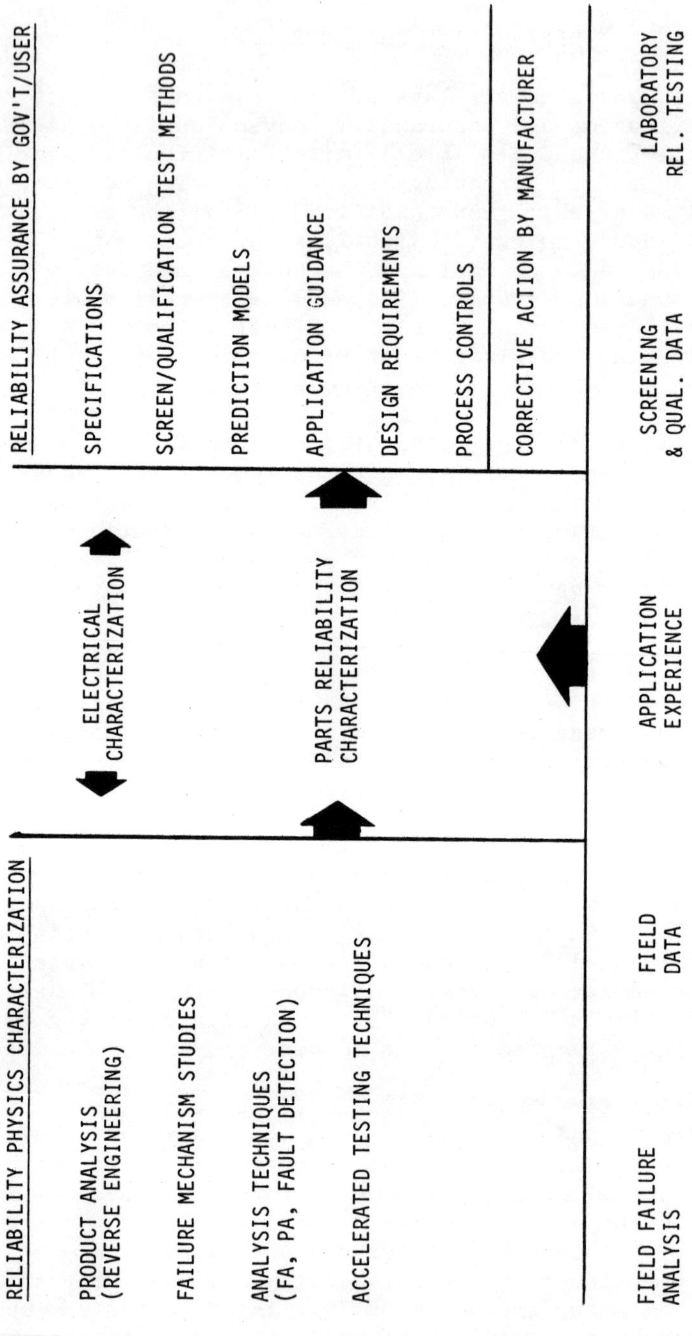

RELIABILITY PHYSICS CHARACTERIZATION

PRODUCT ANALYSIS
(REVERSE ENGINEERING)

FAILURE MECHANISM STUDIES

ANALYSIS TECHNIQUES
(FA, PA, FAULT DETECTION)

ACCELERATED TESTING TECHNIQUES

ELECTRICAL
CHARACTERIZATION

PARTS RELIABILITY
CHARACTERIZATION

RELIABILITY ASSURANCE BY GOV'T/USER

SPECIFICATIONS

SCREEN/QUALIFICATION TEST METHODS

PREDICTION MODELS

APPLICATION GUIDANCE

DESIGN REQUIREMENTS

PROCESS CONTROLS

CORRECTIVE ACTION BY MANUFACTURER

FIELD FAILURE
ANALYSIS

FIELD
DATA

APPLICATION
EXPERIENCE

SCREENING
& QUAL. DATA

LABORATORY
REL. TESTING

Figure 1. Elements of reliability characterization.

Table 1

1977 TECHNICAL STATUS - RELIABILITY CHARACTERIZATION

	Materials	Design	Process	Prediction	R. Assessment	Specs and QRA
SSI	+	+	+	+	+	+
MSI BP	+	+	+	+	+	+
MSI MOS	+	+	X	X	X	+
LS TTL LSI	X	X	X	?	?	X
NMOS LSI	X	X	X	?	?	X
RAD HARD DEVICES	?	X	?	0	0	0
CMOS-SOS	X	X	?	X	X	?
I^2L LSI	X	?	?	0	0	?
GaAs μDEVICES	?	?	0	0	0	0
BUBBLE MEMS.	?	?	X	0	0	0
JOSEPHSON JUCTIONS	0	0	0	0	0	0

Key: + Good shape
X Fair
? Some problems
0 Practically nothing

Table 2

MICROELECTRONICS RELIABILITY CHARACTERIZATION

Reliability of Depositee Glass Improved CVD Techniques	Glass Integrity Critical in Low Poer and Surface Sensitive Devices.
Polymer Coating Study	Can Impose Performance/Reliability Limitations. Coatings Still Desired for Particle Protection.
Epoxy Die Attach Study	Can be Okay. Used Now in Hybrids and SOS Devices.
Thermal Cycle - Thermal Shock	Old Problem Still exists in Plastic Encapsulated Components. Do Not Use Them.
CMOS - CMOS/SOS - MNOS Studies	Inadequate Input Protective Networks. Silicon Island Edge Transistors. MOS Failure Mechanisms in MNOS Devices.
ECL	High Reliability Device But Some Specific Part Types With Design Problems.
Schottky	Very Reliable Technology (Pt-Si Schottky Junction). Susceptible to Static Discharge. Complex Low Power Schottky - Further Study.

FAILURE MECHANISMS

One of the principal objectives of reliability characterization is determining the physical mechanisms which cause device failure. This analysis is done on parts which have failed during various phases of system development and use, or during stress tests designed to evaluate these parts prior to designing them into a system.

A large percentage of field failures are caused by mechanisms involving moisture. Principally, corrosion and surface instabilities are the failure mechanisms and are strongly dependent on the type of package and the environmental conditions during use. Table 3 shows results of moisture analysis related to failure mechanisms in hermetic parts using the quadrupole mass spectrometer. The level of moisture within the package is not always related to a loss of hermeticity but can be traced to the processing and materials used to make the package. Depending on such factors as chip metallization, bond wire type, passivation and glassivation layer quality, the circuit will fail due to local galvanic cell corrosion, deplating of conductor materials during device operation, or surface instabilities due to diffusion of ionic impurities leached from the package sealing glass by the entrapped water vapor.

Table 3

QUADRUPOLE MASS SPECTROMETER ANALYSIS
OF MOISTURE RELATED FAILURE MECHANISMS

Mechanism	Moisture Levels For Failure	Maximum Safe Moisture Levels
Nichrome Resistor Corrosion	0.5 - 1.0%	500 PPM
MOS Surface Inversion	0.5 - 2.0%	200 PPM
Aluminum Corrosion	0.5 - 25%	1000 PPM
Gold Electroplating	0.5 - 15%	1000 PPM

Methods are being sought to eliminate moisture related failure through qualification tests. This can be done using either mass spectrometric or electical measurements on moisture resistive test structures. This is a critical area which deserves further investigation.

Another field failure mechanism often observed on microcircuits sent to Rome Air Development Center (RADC) involves electrical overstress of the interconnect metallization stripes and diffused

Table 4

PROPOSED QUALIFICATION TEST LEVELS FOR SYSTEM TRANSIENTS

Device Technology	Test Level (Volts)	Most Sensitive Path
DTL	50	+ Input to - ground
TTL (54)	55	+ Input to - ground
HTTL (54H)	70	+ Input to - ground
LPTTL (54L)	50	+ Input to - ground
STTL (54S)	25	+ Input to - ground
LSTTL (54LS)	51	+ Input to - ground
ECL	52	+ Output to - output
CMOS	55	+ Vss to - input
NMOS	32	+ Input to - ground
LINEAR	75	Various
I^2L	40	Various
INTERFACE DRIVER	55	+ Input to - ground
INTERFACE RECEIVER	35	Input to Input

junction areas. Recent work in developing test procedures for evaluating the susceptibility of various technologies to "ZAP" failure is summarized in Table 4 [1].

In Table 4 all vaoltages pulsed through 100 ohm source impedance from a 0.10μF capacitor charged to above voltage levels. Pulse decay time equals 10 μsec.

Conductive particle shorting of internal microcircuit interconnection metallization is also of concern in space applications of microcircuits. Under weightless flight conditions, particles of die attach material and lid seal solder may bridge exposed conductors. This mechanism is dependent on the type of package and device layout. The problem can be minimized with proper process controls on package sealing, chip glassivation and device layout. The ability to eliminate mobile conductive particles can be assessed using particle impact noise detection (PIND) tests on sealed packages. In addition, there is considerable interest in coating the internal surfaces of complex hybrid devices with a variety of polymeric materials. Although these coatings can reduce the surface area of exposed concudtors, one must be certain that the polymers are compatible with all other device processes and use conditions.

Table 5

MICROCIRCUIT FAILURE MECHANISMS*

Device Type	Dominant Failure Mechanism	Accelerating Stress**	Apparent Activation Energy	Ref.
CMOS/SOS	V_{TH} Shift Caused by Oxide Charges	V, T	1.1eV	1
CMOS/SOS	V_{TH} Shift Caused by Oxide Charges	V, T	1.1eV	1
CMOS/SOS	Input Leakage Current	V, T	1.1eV	1
CMOS	Sufface Effect	V, T	0.9-1.3eV	2
Linear	Surface Effect	V, T	0.7-3.9eV	2
ECL	Electromigration/High Emitter Resistance	J, T	1.5eV	3
ECL	Electromigration/Base Emitter Shunt	J, T	0.53-0.6eV	3
CMOS	Analysis in Progress	V, T	2.0eV	4
CMOS	Analysis in Progress	V, T	1.9-2.5eV	4
CMOS	Analysis in Progress	V, T	1.9eV	4
TTL	Devices Tested, But an Insufficient Number Failed for Valid Statistics	V, T	--	2
Schottky	Electromigration	J, T	0.56eV	5

*Work sponsored by RADC/RB Griffiss AFB NY
**V = Voltage applied, T = Temperature, J = Current Density

The information about the kinetic behavior of various failure mechanisms is normally generated during stress testing of microcircuits of test structures which incorporate specific parts of device technologies. Table 5 summarizes some results generated on RADC sponsored program. In general, the complementary-metal-oxide semiconductor (CMOS) and linear device technologies were susceptible to surface realted instabilities with activation energies on the order of one electron volt. The high current density technologies such as emitter coupled logic (ECL) and standard series Schottky are more sensitive to electromigration failure. The activation energies for these mechanisms range between 0.53 and 1.5 eV.

Normally this kinetic data is used to develop screening and qualification tests which will remove defective components prior to their use in the field. This is the case with surface sensitive components. A different approach has been used to reduce electromigration related failure.

Since one cannot screen out devices which will fail due to electromigration, sepcifications have been established which limit the maximum design value of current density. This has proven quite effective in the case of aluminum metallized microcircuits. As vendors employ different metallization systems, such as gold or doped aluminum, it may be necessary to develop qualification test procedures which would establish the maximum current density for a particular device type. Questions concerning the most effective use of failure mechanism data will continue to play an important role in future complex microelectronic circuits.

HIGH STRESS TESTING

Failure mechanisms known or suspected to exist in the field are detected and measured in the laboratory using specially tailored high stress testing designed to accelerate these mechanisms, and, hopefully, to establish failure rate curves relatively quickly compared to field data collection. In addition to the usual mechanical and environmental tests, such as centrifuge, vibrarion and shock, salt exposure and seal testing, electrical stress testing usually at elevated temperatures is intended to simulate life use conditions. These tests are listed in Table 6.

ACCELERATED TEST MODELS

Analysis of data from high stress test has been the subject of considerable activity in the last five years. It is important because of its obvious applications to failure rate estimation, prediction at other stress levels, screening and qualification

Table 6

EXAMPLES OF ELECTRICAL STRESS TESTS

Name of Test	Description of Test
Storage	Elevated temperature, with no applied electrical bias
High Temperature Reverse Bias (HTRB)	Reverse bias less than breakdown voltage applied to one or both transistor junctions at elevated temperature
Operating Life	Application of electrical bias and loads to allow operation. Can be dc or pulsed. Generally done at a specified power dissipation.
Electrical/Environmental	Electrical bias stress, combined with 85°C/85% relative humdity, temperature cycle, or pressure cooker.

testing, and product improvement through corrective action. There are two widely accepted (after substantial back-up experimentation) notions of accelerated testing: (1) that temperature provides acceleration of failure rates (or mechanism reaction rates) in conformance with the ancient Arrhenius reaction rate equation which gives the concept of activation energy, and (2) that the failures are distributed lognormally in time, giving the concept of median life (unfortunately, not the failure rate directly) and standard deviation. The Arrhenius equation is shown in Figure 2, where R(T) is the (failure) rate, T is the absolute temperature, and A and B are constants - B being the activation energy. This parameter is of particular interest to reliability physicists since it relates a reliability parameter to a physicochemical or thermodynamic

ARRHENIUS REACTION RATE EQUATION

$$R(T) = e^{A-B/KT}$$

B is interpreted physically as activation energy

Modified Eyring reaction rate equation:

$$R(T,S) = ATe^{-B/KT}\ e^{(C+D/KT)S}$$

Figure 2. Models for accelerated testing.

concept. Indeed, if the reaction rate of a mechanism is linearly related to a device parameter drift rate, for example, the failure mechanism activation energy and, hence, the mechanism itself may be determined. (Some recent results of activation energy degermination through high stress testing of various technology types are summarized in Table 5).

The assumption of the lognornal distribution shown in Figure 3 (Reference McDonnell Douglas) makes data analysis particularly convenient. Plotting cumulative percent failure on normal probability paper vs log of time yields a straight line giving the standard deviation, σ, for the distribution. If more than one mechanism is present, or if the devices have not been properly screened, a straight line may be difficult to obtain. These σ's may be quite variable; however, a σ of 2 is not unusual and may be considered typical. Knowledge of σ is critical for accurate life estimation, since most useful lifetimes are in the tail of life distribution.

Unfortunately, the Arrhenius model does not account for the possible effect of voltage stresses, although it seems intuitively

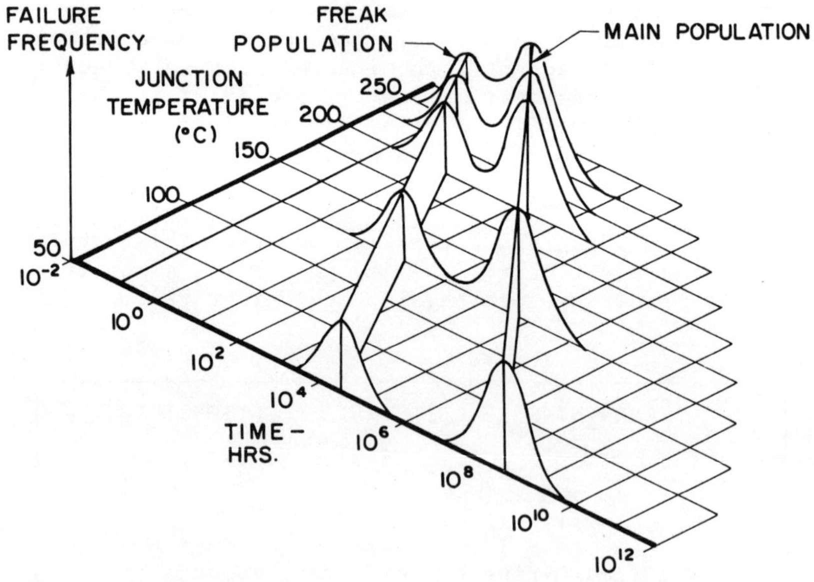

Figure 3. Microcircuit accelerated life testing life-temperature relationships. Ref: RADC TR 76-218, "Evaluation of Microcircuit Accelerated Test Tchniques", McDonnell-Douglas Astronautics Co.-East.

certain that a device tested with full voltage applied ought to perform differently in life than one tested without bias, particularly if the temperature effects of power dissipation are considered. An attempt to model the effects of stresses other than voltage has resulted in the Eyring equation (see Figure 2) where S represents a generalized stress variable. So far, attempts to apply the Eyring equation to such promising technology types as CMOS have not resulted in clear-cut success. The problem may be that there are dramatically different mechanisms or parameteric functions involving, for example, current density, avalanche breakdown, field strength, etc., all simply related to applied voltage. Until these mechanisms are determined, application of the Eyring model may be difficult.

Some of the current problems in accelerated testing and application of this technique to screening and qualification of microcircuits are summarized in Table 7.

Table 7

PROBLEMS IN ACCELERATED TESTING

Uncertainty in E

Variability of σ

Presence of non-accelerable mechanisms

Lack of data on voltage stress dependence

Possible screening damage

PREDICTION

The determination of typical activation energies for various classes of technologies is desired for application to prediction models. One hopes, at least as an initial approximation, that a formula (or model) can be derived for estimating the failure rate of parts for which there are insufficient field or test data. These models may be used, for example, at the early conceptual stages of system formulation, when reliability plans are proposed. The presently used model, subject to periodic updating as new device types and new data become available is as follows:

$$\lambda = \pi_L \pi_Q \pi_P [C_1 \pi_T + C_2 \pi_E]$$

where the failure rate, λ, is related to a learning factor (π_L), a function of product maturity; a quality factor, π_Q, a function of screening and procurement quality level; and a pin factor, π_P, which increases with the pin count (number of connections to the outside world). These factors multiply a sum of contributions to failure related to temperature, π_T, and non-temperature environmental factors, π_E. The constants C_1 and C_2 are very strong functions of complexity, exponentially increasing with gate count above 1,000 gates.

At present, the π_T factor is determined using an Arrhenius rate equation with activation energies of 0.4 eV for bipolar technologies, and 0.7 eV for metal-oxide-semiconductor (MOS) type devices.

STANDARD FAILURE ANALYSIS TECHNIQUES

Most laboratories involved in device failure analysis have concentrated on several standard techniques to characterize the mechanisms which are responsible for microelectronic failure. Proper failure analysis begins with a complete electrical characterization of the part, and then proceeds through optical inspection to the types of procedures outlined in Table 8.

The electron beam techniques of scanning electron microscopy and electron probe microanalysis are truly standards in the field of failure analysis. High resolution structural information can be easily obtained with a scanning electron microscope (SEM). The chemical analysis of microcircuits using X-ray spectroscopy can be performed on the electron probe microanalyzer of a SEM. The data shown in Table 8 is specifically for wavelength dispersive X-ray analysis. Energy dispersive analysis is normally restricted to elements with higher atomic numbers than five and is not as sensitive to trace constituents within the analyzed volume.

The surface selectivity of electron beam generated X-ray data is a function of the primary electron beam energy. For most work in the area of microelectronics, the analyzed volume of material is on the order of a cubic micrometer. To circumvent this limitation, several other charged beam techniques have been developed which analyze a thin surface layer of material.

Auger electron spectroscopy which identifies atoms within the first 50 Angstrom units from the surface of a solid can provide chemical information on thin laminar structures. By combining this electron spectroscopy method with selective sputtering of the surface, it is also possible to study the depth distribution of elements within such materials as deposited glass, metal layers, thermal oxide and the active silicon junction areas. This technique is restricted to relative detectability limits similar to the X-ray spectrochemical methods.

The most sensitive chemical analysis technique is secondary

Table 8

STANDARD FAILURE ANALYSIS TECHNIQUES

Area of Interest	Analytical Technique	Important Features
Device Microstructure	SEM	High Resolution, Large Depth-of-Field. Minimum Sample Preparation
Device Bulk Chemistry	EPMA	Good Spatial Resolution (X-Y) 0.01% Relative Detectability Limits 10^{-15} GM Absolute Det. Limits $Z \geq 5$
Device Surface Chemistry	AES	Good Spatial Resolution (X-Y) Surface Selective (15-50Å) Amenable to Layer Analysis
Chemical Analysis of Layer Structures	SIMS	Fair to Good Spatial Resolution 10Å/SEC profiling Rates No Z Limitations PPB Relative Detectability Limits
Device Surface Electrical Analysis	PS	Fair Spatial Resolution Non-Destructive Surface Inversion Location
Package Ambient Gas Analysis	QMS	Quantitative Analysis of Gas Constituents in Hermetic Devices

ion mass spectrometry although it does require larger analyzed surface areas than Auger electron spectroscopy. The detection of sputtered secondary ions does permit diffusion profile analysis of very low relative concentrations of materials and can provide the failure analyst with information not obtainable with other analytical techniques [2].

In addition to the electron and ion beam techniques, there are a variety of other analytical procedures which have been used for failure analysis. Two methods: (1) photoscanning for device surface analysis, and (2) quadrupole mass spectrometry for package ambient gas analysis, have become relatively standard approaches.

Initial work [3] on the use of a scanned beam of light to investigate surface inversion problems showed the essential non-destructive features of this approach to microcircuit failure analysis. The spatial resolution of this optical technique is suitable for most small scale and medium scale integrated circuit analysis. The major limitation involves accessibility of device regions of interest to external amplifying circuits.

The quadrupole mass analyzer has been adapted to the analysis of the minuscule volumes of gas within hermetic device packages [4]. This technique, which is capable of determining parts-per-million levels of contamination by water vapor and organic materials, is an important adjunct to other failure analysis techniques such as electron probe microanalysis and Auger electron spectroscopy in the study of corrosion and surface-related failure mechanisms.

ADVANCING TECHNOLOGY

As microelectronic devices become more complex and use smaller and thinner active areas, the burden increases on the failure analysis activities to provide detailed structural and physical information.

In the early 1960s, the ability of the microcircuit failure analyst to accurately isolate the physical cause of failure was nearly 100%. In the late 1970s, device geometries easily reach the thousand gate complexity in an area of 20 mm^2. This poses a problem not only in physical fault isolation due to reduced sizes but also an increased difficulty in evaluating dircuit performance, electrically.

Consider a microcircuit which has m inputs and n storage elements. To completely verify a truth table for this device, one must perform 2^{m+n} tests. If we try to test a 4 input gate, we need perform only 16 tests. This could be done in 1.6 microseconds if the electrical test equipment could perform at a 10 MHz signal rate.

Translating this procedure to a 16-bit shift register, the test time increases to 13 milliseconds. A 128-bit random access memory device would require more than 10^{25} years to establish the same truth table verification. The future approach to reliability characterization of advanced microelectronic devices will then depend on more sophisticated testing procedures, as well as the development of new analytical techniques and novel applications of existing ones.

In addition to new approaches to the electrical aspects of failure analysis, there are developments in the physical analysis techniques. The growth of both areas will depend on a mutual feedback of information as is shown in the following examples.

WHAT'S NEW IN FAILURE ANALYSIS

The increased complexity of microelectronic circuits, the electrical test difficulties in fully characterizing these parts and the limited availability of schematic circuits for other than the most organized memory chips have all resulted in challenging boundary conditions on the failure analysis problem.

Computer aided diagnostic procedures are being developed to provide guidance in complex device failure analysis. One task involves reduction in the number of electrical test vectors to reduce the amount of time required to adequately characterize a microelectronic device. Once this is accomplished and several test vectors are established as inoperable conditions, the computer then performs the second task of isolating the faulty circuit element. Depending on the size and sophistication of the computer analysis program, a fault dictionary can be established which could isolate the cause of circuit malfunction to a specific gate or transistor within the device.

By examining the variation in the optical properties of liquid crystal layers due to voltage on the surface of a microelectronic circuit, it is possible to determine both the layout of the active areas of the device and also to isolate the sites of electrical faults [5]. A recent modification of this technique has extended the method for examination of dynamic large scale integrated (LSI) devices at high frequencies [6]. Regulation of the electrical signals which exercise the microcircuit will alter the RMS value of the voltage at various circuit nodes. Depending on this RMS voltage, the liquid crystal material will highlight a chosen location on the device. This is seen in Figure 4, where one stage of a

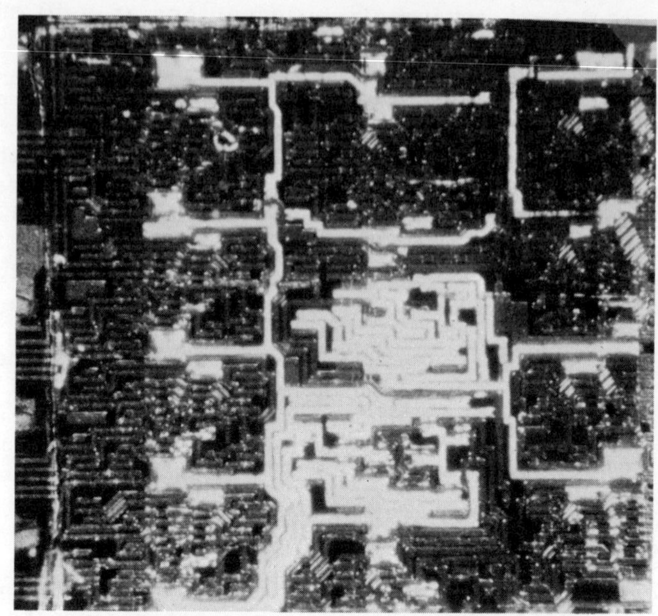

Figure 4. Liquid crystal display of selected area of a 14-stage binary counter circuit.

binary countercircuit is accentuated by the liquid crystal material. Using this abnormal electrical mode of operation, one can establish the layout of complex devices which is representative of normal operation. The technique can also be used to locate failed elements within the device in a manner similar to the earlier version of this technique.

Another optical inspection technique for establishing the electrical characteristics of a microcircuit is called scanning photo-excitation microscopy [7]. This technique uses a scanned pulsed laser beam, selected clocked electrical signals on the device under test, and lock-in amplification of the laser generated specimen photocurrents to establish device layout. Like the liquid crystal techniques, this procedure can be used to examine dynamically operating microcircuits.

Scanning electron microscopy has already been discussed as a standard failure analysis technique. It also has potential for providing detailed electrical information about complex microcircuits using such analysis modes as voltage contrast and electron beam induced current. There are some excellent reference texts [8,9] on these SEM modes of operation and the more fundamental aspects of the instrument.

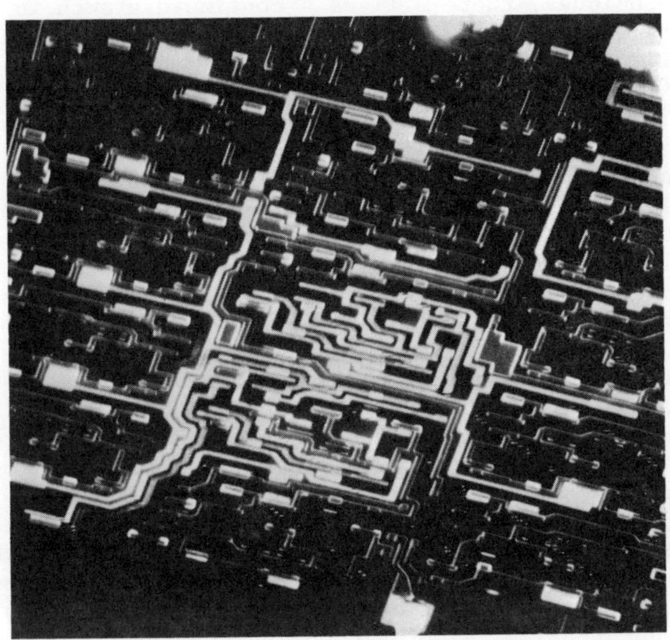

Figure 5. Scanning electron microscope voltage contrast display of identical device area shown in Figure 4.

Recent work [10] has demonstrated the utility of an SEM to make quantitative voltage measurements of selected circuit nodes within an LSI device. Using differential voltage measurement techniques and voltage reference points on the device under test, quantitative measurements can be made over a range of $\pm$ 13 volts with a precision of $\pm$ 100 mV. These types of measurements can be made even through a 2.5 μm deposited glass passivation layer.

Some recent work at RADC has demonstrated the ability of the SEM to display information similar to that obtained with the liquid crystal technique [11]. Figure 5 is an SEM display of the same device region shown in Figure 4. By varying the duty cycle of the counter circuit in a slightly different manner than for liquid crystal analysis, one can accentuate the operating characteristics of a device and do so with higher spatial resolution and greater sensitivity to low values of applied voltage.

These are only a few examples of new developments in failure analysis. Continued work is needed in the development of new approaches to the analysis of submicron regions of microcircuits. In addition, there should be an accelerated application of the existing methods. Through the innovative application of the standard techniques, there will be a continued growth in our understanding of complex microcircuits and their failure mechanisms.

REVERSE ENGINEERING FOR LSI

Reverse engineering is not new. In concept, its application to reliability measurements and failure analysis is shown in Figure 6, where various steps in the process are shown. The difference between this concept and what has been done in the past is in the depth to which it must be carried, especially in the steps beyond the schematic diagram.

In the sense used below, not even the manufacturer of an LSI part, such as a microprocessor, is likely to have a complete schematic diagram. Such parts are commonly developed as a group of previously known and manufactured subsections installed on a single chip, but "optimization" can then merge portions of subsections so that physical separation is lost. The physical proximity can then permit thermal or other sneak interactions which will not appear on an ordinary schematic diagram. This will make the reliability data on the previously known subsection inapplicable. It will render the logical breakdown into subsections inappropriate for test development, since a fundamental premise of such tests is the independence of the subsections. Reliability characterization of such parts is impossible without reverse engineering. The initial conceptual architecture must be discovered along with any

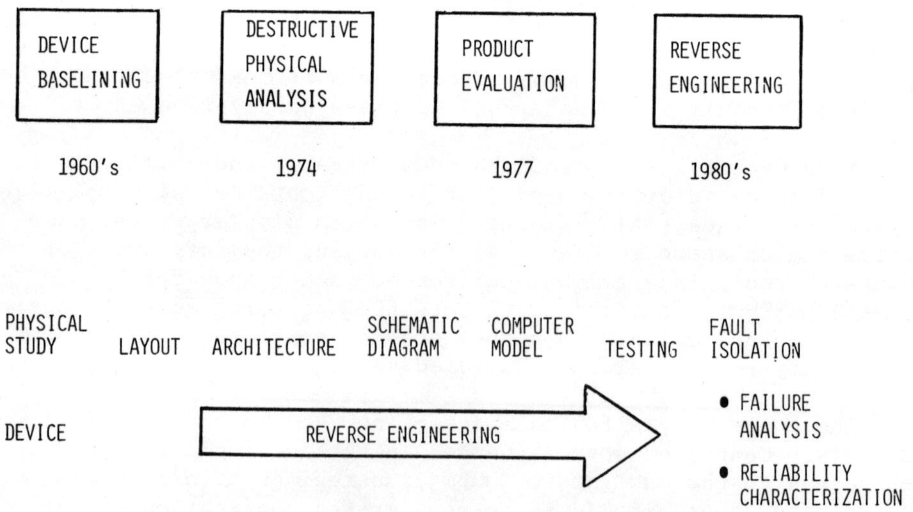

Figure 6. Reverse engineering for LSI.

such unintentional changes or additions. With such a foundation, a short enough set of tests to insure discovery of a failure may be attainable. If not, there is little hope of a valid reliability test since only good luck against large odds would give the time at which a failure occurs.

Reverse engineering also can be an insurance policy. Without it who could know whether a minor change in part design will invalidate an expensive and time consuming reliability characterization already performed? In fact, without reverse engineering on a continuing basis there will be only a small chance of knowing that a change has been made.

Although the cost of reverse engineering will be high, the risk against which assurance is desired is also high: simultaneous failure, at a critical time, by a substantial fraction of nominally identical system circuits as a result of correlated data patterns. The hazard is particularly serious for a system using redundancy since the redundancy is likely to be circumvented by the simultaneous "agreement" on an incorrect result.

CONCLUSIONS

In reviewing reliability characterization of microelectronic devices, we have seen how the methodology has progressed from its rather simple beginning to its present level of sophistication. As device technology has matured, so has the mathematical and physical approach to our understanding of reliability. The answers to many of the problems facing the device analyst are made tractable by the complex electronic technology which generated them. The future success of reliability characterization lies in the development and innovative adaptation of computational and analytical procedures to the description of system and part performance. These activities will then accelerate the distribution of electronic technology not only in the areas of high reliability military systems but also in those commerical applications which form an integral part of our everyday lives.

REFERENCES

1. AF Contract with BDM Corporation, Albuquerque, NM, Contract No. F30602-76-C-0308.

2. RADC TR 77-339, "Microbeam Analysis Techniques for ICs".

3. Haberer, J.R., "Photoresponse Mapping of Semiconductors", Physics of Failure in Electronics, Vol. 5, June 1967.

4. Thomas, R.W., "IC Packages and Hermetically Sealed In Contaminants", Government Microcircuits Application Conference Digest, Sand Diego, CA, 1972.

5. Salvo, C.J., "An Improved Approach to Locating Pinhole Defects in MOS and Bipolar Integrated Circuits Using Liquid Crystals", 14th Annual Proceedings of the Reliability Physics Symposium, Las Vegas, NV, 1976.

6. Burns, D.J., "Microcircuit Analysis Techniques Using Field Effect Liquid Crystals", to be presented at the 1978 Reliability Physics Cymposium, San Diego, CA, 1978.

7. Levy, M., 1977 Reliability Physics Symposium Proceedings, IEEE Catalog No. 77CH1195-7PHY.

8. Wells, O.C., Scanning Electron Microscopy, McGraw-Hill, 1974.

9. Goldstein et al., Practical Scanning Electron Microscopy, Plenum Press, NY, 1975.

10. Touw, T.R. et al., "Practical Techniques for Application of Voltage Contrast to Diagnosis of Integrated Circuits", Scanning Electron Microscopy, 1977, Vol. 1, Chicago, IL, March 1977.

11. Private Communication, J. Bart, RADC, October 1977.

CHAPTER 15

INTEGRATION OF CAD/CAM SYSTEMS FOR PRODUCTION OF STRUCTURAL COMPONENTS

Robert J. Sanderson

Grumman Aerospace Corporation

Bethpage, New York

ABSTRACT

As a result of existing aerospace design and drafting practices, and of existing manufacturing and tooling practices, component mating parts may be fabricated which do not fit properly with each other, producing a situation where product failure portential is established. A typical aerospace vehicle is constructed of various types of mating components such as formed parts, machined parts, stretched parts, honeycomb assemblies, etc. The parts design may be described by diverse means. For instance, the machined part will be completely dimensioned, whereas the formed part will be described by "lofting". Theoretically, these diverse means all originate from one source. However, in actual practice, that source may be interpreted, manipulated, and toleranced as it is used in order to describe the particular parts. In addition, the source is further distorted as the component parts are methodized and tooled. This is especially possible where there is a "family" of tools, such as mock-ups, casts, masters and templates - mainly produced by hand skills - required in order to fabricate the production tools and whereby tolerance is allowed to accumulate between each successive generation within the tool family. The net result of existing design and of existing manufacturing practices may result in a conflict between mating component parts.

With the advent of computer graphics, a technology is afforded whereby product design is truly described by a single source. This source, residing as a mathematical model in a computer, is accessed to discretely describe all mating parts and to generate machine parts to either fabriacte the parts or to fabricate the tools required to make the part. The family of tools is reduced or entire eliminated in the process. The net result is an improvement in component parts quality and in an improved assembly which is more likely to perform its function.

This chapter will describe this technology and will show how it is being implemented to improve product quality.

EXISTING DESIGN AND MANUFACTURING PRACTICES

Types of Components

The typical aerospace vehicle is composed of various types of mating components such as:

* Formed parts for ribs, stringer, frames and brackets.
* Machined parts for major bulkheads, longerons and fittings.
* Stretched parts for exterior skins and ducts.
* Honeycomb assemblies for access panels, doors and equipment shelving.
* Composites for internal and external structural components such as beams, longerons and skin panels.
* Welded parts for built-up sub assemblies, ducting and environmental subsystems.
* Electrical harnesses for all wiring.
* Others such as glass, plastics and fiber glass.

Materials

Materials for the above run the range from the more common aluminum alloys, to titanium alloys, and to boron and epoxy composites. These materials all possess their respective characteristics which must be account for in planning for manufacture.

Means of Components Descriptions

In order to fully describe the components, a variety of methods are employed:

* Formed or flat sheet metal parts are "lofted". That is they are accurately drawn on a stable material. For many years that material was sheet steel or aluminum with a white coated face. For the recent generation of aerospace vehicles, the material is mylar. The drawing is then reproduced for paper

copy, or on metal from which tooling is fabricated. Some dimensioning and tolerancing may also be shown on lofted drawings.

* External skin panels are described similar to sheet metal parts. A full scale mock-up of the area is made. A "master" skin is then formed from a die whose shape was derived from the mock-up, or from a laminate which is cast from the mock-up. In either event, the master skin is then scribed with essential information such as outline of part, chem-mill lines, rivet hold pattern, and mating parts outlines. This master becomes the standard from which all subsequent tooling, and mating parts coordination is derived.
* Tubing may be fully dimensioned, or may be derived from "master tubes" developed from a full scale equipment mock-up, which become the standard to which all subsequent production tubes are formed and inspected. Paper documentation is furnished through schematics and/or with pictures.
* Electrical harnesses are described similarly to tubing.

It is seen, therefore, that components, mating or note, are defined through a variety of means. Each of these means has its own peculiar tolerance at inception and in ultimate fabrication.

Means of Fabrication

Continuing on, these components are fabricated utilizing a mix of equipment and skills from the most elemental such as hand sawing and filing to the most sophisticated such as electron beam welding. The more common examples are:

* Forming - hydro-form (or "rubber" press), drop hammer, stretch.
* Machining - conventional, numerical control, profiling, lathe.
* Welding - resistance, electron beam, inert gas.
* Assembly - hand, automatic, bench, line, ultrasonic riveting.
* Bonding - autoclave.
* Chem-milling - tank.
* Tube bending - hand bending, numerical control bending.
* Electrical - machine assembly, harness hand assembly.
* Composites fabrication - machine lay-up, hand lay-up.

All of this equipment and all of these skills also have their peculiar characteristics and add another possibility for components mismatch. Each process, whether machine or hand, is subject to variation. Such things as machine age, maintenance, operator proficiency, quality surveillance, and hand skills all conspire to produce a satisfactory, marginal, or unsatisfactory component.

Production Tooling

A variety of prototype or production tooling is required in order to fabricate and assemble the components. Such considerations as to numbers of units expected to be produced, rate of production and specific product specifications determine which types of tools will be ordered. There are many different types of which the following are only the main examples:

* Assembly and Sub-assembly Fixtures
* Press Blocks
* Stretch or Hufford Dies
* Autoclave or Platen Press Bond Fixtures
* Welding Fixtures
* Machine Numerical Control Tapes
* Router Fixtures
* Drill and Ream Jigs
* Installation Fixtures

These tools are also constructed using a variety of machine and hand skills. Again, a degree of tolerance is introduced, which will be transmitted to the part or component which the tool is producing.

Tooling Tools

Where particular direct coordination between mating parts of assemblies is required, "tooling" tools may be ordered to affect that coordination. These tools doe not produce parts directly. Their fabrication must necessarily precede the production tooling which they coordinate. As such, while they have been inherently necessary, they add to overall tooling costs, slow down delivery of the production tools, and all another possibility for tolerance accumulation. The major examples are:

* Mock-up Fixtures - Thress dimensional representations of complex curvature surfaces such as exterior skins. Mock-ups are the coordinating media for Stretch or Hufford Dies, Bonding Fixtures, and Skin Drilling and Trimming production tools.
* Masters - Coordinate assembly tools between major sections

or components of the vehicle. Universally required to coordinate tooling which fabricates interchangeable components.

* Templates - Coordinate the family of sheet metal detail tools such as Press Blocks, Press Block Templates, Drill and Router Tools.

Other Considerations

Finally, components and assemblies may have special restraints imposed upon them such as tight tolerances, material and/or heat treat specifications, forming limitations, and interchangeability and replaceability requirements. Consequently, the diverse manufacturing methods and imposed restraints may conspire to produce physical fit problems resulting in product failure potential in spite of existing tight process control and close quality assurance surveillance. This is not to suggest that all, or even many, mating components mismatch with each other. However, when a mismatch problem surfaces, much time and trouble-shooting efforts will be consumed in resolving the problem. Sometimes the problem is not detected for a period. In this case, retrofitting may be called for if the situation is judged to be structurally deficient or is in violations of specifications. Other times, the components are continually "custom" fitted until the problem is corrected. This solution results in excessive labor costs and/or in marginal workmanship.

STRUCTURAL COMPONENTS MISMATCH

Ultimately, poor match of mating components may result in:

* Stress corrosion

* Integral fuel sealing failure

* Aerodynamic drag

With respect to aerodynamic drag, aerodynamic smoothness specifications place tolerances for certain specific areas of the vehicle exterior surfaces. If the components making up the exterior and underlying structure mating surfaces do not fit properly, contour tolerances are exceeded and aerodynamic performance is affected.

Accordingly, a mismatch of mating components may occur in the production of aerospace vehicles. The causes are diversity of means of parts description with their attendant tolerances, and the accumulation of tolerances through the variety of manufacturing processes, production tooling, and tooling tools fabrication. The problem is evident in failure to meet specifications and performance and in excessive labor costs.

THE SOLUTION

Interactive Computer Graphics

If the fit problem is caused by diversity, then an obvious solution would be to narrow or to eliminate the diversities. If a system could be devised whereby the interface surfaces of mating components could be defined excatly the same way, then at least we would be strating off correctly. Furthermore, if that definition could be translated directly to equipment which either produced the part or the tooling for the part, then chances are greatly enhanced for finishing correctly - that is, that the interface surfaces of the mating parts would match correctly.

This is exactly what is happening. There is a burgeoning system which offers a solution. The system defines mating parts surfaces - as well as the entire part-through a math model. The model is pure geometry at the inception. That is, there is no tolerance or interpretation placed upon it at this point. It is a single data base. It is made readily accessible and usable to all subsequent users within the product design and product manufacturing communities. All this is afforded through Interactive Computer Graphics.

What It Is

The system involves the use of computers, computer peripherals, and manufacturing equipment which can be directed by computers. The system is:

* Interactive - The operator, or user, directs and controls the problem and obtains a solution in real time. His input to the system is immediately (anything more than a few seconds is considered slow) digested and is immediately available for editing, erasure, or for the next piece of input to be added.

* Computer - All input and output is manipulated as high-speed calculations and is stored in a large memory bank.

* Graphic - The input to, and the output from the computer is immediately, or on demand, visually displayed to the operator and/or to other interested users via a graphics terminal or "scope".

How It Is Used To Create The Math Model

The math model is essentially "drawn" on the scope and the software and hardware computer systems provide the math model data base. The overall system provides the accessibility and usability required by other engineering and manufacturing disciplines within the company. This overview covers most systems. However, to describe the process in more detail we will look at the software and hardware employed in one of our systems at Grumman.

The trained operator "logs-on" at an IBM 2250 Graphics Terminal (or Emulator). This procedure allows him to retrieve an existing math model for viewing, copying or editing purposes, or to start a new model. All of these operations are performed using an alphanumeric keyboard, for inputting or editing the data; a "light-pen" for signifying to the computer which elements of the model are to be acted upon; and a function-key device which places the system into a variety of many operating modes. With these tools, he is able to do the same things he can do at the drafting table. The difference is that he does them more quickly, more efficiently, and more accurately utilizing the Lockheed developed CADAM (Computer-Graphics Augmented Design and Manufacturing) software package. A sampling of the system modes (or functions) follows:

* Lines - A straight line of any length can be constructed and be immediately displayed on the scope between any two finite locations or between already existing other goemetic elements. This line can at any time be removed from the math model ("erased"), moved, or altered at terminus or either or both ends.
* Circle - A complete circle, arc, or ellipse may similarly be constructed.
* Spline - A general or specific type spline such as a conic may similarly be constructed thru finite points or elements.
* Dimensions - A dimension may be immediately displayed between any selected elements. Witness lines and arrowheads are automatically created. Desired number of decimal places can be signified.

Also, any particular area of the entire model "picture" can be displayed on the scope at any time. The area can be "viewed" in any size. This capability allows viewing all or a large part of the entire picture in a smaller size, or "homing" in on a particular area in a large size.

An analysis capability allows instant interrogation of the model for true lengths, angles, locations, and other descriptive geometry data which is normally performed manually at the drafting table and may be extremely time consuming.

Another capability allows "overlaying" of different models for comparison, or for "piecing" together for a larger picture.

All of the usual drafting conventions can symbols such as notes, balloons, arrows, and deltas can be quickly constructed.

Details may be created or may be copied from other stored models. Examples are bushings, fasteners, standard clamps, and company drafting forms.

In short, a complete design may be constructed at the scope as a stored math model. Paper output for the model at Grumman is presently obtained by:

* Flat Bed Plotter - A pen on vellum or mylar accurate drawing made to any scale at a reasonably fast rate (up to about 40" per second).

* Electrostatic Plotter - An extremely fast output whereby all data is "dumped" as a series of extremely close points on an unwinding roll of paper at a rate of approximately 3/4" per second.

In addition to the design and drafting capability, another mode - Numerical Control - allows the operator to generate an N.C. program at the scope. Using the light-pen and keyboards he can input machine commands such as feed, speed, tolerance, and cutter shape, and construct a cutting path around the displayed math model of a part. A "replay" mode permits checking and/or editing the program by "displaying" the cutter motion on the scope, and by displaying pertinent data. The completed program may be post-processed for a particular machine tool at the scope. Output is then obtained as a computer listing of the pertinent program data, and as the actual N. C. tape to machine the part. In the process, calculations normally performed by the N. C. Programmer are automaticaly performed by the computer. The computer "knows" where a particular surface is with respect to all other surfaces. The programmer does not have to calculate and manually record the location.

Applying it To Improve Mating Parts Match

Grumman is implementing Interactive Computer Graphics within the Aerospace Corporation. In 1973, the Corporation secured a "stand-along" graphics system which has been used since that time for mainly electrical circuits definition, and also for schedules and charts preparation. In 1974, and ad hoc committee, after researching the graphics subject, recommended that the Corporation secure the Lockheed software package and that it be run on our

Grumman Data Systems Corporation IBM 370 Main Frame, using 2250 type Graphics Terminals as input media. The program was given department status. Since that beginning, many people from various functional areas within Grumman have been trained and are integrating graphics within their respective areas. It is planned to have a complete system in place for a major program. Meanwhile, the system has been used on smaller contract, and for revisions to existing large ones.

With respect to using graphics to improve component parts fit, the following scenario is developing:

Vehicle contour definitions - or "wetted" areas - are described geometrically using a particular type of conic form of the various areas or "patches". These patches are batch loaded into the computer. "Slices" of fuselage station, water-line, or canted plane cuts are derived. These operations are performed with company developed software programs. The results are then "passed" to the previously mentioned CADAM progra. The slices are then "filled" in with all the design details such as webs, stiffeners, fittings, rivets, and other hardware. It is here that the single data base is established.

As these components are developing, they are stored in the computer for subsequent use. All pertinent geometrical data is created in a discrete identical manner which will reside in the computer as pure math numbers and will leave no room for interpretation as it is used. We do not now have the myriad means of components descriptions as related earlier in this text. Most importantly, mating parts will share the same discrete data for the description fo their interfaces. Paper drawings may be obtained from the plotters previously mentioned.

Manufacturing, having the task of converting the math model into physical shapes and assemblies, now accesses the data at the scope in order to do this. N. C. programs are generated at the scope at this time for machined parts, and for some production tooling. Although many aerospace vehicle machine parts have been fabricated by N. C. equipment for the better part of twenty years, the advent of interactive computer graphics adds a new look. The part description resides in the computer as math data, and the N. C. program is derived from that data. Similarly, N. C. programs are generated to fabricate production tooling for sheet metal parts. Tubing has been automatically formed for some years on N. C. machinery. However, the "model" was generally a "digitized" hand shaped master which was fashioned from an equipment mock-up or from an actual vehicle. We are progressing into the era when both the tubing will be defined, and the N. C. program to shape it will be generated at the scope.

Tool design is accomplished at the scope around accessed engineering information. Large advantages are gained over traditional tool design drafting methods in that:

* Engineering data is immediately available to tool design.
* Tool design is completed "around: accessed engineering data.
* All tool design information necessary for dimensioning, and descriptive geometry calculations are handled automatically.
* Tool design drafting time is reduced.

Finally, Quality Control is able to use the math model in order to generate N. C. tapes to check both parts and tools which were derivatives of the math model. At this time, this piece of the total picture is at the planning stage. However, it is inevitable that as computers, and especially interactive computer graphics, are used in the design and manufacturing processes, Qaulity Control will also use the same data to accomplish its tasks. The integration of CAD/CAM systems for production of structural components will then be complete. And not coincidentally, the mating structural components proper fit is assured.

CHAPTER 16

RELIABILITY ASSURANCE OF AIRCRAFT STRUCTURES

Herbert F. Hardrath

NASA-Langley Research Center

Hampton, Virginia

ABSTRACT

Airworthiness certification requirements recently adopted by the U. S. Air Force rely heavily on damage tolerance to assure in-service reliability of airframes. FAA requirements are currently undergoing revision, and will also depend on damage tolerance characteristics of airframe structures. Two major impacts of these developments are evident. First, designers and operators will pay more attention to fatigue crack propagation rates and progressive deterioration of residual static strength than heretofore, and secondly, inspection will be an integral and acknowledged feature of continued reliability in service.

INTRODUCTION

Commercial aircraft enjoy a very high safety record compared with the records of competing modes of travel. This record is strongly dependent upon achieving a high degree of reliability in the aircraft structure and other systems. This reliability must be provided and maintained in a vehicle that must be as light as possible to minimize fuel consumption; that must operate for very long times to assure a financial return on a large initial investment; and that must operate in any atmospheric environment found on earth. This combination of requirements forces aircraft structures to be designed more precisely than those of other frequently used transporation vehicles and civil structures.

The purpose of this chapter is to portray the principles employed to assure this high degree of reliability. As may be expected, the procedures used were developed in an evolutionary manner, are periodically reviewed and are updated as developing technology permits. Recent changes to military and civil airworthiness requirements are discussed and new considerations involved in the certification of composite structures are outlined.

A STATISTICAL APPROACH

Discussion of reliability almost inevitably involve reference to, if not major reliance on, statistical assessments. The well-known bell-shaped probability density function shown in Figure 1 is the usual focus of such assessments. The foremost life-limiting mechanism for aircraft is fatigue failure resulting from the repeated loads encountered in service. Corrosion is another concern, but it usually is not included quantitatively in life prediction analyses. In Figure 1, the probability of encountering a significant fatigue problem during aircraft service is plotted against the life time (usually measured in flight hours for aircraft), at which such a problem is encountered.

In some industries, for example the electronics industry, large numbers of components are subjected to simulated environments to define this curve with a high degree of precision. Unfortunately, aircraft structures and their simulated service tests are much too expensive to allow such an approach. Only a few investigations have been undertaken systematically to at least estimate such curves for specific instances.

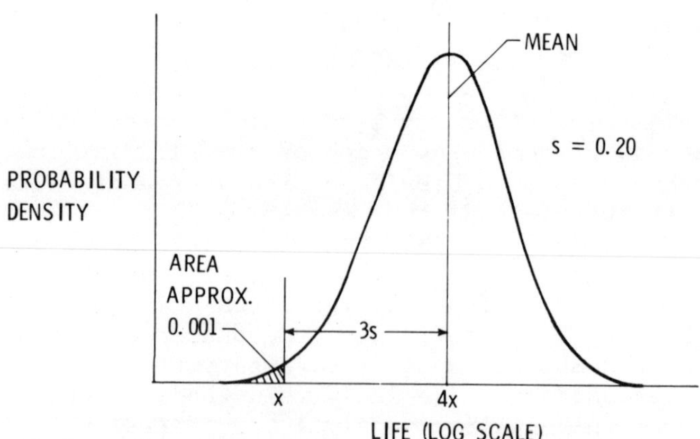

Figure 1. Distribution of failures in identical tests.

For example, the Australian Aeronautical Research Laboratories [1] performed 92 fatigue tests of F-51 fighter airframes during the 1950's. Analysis of those tests results indicated that the standard deviation, s, of the logarithm of life was approximately 0.2. (A log-normal distribution of lives had been assumed.) Debate continues over the appropriate distribution function and the constants required to predict behavior in aircraft stuctures.

In the absence of more definitive information, Figure 1 has been drawn using a log scale for life and s = 0.2. This curve implies that structures that are as nearly identical as is achievable in practice will have a distribution of lives as shown if they are tested under identical fatigue loading conditions.

The shaded area under the left tail of the curve represents a hypothetically acceptable failure rate of .001, or 1 aircraft in a fleet of 1000. The life at which this failure rate is achieved corresponds approximately to the mean life minus three standard deviations, as shown, and, for the example at hand, that life is one-fourth of the mean life. The obvious implication of the foregoing is that aircraft would be highly reliable if discarded at 25 percent of their mean lives. Several factors prevent practical use of that suggestion. First, one never has the luxury of establishing the mean life through multiple tests. Thus, a larger "factor" must be applied to relate the "safe life" to an observed life. Further, fatigue tests are generally conducted indoors, but aircraft are subjected to moisture, industrial effluents, exhaust gases, galley and restroom drippings, microbial wastes, and other hostile environments that degrade life. For economic reasons, compromises are necessarily made to simulate loading conditions and their distributions. As a result, certain parts of the structure may not have been interrogated adequately even though a complete structure was tested. All of the foregoing considerations are for tests conducted under the same fatigue load schedule. Particularly for military aircraft, the actual loadings experienced in service can vary greatly among the aircraft of a given type and all these can be very different from the loadings applied in the tests. Consequently, the life factor would have to approach 100 if one were to rely completely on the statistical considerations outlined. The economic impact of such a policy would be that the average aircraft would be discarded when it still possessed 99% of its useful life! Obviously, some more economical solution is required.

The exclusive use of the statistical approach is rendered even less viable if the history of service failures is examined. Troughton and Harpur [2] reviewed results of full-scale fatigue tests and service fatigue failures encountered in a large number of British military and civil aircraft that had been subjected to full-scale fatigue tests. They compared the life when a given defect was discovered in service with that when the same defect was found during the full-scale fatigue test. To be "safe" one would prefer to have encountered the defect earlier in test than in service. However, almost all data indicate failures in service occurred earlier than in test. About 80% of the service failures occurred earlier than one-fourth of the test life suggested in the foregoing criterion. Furthermore, about one-half the service failures were not identified in the fatigue test! The authors [2] presented an analysis of the probable reasons for these disconcerting discrepancies. The observiations presented earlier are dominant among the reasons. Similar analyses have not been published for American-built aircraft. However, the Federal Aviation Administration publishes an annual report [3] that describes all Airworthiness Directives in force for commercial aircraft bearing FAA Airworthiness Certificates. A cursory analysis of such directives listed in the 1974 issue of that document reveals that about two-thirds of the cases that are attributable to fatigue cracking problems require corrective action before the aircraft has been flown 5000 hours. The 5000-hour life was arbitrarily chosen because it represents a small portion of th- expected service life of civil transport aircraft. (Commercial aircraft are commonly utilized for over 50,000 hours.) The conditions in question are those that are judged to be potentially critical unless corrected. Of course, the data are for a wide assortment of aircraft, some of which have seen as many as 40 years of service.

Form the foregoing, the statistical approach does not provide adequate reliability within the current state of the art, at least not without prohibitive seight penalties. Consequently, an alternate approach must be employed.

CURRENT AIRWORTHINESS SPECIFICATIONS

For more than two decades, the airworthiness of American civil transport aircraft structures has been certified almost exclusively on the basis that the structures were demonstrated to be "fail-safe". The applicable regulation [4] requires the applicant to show "by analysis, tests, or both, that catastrophic failure or excessive structural deformation, that could adversely affect the flight characteristics of the airplane, are not probable after fatigue failure or obvious partial failure of a single principal structural element. After these types of failure of a single principal structural element, the remaining structure must be able to withstand static loads" that are specified in some detail.

In effect, this regulation frankly admits that failures may be encountered in service, but requires that sufficient strength be maintained in spite of such failures. The very good safety level enjoyed by current aircraft in spite of the rather disconcerting statistics outlined in the previous section is attributable to the good success of this regulation. Fail-safety is usually achieved by selecting "tough" materials and be designing redundant structural configurations.

This regulation will be revised in early 1978. The new regulation will require that damage escaping detection at one inspection will not grow to critical size before the structure is inspected again. The present fail safe requirement is retained to assure adequate strength in the presence of more severe damage regardless of its source. Deleterious environmental effects must also be accounted for. Those parts of an aircraft that cannot be designed to the damage tolerant or fail-safe must be shown to have very long lives. However, this option is open to only those specific components.

Aircraft manufacturers, of course, have a responsibility to their customers to provide a long useful life in their aircraft. Consequently, designs are carefully scrutinized to control high stress concentrations and a full-scale fatigue test is usually performed to check the design. In spite of the lack of precision of such tests described earlier, these tests frequently identify potential trouble spots. The manufacturer then enhances his product by retrofitting existing aircraft and modifying the design of aircraft not yet completed. Sometimes the manufacturer and airline negotiate a warranty on trouble-free life and on a prorated sharing of repair costs. The U. S. Air Force adopted a related specification [5] in 1974. It includes most of the essential parts of the FAA requirement, but is much more specific in several details. For example, the manufacturer is required to assume an initial flaw exists at eahc potentially sensitive spot in the structure. The size of the flaw is based upon the sensitivity of inspection techniques employed during manufacture. This flaw must then be shown not to grow to dangerous size, within an inspection interval. The specification includes a complex matrix of types of inspection, loads that must be supported during an inspection interval, and generic structural characteristics, all of which govern how the specification is to be applied in each instance. A second specification [6] requires the manufacturer to demonstrate that the cost to repair normal service-induced damage will not exceed prescribed levels. The Navy and Army are planning, but have not yet adopted, similar specifications.

IMPACT OF AIRWORTHINESS SPECIFICATIONS

Because of the emphasis placed on rates of flaw propagation and residual static strength, the new specifications are forcing aircraft designers to consider these properties much more carefully than was done earlier. Relatively less emphasis is placed on fatigue analysis by S-N curves, cumulative damage, and similar procedures.

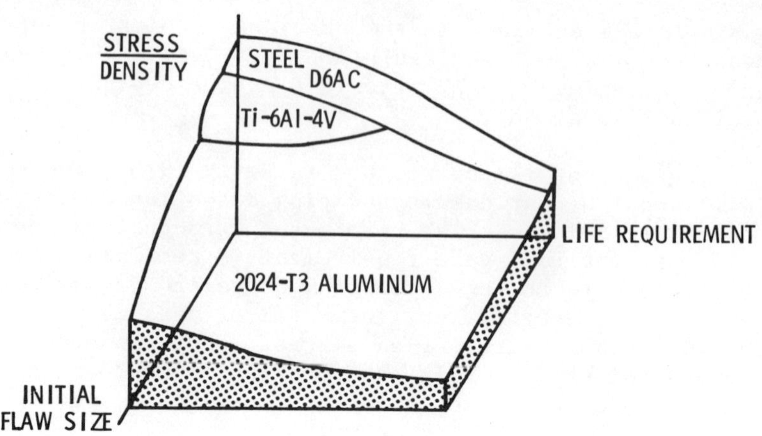

Figure 2. Relative efficiencies of some aircraft materials.

Materials are selected which have relatively low rates of fatigue crack propagation under service loadings. The procedure for choosing the most effective material is shown conceptually in Figure 2 [7]. The vertical axis of the 3-dimensional diagram is the ratio of design stress (for some critical load condition) to the density of the material. This parameter measures the "efficienc of the structural design and is a critical design consideration. Th right hand axis is the life required or the inspection interval. Th left hand axis is the flaw size defined by the sensitivity of inspec tions utilized. The curved surface represents the locus of points at which one of the three materials meets the life requirement assuming a given initial flaw size and does so with a higher stress-density ratio than the other two candidate materials. The three candidate materials were chosen arbitrarily for purposes of this illustration. The figure illustrates that the modest-strength 2024-T aluminum alloy is superior to the other two candidates for most practical combinations of flaw size and life requirement. The titanium alloy and steel are more efficient only when very small flaws (less than one or two millimeters) are detectable with high confidence. Higher-strength, but less damage tolerant, aluminum alloys do not have superiority anywhere within the range of this analysis. Primarily for this reason, these alloys are rarely used in fatigue-prone areas of aircraft structures.

Another practice that enhances damage tolerance is to keep the design stresses low. Although this practice results directly in higher structural mass, the compromise is made in the interest of safety. Fortunately, an enhanced fatigue life is automatically a by-product.

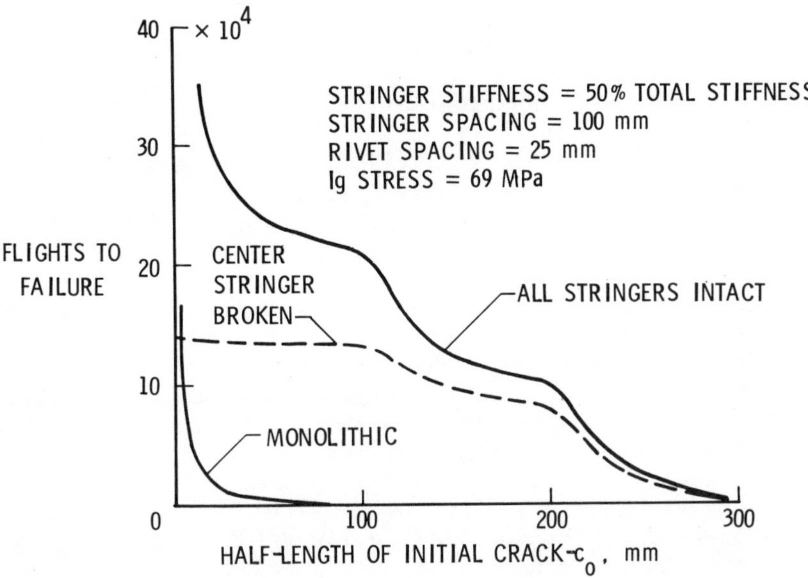

Figure 3. Crack propagation life for 7075-T6 aluminum alloy panels.

Because structural redundancy can provide added protection against uncontrolled flaw propagation, it is usually provided in commercial aircraft. To assess its efficacy, improved analyses are being developed that allow the designer to calculate local stress intensities for cracks that might occur. These stress intensities are a necessary step in predicting rates of crack growth and residual strength. Figure 3 illustrates an example problem solved by such an analysis [8]. For the example, behavior under the expected load spectrum for a point on the wing of a commercial airliner was calculated. The configuration chosen was a flat panel stiffened at 100 mm spacing with equal-sized stringers having an aggregate stiffness equal to half the total stiffness of the panel. The material chosen was 7075-T6 aluminum alloy; the stress was 69 MPa (10 ksi, a rather low value for current aircraft structures) per unit load factor; and the stiffeners were viteted to the sheet at 25 mm spacing. All these parameters have important influences on the predicted behavior. The diagram shows the number of typical flights that may be flown safely (vertical axis) after a specified crack (horizontal axis) is present. For this example, the crack was assumed to be centered over the middle stringer of the array; however, other configurations may be analyzed.

The lower curve illustrates that a very limited life is available for monolithic panels. In such structures, no crack inhibiting or arresting features are available and flaws propagate relatively rapidly. In contrast, the upper curve shows that structures with riveted stringers have an order of magnitude more life with a given crack present, and can tolerate much longer flaws (several stringer spacings) without failure. The dashed curve illustrates the state of affairs if the center stringer were also broken. An intermediate life is available even though the load from the broken stringer must now be shared by neighboring structure Analysis capability such as illustrated is developing at a rapid rate and is becoming a critical design tool.

Because the potential presence of flaws is becoming acknowledged, the operator of an aircraft must be able to inspect for such flaws in order that they may be identified well before they reach critical size and before cost of their repair becomes prohibitive. To this end, the manufacturer deliberately designs his structure to be inspectable and advises the operator on how and where to look for cracks. Inspectability is enhanced by providing appropriate access hatches so that the inspector or his instruments can reach critical parts. Freuqently, structures are deliberatly designed so that the most likely failures occur on the exterior surfaces of the structure where inspection is obviously easier and more dependable.

Unfortunately, redundant and readily inspectable structures inherently imply more parts, attendant higher manufacturing cost, and new stress concentrations that invite new fatigue failures. The trade-offs have not been demonstrated quantitatively, and are influenced strongly by individual judgments, company policy, and past experience.

RELIABILITY OF INSPECTIONS

To design and certificate a structure with flaw behavior in mind, the sensitivity and reliability of inspections must be considered quantitatively. Depending upon many particular configurational constraints and likely flaw locations, one of several inspection procedures can be of interest. Among the most preferred and "reliable" is visual inspection. However, flaws must usually be at least 5 mm long to be detected by this method. Many situations require better sensitivity and adaptability to remote areas that preclude visual examination. In metallic structures, eddy current procedures are used frequently. This method is more tedious, expecially when large numbers of stations are to be interrogated, but somewhat smaller cracks may be identified.

Figure 4 [9] presents results of a study in which aircraft parts were inspected in situ by two consecutive teams using eddy current techniques. The parts were then examined more critically by laboratory methods, including destructive examinations, to

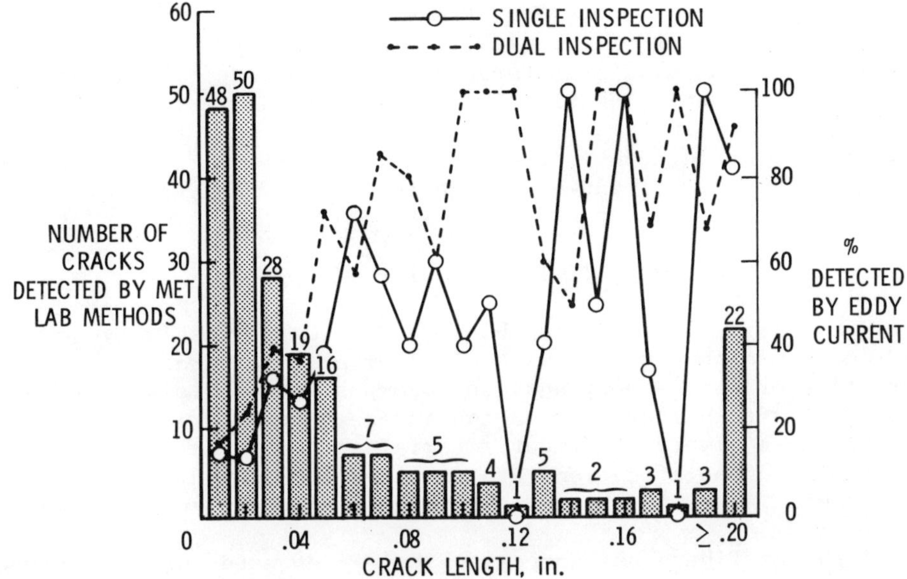

Figure 4. Crack detection reliability.

define the actual crack population more accurately. The bars indicate the numbers of cracks found in each of 19 crack length classes up to 5 mm (.2 inch), and a final bar for all cracks ≥ 5 mm. The symbols, connected by straight lines, represent the percentage of the cracks present that were also found by eddy current. As expected, the reliability of detection is rather poor for cracks shorter than 1 mm, averages about 50% for cracks between 1 and 5 mm and seems to be limited to about 80% for larger cracks. The impacts of such data are several. The sensitivity and reliability of the inspection should be improved; inspections must be performed more frequently than desired; or the initial flaw size must be assumed to be large enough to allow for the uncertainties. All these prospects contribute to higher costs, higher structural mass and/or undesirable schedule impacts. Each competing inspection tool needs to be evaluated in a similar manner to help select the optimum inspection system and schedule.

As indicated earlier, accessability for inspections is a factor that affects the reliability of inspections. A rationale [10-]2] has been developed by which a strategy based on sensitivity and accessability can be developed and its reliability can be estimated. Although such concepts are not yet developed to the point where reliability of an aircraft can be evaluated quantitatively in a statistical sense, the calculations are useful in developing optimum inspection procedures. Present inspection procedures are based primarily on past experience. An effective system exists through which the airlines, the manufacturers, and FAA exchange information on newly identified trouble spots in particular aircraft. This information is extremely useful when other similar aircraft are inspected. Continued vigilance by all three parties is vital to the continuance of high reliability in service.

RELIABILITY IN COMPOSITE STRUCTURES

The foregoing procedures have developed progressively over four decades of experience with aluminum alloy aircraft structures. Currently, graphite-epoxy and other composite materials are being applied in aircraft structures and future aircraft will probably utilize these highly efficient materials extensively. The present airworthiness specifications are expected to apply directly, but some special means may have to ve developed to apply them.

Although fiberglass composites have been used for many years for radomes and control surfaces of aircraft, these applications have almost always been in components that were not critical to safety of flight. More recently, advanced composites made with boron, graphite, or Kevlar fibers and epoxy or aluminum matrices have been developed. Flight hardware has been constructed successfully and a growing number of components are in routine use. Significant components made of these composities are in regular production for several military aircraft types.

General comments that can be made about the serviceability of composite structures follow. Their mass is generally 10 to 25% less than that of the equivalent aluminum alloy structure. As technology develops and usage increases, the cost of construction is expected to be competitive. On the whole, resistance to fatigue failure is much better than for aluminum alloys. However, the mode of fatigue failures is very different and residual strength of components containing significant flaws is somewhat poorer than for the best aluminum alloys. These last characteristics will introduce new considerations into the design and certification of composite structures.

Laboratory fatigue studies [13] of boron-epoxy composites have shown that matrix cracking is usually the first step in the fatigue failure process. The scanning electron microscope pictures in Figure 5 show this behavior for a boron-epoxy composite specimen. The specimen was cut from an 8-ply laminate having its plies laid to produce quasi-isotropic properties. The specimen had a central hole (diameter was 1/6 the width of the specimen) and was subjected to 10^7 cycles of repeated tension without catastrophic failure. The

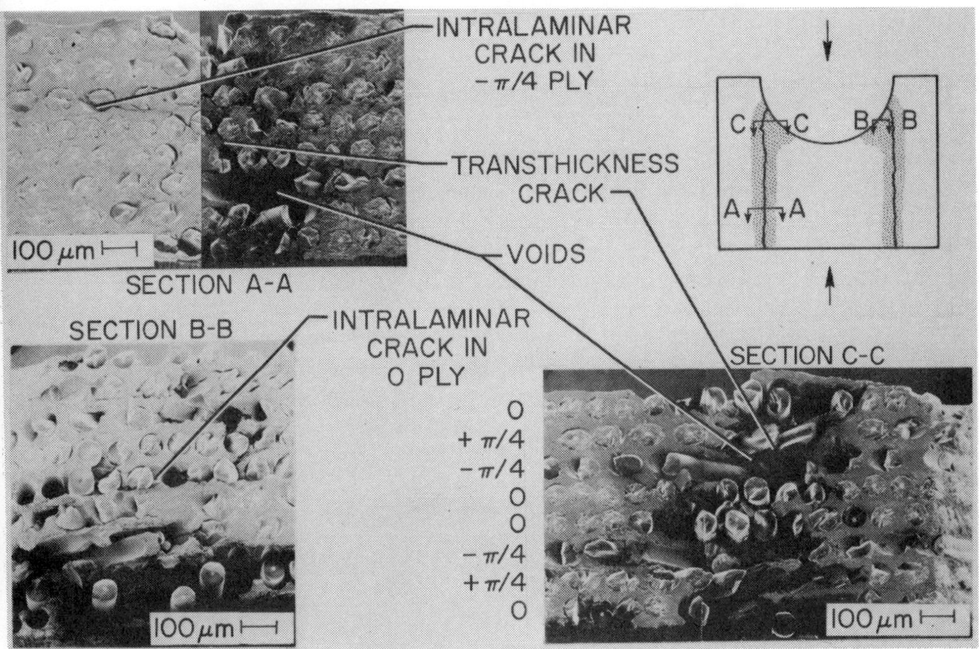

Figure 5. Matrix damage in a $(90/\underline{+}45/0)_s$ laminate in 10^7 tension loads.

specimen was deliberately cut to allow microscopic examination. As seen in the micrographs, the matrix is cracked between fibers of a given ply and usually in the plies oriented at $\underline{+}$ 45° to the axis of loading. X-ray inspections revealed that only a few fibers in these same laminae had failed at this stage of the test.

Similar specimens were repeatedly loaded in compression with similar results. Once such matrix cracking occurs in a compression fatigue test, fibers are free to buckle locally and (because they are quite brittle) eventually to break. For a given applied strain amplitude, a compression fatigue specimen is likely to fail at a shorter life than a tension specimen. Depending upon the ply orientation, the flaw propagation may be interlaminar, intralaminar, axial, transverse or a mixture of these. These circumstances complicate the inspection of composites, expecially because delamination and even fiber breakage of interior plies are not usually detectable visually, nor by X-rays nor by eddy current. The currently favored tool is C-scan ultrasonic inspection. This process appears well suited for identifying small delaminations, but experience must be built up to assure appropriate interpretation of the signals generated. X-rays are unsuitable for most graphite composite inspections because the material is transparent to the rays. Questionable areas can be made visible by application of solutions that are opaque to the rays, but this technique works only for cases where the cracked matrix is vented to an exterior surface.

Studies are underway to define the deleterious effects of accidental damage from a variety of causes. Preliminary results indicate that residual strength, particularly in compressively loaded panels, can be reduced quite significantly by moderate damage [14]. Softening strips and other means for enhancing tolerance to such damage are being developed, but the design of such schemes has not been quantified at this date.

CONCLUDING REMARKS

The foregoing review has outlined the current procedures for assuring high reliability in aircraft structures. Design for damage tolerance and pursuit of a viglant inspection system are the primary means by which current structures achieve a very high safety record. Calssical statistical assessments could not match this reliability without prohibitive economic and weight penalties. Advanced composite materials offer a very attractive increment on structural efficiency, but introduce a new dimension in structural design and inspection to assure reliability corresponding to that which is common in metallic airframes.

REFERENCES

1. Payne, A.O.,"Determination of the Fatigue Resistance of Aircraft Wings by Full-Scale Testing", in Full-Scale Fatigue Testing of Aircraft Structures, F.J. Plantema and J. Schijve, Eds., Pergamon Press, New York, 1961, pp 76-132

2. Harpur, N.F. and Troughton, A.J., "The Value of Full-Scale Fatigue Testing", in Fatigue Design Procedures, E. Gassner and W. Schutz, Eds., Pergamon Press, New York, 1969, pp 343-75.

3. Anon., Summary of Airworthiness Directives for Larger Aircraft: Federal Aviation Regulations, Part 30, 1974, Federal Aviation Administration, Oklahoma City, OK.

4. Anon., Code of Federal Regulations, Title 14, Aeronautics and Space, Part 25, 1977, U. S. Government Printing Office, Washington, D.C.

5. Anon., Airplane Damage Tolerance Requirements, Military Specification MIL-A-83444 (USAF), 1974, U. S. Government Printing Office, Washington, D.C.

6. Anon., Airplane Structural Ground Tests, Military Specification, MIL-A-008867B (USAF), 1975, U. S. Government Printing Office, Washington, D.C.

7. Elber, W. and Davidson, J.R., A Materials Selection Method Based on Material Properties and Operating Parameters, NASA TN D-7221, April 1973, National Aeronautics and Space Administration, Washington, D.C.

8. Hardrath, H.F., "Fracture Mechanics", J. of Aircraft, Vol. 11, No. 6, June 1974.

9. McCarthy, J.F., Tiffany, C.F. and Orringer, O.,"The Application of Fracture Mechanics to Decisions on Structural Modifications of Existing Aircraft Fleets", in Case Studies in Fracture Mechanics, T.P. Rich and D.J. Cartwright, Eds., U. S. Army Materials and Mechanics Research Center, Watertown, MA, 1977.

10. Davidson, J.R., "Reliability After Inspection", in Fatigue of Composite Materials, ASTM STP 569, American Society for Testing and Materials, 1975, pp 323-34.

11. Davidson, J.R., "Reliability and Structural Integrity", in Recent Advances in Engineering Science, Vol. 7, Scientific Publishers, Inc., 1977, pp 387-98.

12. Davidson, J.R., "Rationale for Structural Inspections", Aircraft Safety and Operating Problems, NASA SP-416, October 1976.

13. Roderick, G.L. and Witcomb, J.D., "Fatigue Damage of Notched Boron/Epoxy Laminates Under Constant-Amplitude Loading", in Fatigue of Filamentary Composite Materials, ASTM STP 636, American Society for Testing and Materials, 1977, pp 73-88/

14. Rhodes, M.D., Williams, J.C. and Starres, J.H., Jr., Effect of Low-Velocity Impact Damage in Compression Strength of Graphite/Epoxy Hat-Stiffened Panels, NASA TN D 8411, 1977.

CHAPTER 17

NEUTRON RADIOGRAPHY UTILIZING SELECTED ENERGY INTERACTIONS

John J. Antal

Army Materials and Mechanics Research Center

Watertown, Massachusetts

ABSTRACT

We intended in this chapter to describe the work of the Materials Sciences Division of AMMRC in attempting to bring new methods of materials characterization to the fore. I wanted to talk about neutron radiography, but earlier in the meeting it became clear that we were in trouble, not having brought with us a single visual aid showing a crack or other mechanical failure. By the second day of the meeting we were in even deeper trouble, not having a slide showing incipient stress corrosion.

But the program soon unfolded a fine new picture of the status of fracture mechanics. It was good to see theory accept the idea of macroscopic defects in real material, and to see the producers of this material striving to characterize better the materials they can provide rather than attempting to provide the perfect materials theory would like to demand. The non-destructive testing community then expressed their desire to contribute at the design stage so that measurable minimum defect sizes would always be specified.

If this approach is accepted it will undoubtedly follow that the designer will soon be pressuring the NDT community to refine its technologies to provide even better resolution of defects to satisfy the designer's needs. And at last we can see where the work we are attempting to do in the area of nondestructive materials characterization might be of interest to the reader. Two particular developing techniques will be reviewed in this chapter: fission neutron radiography and subthermal neutron radiography.

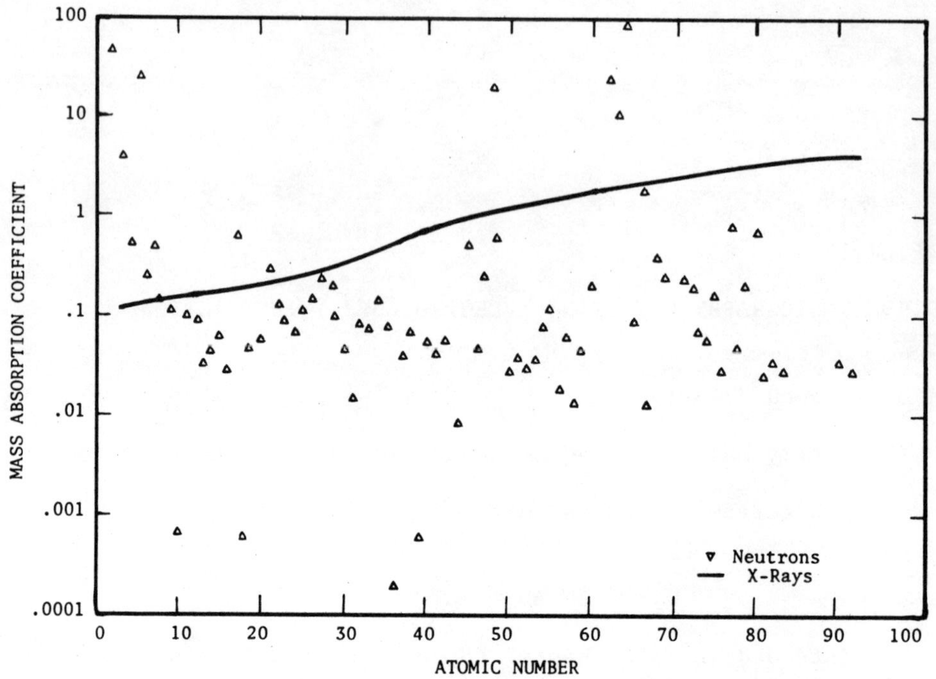

Figure 1. Mass absorption coefficient for thermal neutrons and 120 KeV X-rays vs atomic number for 80 elements having their natural isotopic abundance.

NEUTRON RADIOGRAPHY

Normally neutron radiography refers to a nondestructive examination technique executed in a manner identical to X-ray radiography but employing thermal neutrons as the penetrating radiation. Thermal neutrons are those having an average energy of 0.025 electron volts. Neutrons are able to penetrate most materials readily even at these low energies because they are not electrically charged and do not interact with the multitude of electrons present in materials. X-rays, on the other hand, do so interact and thus require 150 Kilovolts of energy or more in radiography applications. Neutrons do interact, often strongly, with the nuclei of atoms in materials and with the atomic lattice of materials, thereby providing selective absorption and scattering characteristics for contrast in radiography.

These differences between neutron and X-ray interactions with materials usually result in neutron radiography becoming a technique which is complimentary to x-radiography in a particular situation.

Figure 1 illustrates one aspect of the differences between neutrons and X-rays which is important to the radiography of items composed of a variety of materials. Presented in Figure 1 is the relationship between the mass absorption coefficient and the atomic number of most of the elements with which we commonly associate. X-rays interact with materials in a predictable manner where the interaction becomes stronger as the atomic number increases. Neutrons, on the other hand, appear to follow no readily perceived pattern except perhaps that they interact strongly with the group of lightest elements. Often of great interest in neutron radiography are the very strong absorbing elements: boron, cadmium, and most of the rare earths. Always of great interest is the ability of hydrogen to react strongly with neutrons.

Thermal neutron radiography is not complimentary to x-radiography in all instances. For example, the ability of both these radiations to penetrate iron, our most commonly encountered material, is about equal. In an attempt to fill this and other holes in the radiography spectrum, we are looking into the use of neutrons of both very high and very low energies as radiographic radiations.

Neutrons are most often obtained from fission sources which provide a spectrum of neutrons with an average energy of about 1.5 million electron volts (McV). These neutrons are moderated in energy by passing through large amounts of certain materials wherein they attain the energy of the thermal neutrons commonly employed in neutron radiography. If the fission neutrons themselves could be eliminated from the system and a higher intensity of a much more penetrating radition would be available for examining thick sections. But the usefulness of this technique is dependent upon the physics of the interactions of these neutrons with materials and the data presented in Figure 2, the analog to Figure 1, must be considered.

Data for the lighter elements has been separated from the data for the heavier elements in Figure 2 in order to avoid confusion. In order to gain contrast in neutron radiography adjacent materials should contain elements whose cross sections are widely separated from each other in the figure. This is not too often the case for fission neutrons, so fission neutron radiography would be expected to have limited application. Yet in the region of lighter elements the cross section diverges nicely at 1 MeV indicating fair contrast for these elements relative to the remainder, and cadmium is particularly more transparent to fission neutrons than it is to either thermal neutrons or X-rays.

At the opposite extreme of low energies, we find subthermal neutrons, those defined as having energies below 0.004 electron volts. These neutrons have very interesting interaction properties with materials because they are at the limit of energy where they

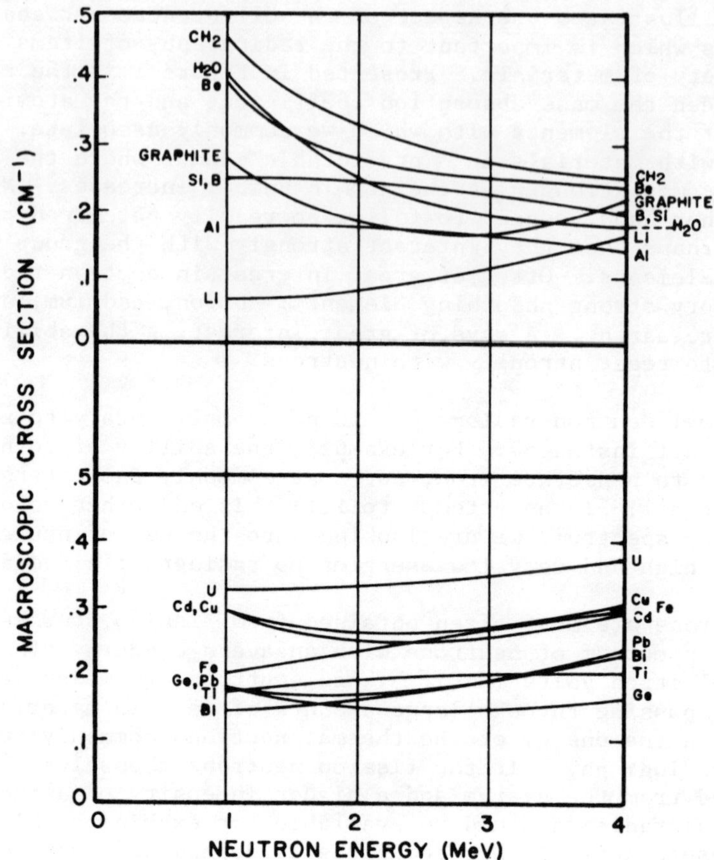

Figure 2. The probability of interaction by neutrons of high energy neutrons with a variety of materials. The lighter elements are in the upper graph and the heavier elements in the lower graph. The data have been drastically averaged to smooth over numerous resonance peaks occurring throughout the energy range shown.

might scatter from materials through diffraction processes. Thus subthermal neutrons may be scattered strongly by the lattice structure of a certain crystalline material and might pass uninhibited through an adjacent crystalline material. What is important to subthermal neutron radiography then is the exact crystal structure of the materials being examined. This takes such a variety of forms that it is not easily presented like the data of Figures 1 and 2. Each combination of materials must be examined individually to assess the usefulness of subthermal neutron radiography to a particular application. One factor of great interest is that the crystal structure of iron is such that diffraction scattering does not occur for subthermal neutrons and iron is noticeably more transparent to these neutrons than to thermal neutrons.

NEUTRON SOURCES

The utility of neutron radiography is sometimes determined by the nature of the source of neutrons. The best source of neutrons for radiography is the nuclear reactor. Radiography at this type of source is carried out in a routine manner similar to X-ray radiography. The intensity of this source of neutrons is such that exposure times are similar to those for x-radiography. The same X-ray film is generally used to record the radiograph, with the addition of a screen to convert the detected neutron to gamma or electron radiation which is able to sensitize the film. Because of the complexity of the nuclear reactor, samples to be radiographed must be of a sort that are easily carried to the source.

The other major source of neutrons for radiography is californium-252, a man-made radioactive isotope which is self-fissioning. The advantage of this source is that it is reasonably portable and requires very few man-hours of labor for operation during an exposure. The disadvantage of this source is that the intensity of neutrons it produces from a reasonably-sized source is low relative to that from a nuclear reactor. High intensities of unwanted gamma-rays are always emitted from neutron sources, which means that the need to provide heavy personnel shielding limits the size of source which can be considered to be "portable". Thus, even though the radioactive californium source is itself always very portable, in practice only a source small relative to a nuclear reactor is considered for use.

Californium-252 has become available in quantity only in recent years and our programs at AMMRC are directed to exploring the use of this isotope as a radiographic source.

Subthermal neutrons are not produced by a special source, but are present in the emissions from any neutron source which is heavily moderated. Subthermal neutrons are segregated from a thermal neutron beam by passing the beam through a filter. The most common filter material is polycrystalline beryllium maintained at a low temperature. The subthermal neutrons are in the "tail" of the thermal neutron spectrum and as might be expected, their intensity from any neutron source is low. In fact, this is the major factor which is retarding the application of subthermal neutron radiography at the present time.

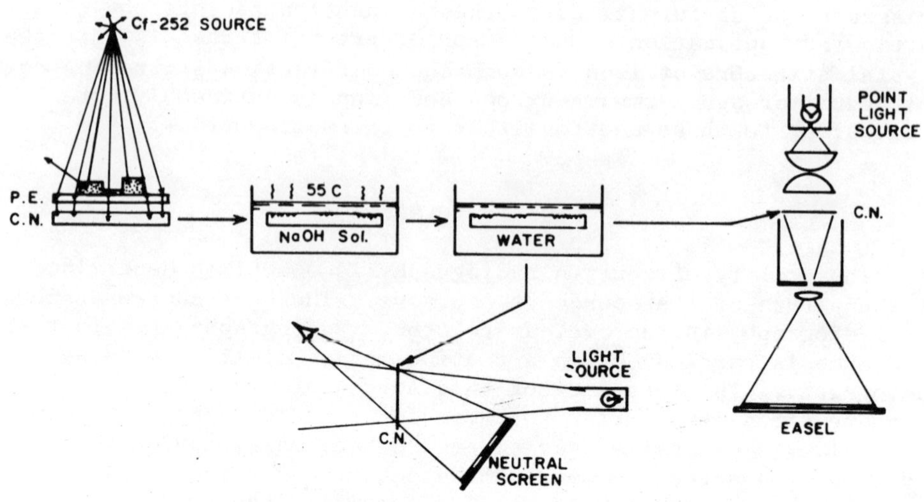

Figure 3. Scheme for producing fission neutron radiographs utilizing californium-252. The object being radiographed is shown in dotted cross section below the source. The polyethylene converter sheet (P.E.) is about 5 mils thick. The cellulose nitrate (C.N.) detector sheet is about 1 mil in thickness. Etching in a hot sodium hydroxide solution for about 10 minutes and rinsing in water are the only processing required and are accomplished in ordinary room lighting. The dried cellulose nitrate film appear to the eye as a negative with a frosted image and may be printed in an enlarger set for dark-field illumination of the negative, or viewed by eye against a neutral background.

FISSION NEUTRON RADIOGRAPHY

Using californium-252 as a source, fission neutron radiography could be carried out in a manner identical to that of x-radiography with radioactive sources. The problem is to find an efficient detector for these penetrating high energy neutrons. Several approaches to imaging high energy neutrons have been taken in the past. Our approach was to allow the fast neutrons to impinge upon a polyethylene sheet which acts as a neutron-to-proton converter. When neutrons collide with the hydrogen nuclei in this material they are ejected as protons through the opposite face of the sheet. The protons are then detected in a sheet of cellulose nitrate which backs up the polyethylene sheet. Upon entering the cellulose nitrate sheet, the charged protons literally plow through the material, dislodging many atoms from their normal sites along the

Figure 4. Photograph of items for which fission neutron radiographs are presented in following figures. The item at the top is a 4 kw transmitting tetrode with a ceramic body. The item at the bottom center is a 6-foot flexible tape in a steel case, and the other two items are electrical potentiometers.

line of travel. After the proton dissipates its energy in the material, a damage "spike" of disordered atoms remains. The cellulose nitrate film is processed in a caustic solution which chemically etches the film surface, but etches the area around and through the small radiation-damaged volume at a much more rapid rate. The result is the development of large etch pits in the film surface at the point of proton impact which can be readily viewed under a microscope. The whole process is described in Figure 3.

After processing in the etching solution, the amount of light scattered by the numerous tiny etch pits is sufficient that the variations in etch-pit density form an image when viewed properly. Placed in the negative carrier of an enlarger set for dark field

Figure 5. Fission neutron radiograph of the cathode structure of a 4-kw transmitting tetrode.

imaging, prints are readily made, or the etched film may be viewed by eye optimally if arranged as shown at the bottom of Figure 3.

The fission neutron radiography system described here is not very sensitive as presently developed, but it does work. As evidence of this, Figure 4 is a photograph of several items which we have radiographed in the manner described with fission neutrons from californium-252. The actual radiographs make up Figures 5, 6, 7 and 8. The radiographs were made with each item 20 cm from the californium source. The neutron fluence was 2 x 10^{11} neutrons per cm^2, accumulated in an exposure over a week-end. Because the etch pits are extremely small, the resolution obtained is very good.

Our work has concentrated on producing cellulose nitrate films of suitable characteristics. Cellulose nitrate procured from commercial firms has not been satisfactory for this radiography use. We are continuing to develop films of higher sensitivity in hopes of making fission neutron radiography a practical technique of nondestructive examination. The detection system has the advantage of being very simple and inexpensive. It is also totally insensitive to gamma-rays and so may be used to radiograph radioactive materials.

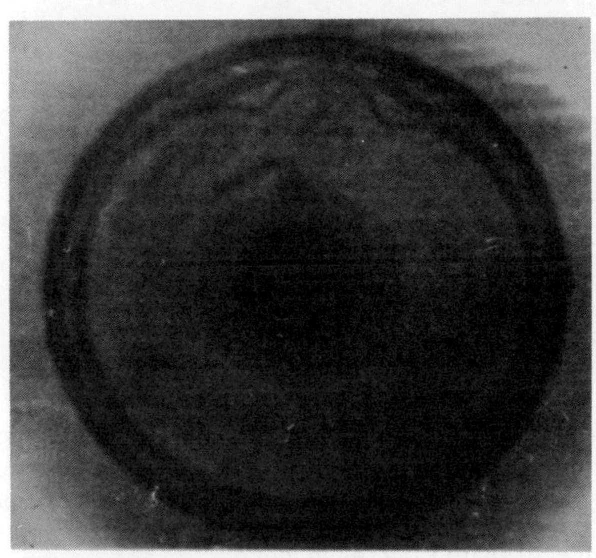

Figure 6. Fission neutron radiograph of two electronic potentiometers. The item on the left is of the carbon film variety and has a plastic knob, two nuts and a washer on its shaft. The item on the right is wire-bound and is viewed along the direction of the shaft.

Figure 7. Six-foot steel tape in its case radiographed with fission neutrons. The individual turns of the return spring (5 mils thickness) are well-resolved. Non-parallel radiation from this point source gives the appearance of blurring seen here.

SUBTHERMAL NEUTRON RADIOGRAPHY

We are not yet able to take radiographs with subthermal neutrons from californium-252. As noted earlier, the intensities available are low and this is especially true where californium-252 is involved A procedure which has successfully increased the neutron intensity obtained from reactor sources is to lower the temperature of the moderator material. Neutrons come to a hynamic equilibrium in moderator materials at the temperature of the moderator. Thus if the moderator temperature is lowered, so is the average neutron energy. As one example, the HERALD reactor in Aldermaston, England, has installed a low temperature facility which cools a small amount ofhydrogen and deuterium moderator mixture near the reactor core to a temperature of -258 C. This successfully produces a flux of 10^6 subthermal neutrons/cm^2/sec at a beam port where radiography is performed routinely. It is a very complex facility and it requires a very large amount of refrigeration primarily to combat the heating effect of the gamma-ray energy absorbed in the components of the facility. This source will, of course, never be portable, but a similar arrangement for a small californium source might have a

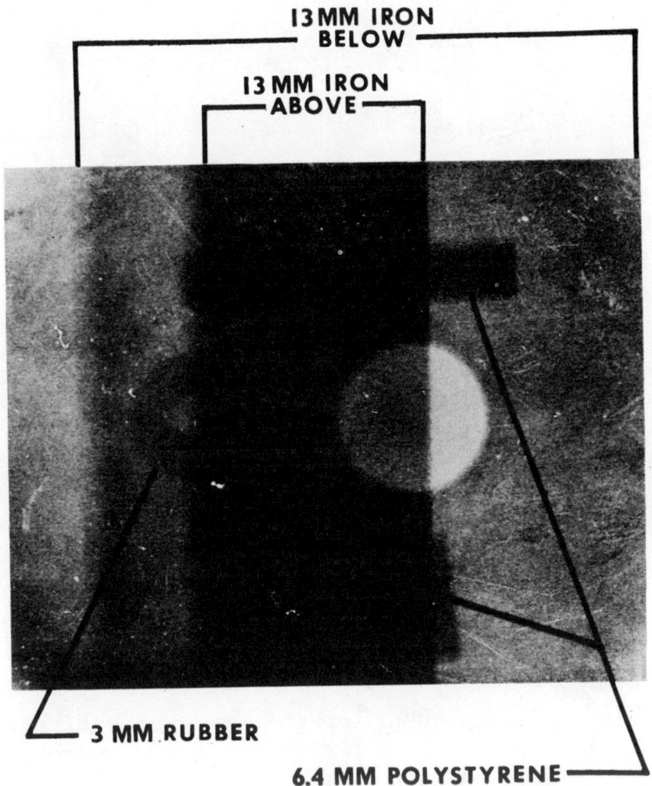

Figure 8. An illustration of the imaging ability of fission neutrons. The more opaque items contain an abundance of hydrogen. They are positioned between two mild steel plates which appear to be quite transparent. The lower steel plate has a hole through its center.

much greater chance of becoming portable.

Whereas in the nuclear reactor setup it is possible only to cool a small amount of moderator, with a tiny californium source (ours is 1-1/4 inches long and 3/8 inches in diameter) the source can be placed entirely within the small moderator and the whole assembly cooled with a very small amount of refrigeration.

We have been working toward this end through a series of investigations being carried out in our laboratories. The moderator selected was a 5-inch diameter paraffin sphere which is easily and inexpensively fabricated. The californium source is placed at the center of the sphere which resides in a polyform container. A one

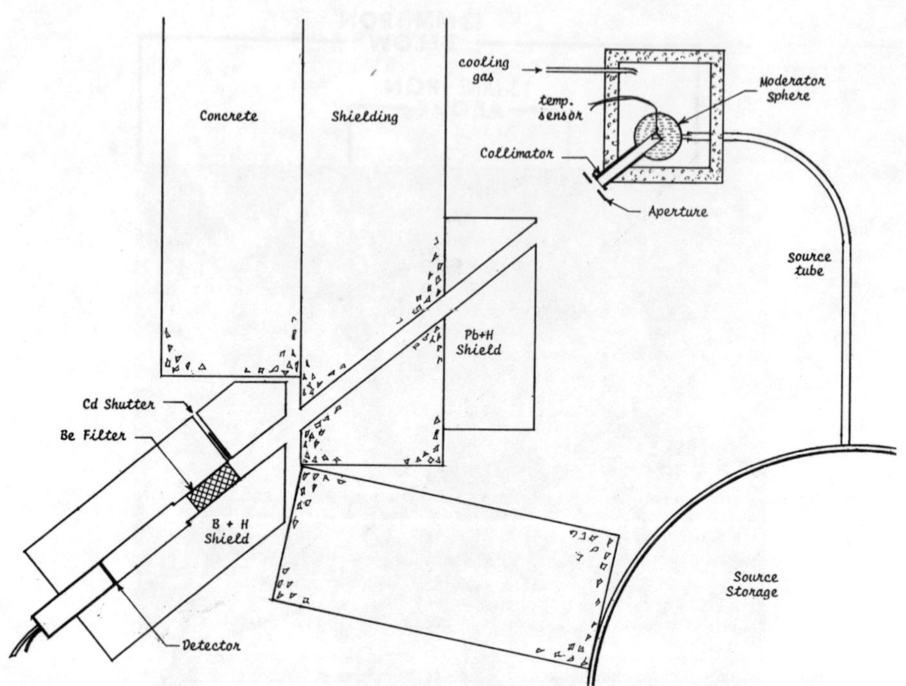

Figure 9. Experimental setup for the study of subthermal neutrons from a californium-252 source. The source is pushed into the moderator sphere by a cable in the source tube from the opposite side of the source storage container. The collimator is so designed that the detector sees only the base of the reentrant hole in the moderator sphere.

inch diameter hole through the sphere just above its equator provides for the insertion of various moderator materials in plug form and provides a re-entrant hole from which the subthermal neutrons are emitted. The assembly is cooled by evaporating liquid nitrogen and can reach a low temperature of -195 C. A drawing of the laboratory setup is shown in Figure 9. The source is attached to a cable and may be removed to a safe storage place at any time to allow for alteration of the experimental setup. At present the beryllium filter is not cooled below room temperature, but it would improve the beam flux if this were done.

The setup pictured in Figure 9 is not used for radiography; the collimator is incorrectly configured for that purpose. The subthermal neutron beam is simply monitored by a sensitive detection system to provide data on improvements in beam intensity. Figure 10 shows some interesting data which has been obtained with this apparatus. It shows the increase in subthermal neutron intensity which is obtained when the temperature of the system is lowered from room temperature to that of liquid nitrogen. Data is given for two different moderator materials and indicates that a 4-1/2 times increase in intensity can be obtained by cooling in this simple manner. Also note that the intensity is increasing sharply at the lowest temperature, indicating that it would be desirable to go to even lower temperatures.

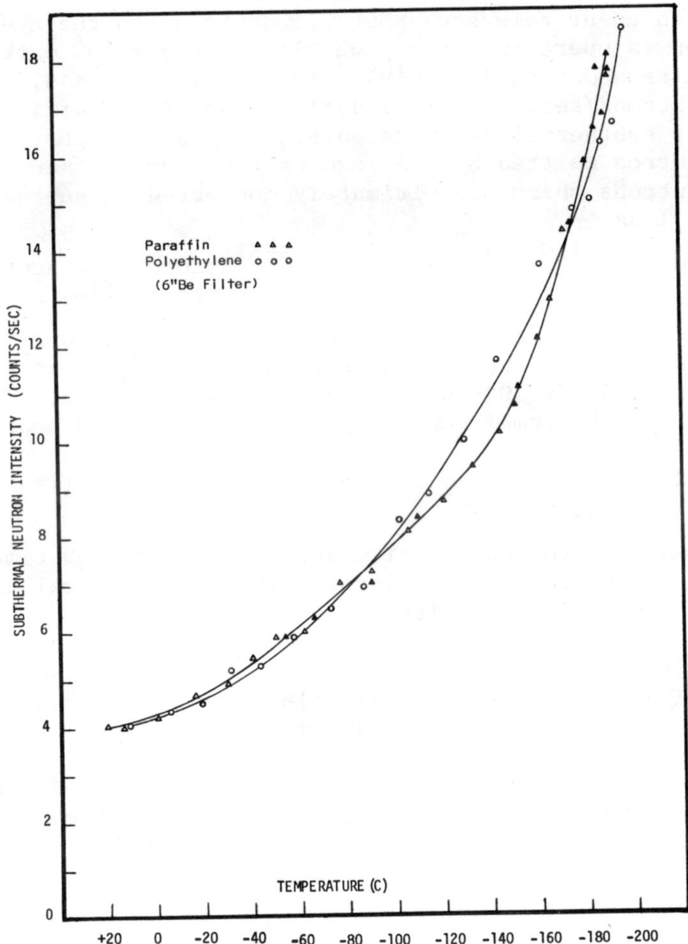

Figure 10. Subthermal neutron intensity as a function of moderator sphere temperature for two organic solid moderators. The large increase in intensity upon cooling is evident.

Since we are interested in this arrangement as a source of subthermal neutrons, it is interesting to estimate the efficiency of subthermal neutron production relative to that of a working reactor neutron source. The HERALD low temperature facility has a hyrodgen-deuterium moderator in liquid form at a temperature of -258 C. With the reactor operating at 5 megawatts of power, 1.0×10^6 subthermal neutrons/cm^2/sec issue from a particular beam port.

The energy released by one fission event is about 200 MeV, which is 3.2×10^{-11} watt-seconds. That is, 1 watt of reactor power requires that 3.1×10^{10} fissions/sec take place. Since each fission event releases about 2.5 neutrons on the average, and the reactor is operating at a power level of 5×10^6 watts, this reactor emits about $(3.1 \times 10^{10}) \times (5 \times 10^6) \times 2.5 = 3.9 \times 10^{17}$ fission neutrons/sec. We might define a practical efficiency as the flux of subthermal neutrons obtained at a beam port for each fission neutron emitted by the reactor fuel core, since it is the fission neutrons which are ultimately converted to subthermal neutrons. Thus

$$\text{Eff(Reactor} = \frac{1.0 \times 10^6}{3.9 \times 10^{17}} = 2.6 \times 10^{-11} \text{ subthermal neutrons per fission neutron}$$

A 1.5 milligram californium-252 source emits 3.6×10^9 fission neutrons per second, from which we have obtained in our laboratory a beam flux of 3.1×10^{-8} subthermal neutrons/cm^2/sec. The efficiency of subthermal neutron production for this source is

$$\text{Eff(Cf-252)} = \frac{11}{3.6 \times .10^9} \text{ subthermal neutrons per fission neutron}$$

The efficiency of the californium source is some 1500 times greater than the reactor source in the production of subthermal neutrons in a practical operating facility.

The HERALD reactor has both thermal neutron and subthermal neutron radiography services available on a routine basis. They recently noted that 80% of the radiography service provided to the British Defense Establishment is accomplished with subthermal neutrons. So we find much to encourage us to continue our efforts in attempting to apply extreme-energy neutrons to the radiography of materials.

CHAPTER 18

THE APPLICATION OF MODERN ANALYTICAL INSTRUMENTATION TECHNIQUES TO THE ANALYSIS OF EXPLOSIVE MIXES

F. C. Burns, H. F. Priest and G. L. Priest

Army Materials and Mechanics Research Center

Watertown, Massachusetts

Upgrading and modernization of munition manufacturing facilites to improve production capabilities and to increase cost effectiveness are under way at several Army installations. It is a requirement that each batch of blended explosive mixture be analyzed before loading into detonator and related items to insure that they are properly mixed. Rates of production are being increased considerably, placing new requirements on analytical procedures. The traditional methods consisting of consecutive extractions with a variety of solvents are cumbersome and much too slow, requiring four hours to analyze a single batch of explosive mix. The batches are very small because of the extreme sensitivity of this class of explosives, making it necessary either to have many batches being stored in explosive facilities while awaiting analysis or to reduce the time for the analysis (of each batch) to not more than thirty minutes. The latter is a much more desirable and cost effective approach.

R.F.Q.s for an automated analysis system requiring 30 minutes per test were sent to approximately 46 commercial vendors, none of whom responded. This is not surprising because it is a highly specialized problem requiring a combination of techniques.

The composition of a typical explosive mix and one which is extensively used is shown in Table 1.

This is probably the most difficult composition to analyze, but we believe that a combination of X-ray and 14 MeV neutron activation analysis will provide adequate results. Preliminary

Table 1

COMPOSITION OF TYPICAL EXPLOSIVE MIX

Lead Styphnate	$(Pb\ C_6\ H_3\ N_3\ O_9)$
Tetracene	$(C_2\ H_8\ N_{10}\ O)$
Antimony Sulphide	$(Sb_2\ S_3)$
Barium Nitrate	$[Ba(NO_3)_2]$
Lead Azide (Dextrin)	$[(N_6\ Pb)\ (C_6\ H_{10}\ O_5)]$

feasibility studies have been initiated to confirm our hypothesis.

Using the explosive mix shown in Table 1, we have proposed the following two step approach: (1) X-ray spectroscopy, either dispersive or non-dispersive, but probably the latter, with an isotopic source, will provide precise analysis with an accuracy of ca. $\pm$ 0.1% for the lead, barium and antimony and will give directly the concentrations of antimony sulfide and barium nitrate.

Furthermore, if the ratios of lead, barium, and antimony are incorrect then it is obvious that the mix is either not properly constituted or not properly mixed.

(2) Non-destructive 14 MeV Neutron Activation Analysis will be used to determine the oxygen and the nitrogen concentration in the explosive mix. The unique 14 MeV neutron activation analysis system at AMMRC is being used for the feasibility studies, but commercial units adequate for the primer analyses are available.

The AMMRC system is made up of three subsystems: irradiation and transfer, uncapping, and counting [1].

The determination of oxygen is an interim check. If the total oxygen content is correct, then it is a fair assumption that this, coupled with correct lead, barium, and antimony ratios, indicates a satisfactory explosive mix. However, if each component needs to be determined precisely, then neutron activation analysis for nitrogen gives the necessary data for setting up three simultaneous equations which can be solved for each of the remaining constituents (Appendix I).

To meet the 30 minute batch analysis deadline, automation using micro processors and computer read out will be part of the final systems to be installed at the various installations.

APPENDIX I

From X-ray Spectroscopy

Quantity of lead = P
Quantity of barium = R
Quantity of antimony = S

From Neutron Activation Analysis

Quantity of Oxygen = T
Quantity of Nitrogen = U

From the quantity of Barium, R, calculate the quantity of Barium Nitrate in the mix and the quantity of nitrogen and oxygen in that amount of $BA(NO_3)_2$.

From the quantity of antimony calculate the quantity of Sb_2S_3 in the mix.

Using the following quantitative conversion factors based on formula weights

$$g = \frac{0}{PbSt} \qquad h = \frac{0}{Tet.} \qquad i = \frac{0}{N_6P_6 + Dex.}$$

$$k = \frac{N}{PbSt} \qquad i = \frac{N}{Tet.} \qquad m = \frac{N}{N_6Pb + Dex.}$$

$$o = \frac{Pb}{PbSt} \qquad q = \frac{Pb}{N_6Pb + Dex.}$$

a = Quantity of lead styphnate (PbSt)
b = Quantity of Tetracene (Tet.)
c = Quantity of Dextrinated Lead Azide (N_6Pb + Dex.)

$$ag + bh + ci = [T - 0 \text{ from } Ba(NO_3)_2]$$

$$ak + bl + cm = [U - N \text{ from } Ba(NO_3)_2]$$

$$as + cq = P$$

These three equations in three unknowns can now be solved for a, b, and c which now gives a complete analysis for the explosive mix.

REFERENCES

1. Burns, F.C., Priest, H.F. and Priest, G.L., Analytical Chemistry, Vol. 42, April 1970, pp 499-503.

INDEX